Defects in
Optoelectronic Materials

Optoelectronic Properties of Semiconductors and Superlattices

A series edited by *M. O. Manasreh*, Dept. of Electrical and Computer Engineering University of New Mexico, Albuquerque, USA

See the back of this book for other titles in Optoelectronic Properties of Semiconductors and Superlattices.

Defects in Optoelectronic Materials

Edited by

Kazumi Wada

Massachusetts Institute of Technology
Cambridge, Massachusetts, USA

and

Stella W. Pang

University of Michigan
Ann Arbor, Michigan, USA

Gordon and Breach Science Publishers

Australia • Canada • France • Germany • India •
Japan • Luxembourg • Malaysia • The Netherlands •
Russia • Singapore • Switzerland

Amsteldijk 166
1st Floor
1079 LH Amsterdam
The Netherlands

British Library Cataloguing in Publication Data

ISBN 90-5699-714-9

CONTENTS

ABOUT THE SERIES

The series *Optoelectronic Properties of Semiconductors and Superlattices* provides a forum for the latest research in optoelectronic properties of semiconductor quantum wells, superlattices, and related materials. It features a balance between original theoretical and experimental research in basic physics, device physics, novel materials and quantum structures, processing, and systems—bearing in mind the transformation of research into products and services related to dual-use applications. The following sub-fields, as well as others at the cutting edge of research in this field, will be addressed: long wavelength infrared detectors, photodetectors (MWIR–visible–UV), infrared sources, vertical cavity surface-emitting lasers, wide-band gap materials (including blue-green lasers and LEDs), narrow-band gap materials and structures, low-dimensional systems in semiconductors, strained quantum wells and superlattices, ultrahigh-speed optoelectronics, and novel materials and devices.

The main objective of this book series is to provide readers with a basic understanding of new developments in recent research on optoelectronic properties of semiconductor quantum wells and superlattices. The volumes in this series are written for advanced graduate students majoring in solid state physics, electrical engineering, and materials science and engineering, as well as researchers involved in the field of semiconductor materials, growth, processing, and devices.

FOREWORD

Defects in Optoelectronic Materials contains a valuable collection of reports that are an excellent demonstration of the value of basic research in the semiconductor industry. The coverage ranges from fundamental physics to its applications in devices and processing. Each chapter gives vivid examples of how the understanding of materials behavior has enabled the introduction of new materials and the surmounting of perceived processing limits. Defects in solids are an ideal subject to demonstrate this value chain. The defect, though determinant in the function of an electronic material, is not an independent species, but is dependent in its influence on the response of the host to its existence. New phenomena are encountered for the first time; perceived as barriers to implementation of a technology; understood; and later employed in commercial success.

Section one describes the behavior of point defects in the bulk and near surfaces. Walukiewicz develops the concept of the chemical potential of charge carriers in the free energy of the electronic materials system. In compound semiconductors this chemical potential can be the driving force for bond reconstruction and sublattice site exchange to give rise to amphoteric doping behavior. This understanding has been key to the current utility of wide band gap semiconductors and the process challenge of hydrogen passivation of dopants. Wada provides an approach to dealing with the surface as a defect and its interaction with point defects in the bulk. Since processing is usually initiated or terminated at a surface, the slow diffusing species can accummulate in near-surface regions and homogeneously precipitate in the form of dislocation loops, voids and gaseous bubbles.

Section three addresses plasma process-induced damage. Hu and Chen give a thorough review of both the chemical and defect contributions to the etch process. They give insight into defect engineering for dry etch applications with discussions of defect generation, in-diffusion and interaction with microstructure. Pearton and Shul follow with excellent case studies of engineering applications of dry etch processing to wide band gap semiconductors. Pang presents the key issues in dry etch process design and the related approaches to defect control.

Section six addresses electrical contact formation. Okumura presents the fundamentals of metal semiconductor reactions for Schottky barrier and ohmic contact formation. The roles of interfacial phases, defects and material work functions are related to contact performance. In particular, the real-world issue of interface inhomogeneity is considered. Auret gives a thorough review of charged particle displacement damage in III-V compound semiconductors. The electrical properties of defects are covered in detail including carrier compensation in GaAs and the defect state parameters of point defects.

Section eight addresses the role of defects in device behavior. Shinozuka provides a fundamental approach to the metastability of defect configuration within a host in terms of bond reconstruction, thermodynamic equilibria and kinetic effects for both open and closed systems. This basis is then cast in the context of device operation and reliability. Wada and Fushimi describe the tendency of injection mode, compound semiconductor devices to degrade under operation. Heterostructure bipolar transistors and double heterostructure lasers are considered and defect engineering approaches such as dopant selection are described.

This book is a valuable resource for both academic and commercial environments. The reader has here a well-constructed reference for fundamentals and a handbook for applications.

L. C. Kimerling
Massachusetts Institute of Technology

PREFACE

This book will be of interest to device process engineers, defect physicists, and students. To device process engineers, this book explains possible origins of problems arising during materials processing based on current defect physics and provides fundamental ideas to control them. To defect physicists, this book highlights the problems concurrently challenging device process engineers. To students, this book introduce defect issues in current industries related to materials science and engineering.

The scope of this book ranges from anomalous phenomena in III-V and II-VI compound semiconductor growth and processing to device degradation. They are described by the defects involved and by defect behavior under electronic excitation of materials. Defects in device processing are described. Device processing covered consists of crystal growth, plasma processing and ion-implantation. Fundamentals of defect incorporated during crystal growth are described, plasma-process induced defects are explained and characterized, and defects introduced by radiation damage of electron beams and their characterization methods are shown. Materials analyzed are Si and III-V compound semiconductors including group III nitrides. Defects generated during device operation are described. Device degradation is a serious problem in future integrated circuits development. Current understanding of instability and defect reaction under electronic excitation (i.e., device operation) is presented and device degradation of GaAs/AlGaAs heterojunction bipolar transistors is described. With all these chapters, this book provides a perspective look at a range of defect issues in materials and devices.

Finally, we thank all of the authors who have put together the well-organized chapters for this book. We are also indebted to the reviewers for their helpful comments and suggestions. The editorial help from Concetta Seminara-Kennedy at Gordon and Breach Science Publishers is also greatly appreciated.

Kazumi Wada, Cambridge, Massachusetts

Stella W. Pang, Ann Arbor, Michigan
February 1999

CHAPTER 1

Saturation of Free Carrier Concentration in Semiconductors

W. WALUKIEWICZ

Materials Sciences Division, Lawrence Berkeley National Laboratory
1 Cyclotron Rd. Berkeley, CA 94720

1. INTRODUCTION

Ability to control the type and magnitude of conductivity is the most important and unique feature that distinguishes semiconductors from all the other solid state materials. Practically all applications of semiconductor materials rely on the availability of reproducible and well controlled doping. In principle, doping can be achieved by introduction of any point or extended defects that can contribute free electrons to the conduction band or free holes to the valence band. However, with few exceptions, in all practical cases doping is realized only by the intentional incorporation of foreign atoms that substitute host crystal lattice sites. It has been realized very early that many of the large variety of semiconductor materials are difficult to dope. The problem has been especially severe in wide-gap semiconductors where in many instances n- or p-type doping cannot be achieved at all, significantly limiting the range of applications of these materials [1–5].

Over the years several methods have been developed to dope semiconductor materials. Dopants can be introduced during the growth. They can also be incorporated during post-growth processing via ion implantation, diffusion and neutron transmutation doping. There are several reasons that can prevent a semiconductor material from being efficiently doped. In some instances foreign atoms cannot be incorporated in large enough concentrations into the semiconductor crystal. In other cases dopants incorporated into the crystal lattice tend to form electrically inactive defects by substituting the wrong sites or by forming precipitates. Finally, even if dopants are incorporated on the proper sites their electrical activity can be reduced or completely eliminated by compensation and/or passivation by other dopants or native defects.

Although the studies of doping issues have a long history, the problems of the limitations in electrical activity of dopants have become even more important in new emerging device technologies that put stringent demands on a nanoscale control of electronic and structural properties of semiconductor materials. Such devices require preparation of small size structures with very high doping levels and abrupt doping and composition profiles. The limits of the maximum doping levels and the question of the stability of dopant and compositional profiles are becoming central issues for the new device technologies. The doping issues in epitaxially grown III–V compound semiconductors have been reviewed in a recently published book [6].

The past few years have witnessed spectacular progress in development of a new generation of short wavelength optoelectronic devices based on group III nitrides [7–10] and wide gap II–VI semiconductors [11–14]. In both cases the progress was made possible by discovering more efficient ways to activate acceptor impurities in these material systems. Despite this progress the high resistance of p type layers is still a major hurdle in the development of the devices requiring high current injection levels. There have been numerous attempts to understand the maximum doping limits in semiconductors. Most of these were aimed at explaining limitations imposed on a specific dopant in a specific semiconductor. Thus, it has been argued that in the case of amphoteric impurities in III–V compounds, doping is limited by the impurities occupying both acceptor and donor sites. Redistribution of impurities can also lead to limitations of the maximum doping level in the materials with impurity diffusion strongly depending on the Fermi energy [6]. Since in most instances diffusion occurs only in the presence of defects this mechanism is closely related to the more general concept of compensation by lattice defects. It has been proposed recently that in semiconductors with the surface Fermi energy pinned in the bandgap, the electrostatic repulsion between charged impurities can lead to limitations on the maximum concentrations of dopants [15].

Formation of new stable solid phases involving dopant atoms can be a severe limitation in achieving high doping levels. This limitation depends on the chemical identity of the dopants and the host lattice elements and may be critical in the cases where only a limited number of potential dopants is available [16, 17].

Passivation of donor and acceptor impurities by highly mobile impurities is another major mechanism limiting the electrical activity of dopants. Hydrogen, Lithium Fluorine and Copper are known to passivate intentionally introduced dopants in semiconductors. Hydrogen has been an especially extensively studied impurity as it is a common contaminant in all the growth techniques involving metalorganic precursors [18, 19]. In some cases hydrogen can be removed during a post-growth annealing. Magnesium doped p-type GaN is frequently obtained by thermal annealing of MOCVD grown, hydrogen passivated films [20, 21]. However in other instances, as the case of N doped ZnSe, hydrogen is too tightly bound to the N acceptors and cannot be removed by a thermal annealing cycle [22]. This eliminates metalorganic precursor based epitaxial techniques as methods to grow a p-type ZnSe [23].

In this chapter, the saturation of the doping levels in semiconductors will be discussed in terms of the amphoteric defect model (ADM). In recent years, the model has been successfully applied to numerous doping related phenomena in semiconductors. It has been used to explain doping induced suppression of dislocation formation [24] as well as impurity segregation [25, 26] and interdiffusion [27] in semiconductor superlattices. We will show that ADM provides a simple phenomenological rule capable of predicting trends in the doping behavior of a large variety of semiconductor systems.

2. THE AMPHOTERIC DEFECT MODEL

Preparation of semiconductor materials unavoidably leads to the incorporation of crystal lattice defects. The defects may take the form of isolated point imperfections or they may interact with each other forming extended defects. It is obvious that a proper treatment of defect formation and defect interactions is an immensely complex problem. Incorporation of dopants will complicate it even further, making any effort to consider it in its full generality extremely difficult. This is why in most cases, the attempts to describe incorporation and activation of dopants have only been qualitative. To obtain a more quantitative description, a large number of approximations and simplifications need to be introduced, making such determinations much less reliable. Here we introduce a simple phenomenological description of defects and their interactions with dopants.

All the point defects and dopants can be divided into two classes; delocalized shallow dopants and highly localized defects and dopants.

Shallow hydrogenic donors and acceptors belong to the first class. Their wave functions are delocalized and formed mostly out of the states close to the conduction band minimum or the valence band maximum. As a result the energy levels of these dopants are intimately associated with the respective band edges, conduction band for donors and the valence band for acceptors. In general the energy levels will follow the respective band edges when the locations of the edges change due to external perturbation such as hydrostatic pressure or changing alloy composition.

In contrast, wave functions of highly localized defects or dopants cannot be associated with any specific extremum but are rather formed from all the extended states in the Brillouin zone with the largest contribution coming from the regions of large density of states in the conduction and the valence band. Consequently the energy levels of such defects or dopants are entirely insensitive to the location of the low density of states at the conduction and valence band edges. It has been shown that transition metal impurities with their highly localized d shells belong to this class of dopants [28, 29]. The insensitivity of the transition metal energy levels to the position of local band extrema has led to the concept of using these levels as energy references to determine the band offsets in III–V and II–VI compounds [29] and the band edge deformation potentials in GaAs and InP [30].

A good illustration of the difference in the behavior of delocalized and highly localized centers is represented by donor dopants in compound semiconductors. When these dopants substitute proper sites they form shallow hydrogenic levels. However they can also exist in a highly localized DX center configuration [31–33] in which the center is occupied by two electrons [34–36]. Formation of DX centers is associated with a large lattice relaxation that involves displacements of the impurity and/or the nearest neighbor host lattice atoms [34]. The difference in the dependence of these two types of states on external hydrostatic pressure and alloy composition is used to change the ground state of such a system by changing the relative location of the DX center and the shallow donor energy levels [35, 37].

Properties of intentionally introduced dopants have been studied extensively and are quite well understood. Much less is known about native point defects. The main reason for this is the difficulty of identifying specific defects. In fact, with few exceptions, the energy levels of most native point defects are not known. Measurements of the pressure dependence of the energy levels of anion antisites defects in GaAs [38] and InP indicate [39] that they are highly localized and have pressure dependencies similar to the transition metal impurities.

Even more compelling evidence for the localized nature of the native defects has been provided by studies of semiconductor materials heavily damaged with high energy particles [40–46]. It has been found that for sufficiently high damage density i.e. when the properties of the material are

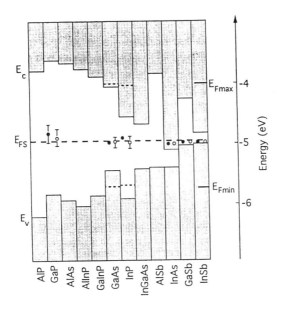

Fig. 1. Band offsets and the Fermi level stabilization energy (E_{FS}) in III–V compounds. The energy is measured relative to the vacuum level. The filled circles represent stabilized Fermi energies in heavily damaged, high energy particle irradiated materials. The open circles correspond to the location of the Fermi energy on pinned semiconductor surfaces and at metal/semiconductor interfaces. The dashed lines show the location of the Fermi energy for a maximum equilibrium n- and p-type doping in GaAs and InP.

fully controlled by native defects, the Fermi energy stabilizes at certain level and becomes insensitive to further damage. The location of this Fermi level stabilization energy, E_{FS}, does not depend on the type or the doping level of the original material and therefore is considered to be an intrinsic property of a given material. As is shown in Figure 1 the Fermi level stabilization energies for different III–V semiconductors line up across semiconductor interfaces and are located approximately at a constant energy of about 4.9 eV below the vacuum level [47]. This is a clear indication that the native defect states determining the electrical characteristics of heavily damaged materials are of highly localized nature. As is seen in Figure 1 the location of the stabilized Fermi energy in heavily damaged III–V semiconductors is in a good agreement with the Fermi level pinning position observed at the metal/semiconductor interfaces [48]. This finding strongly supports the assertion that the same defects are responsible for the stabilization of the Fermi energy in both cases.

The mechanism explaining the defect induced stabilization of the Fermi energy is based on the concept of amphoteric native defects. The stabilization of the Fermi energy can be understood if we assume that the type of defects formed during high energy particle irradiation or metal deposition on the semiconductor surface depends on the location of the Fermi energy with respect to E_{FS}. For Fermi energy $E_F > E_{FS}$ ($E_F < E_{FS}$) acceptor-like (donor-like) defects are predominantly formed resulting in a shift of the Fermi energy towards E_{FS}. Consequently, the condition $E_F = E_{FS}$ is defined as the situation where the donor and acceptor like defects are incorporated at such rates that they perfectly compensate each other leaving the Fermi energy unchanged.

Such an amphoteric behavior of simple native defects is supported by theoretical calculations that show that depending on the location of the Fermi energy vacancy like defects can acquire either negative or positive charge acting as acceptors or donors, respectively. In the case of GaAs it was shown that both gallium and arsenic vacancies can undergo the amphoteric transformation [49]. For example, as is shown in Figure 2 V_{Ga} is a triple acceptor for $E_F > E_v + 0.6$ eV. However for lower Fermi energies this configuration is unstable and the vacancy undergoes a relaxation in which one of the first neighbor As atoms moves towards the vacant Ga site. The transformation is schematically represented by the reaction,

$$V_{Ga} \Leftrightarrow (V_{As} + As_{Ga}) \tag{1}$$

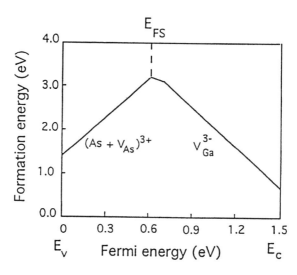

Fig. 2. Formation energy of a gallium vacancy and the related donor defect as function of the Fermi energy in the GaAs band gap [49, 50].

In arsenic rich GaAs the calculated formation energy of V_{Ga} is below 1 eV for E_F at the conduction band edge [50]

A similar amphoteric behavior is also predicted for V_{As} where the transformation is given by the reaction [49],

$$V_{As} \Leftrightarrow (Ga_{As} + V_{Ga}) \qquad (2)$$

In this case the V_{As} donor that is stable in GaAs with E_F larger than about $E_v + 0.8$ eV and transforms to an acceptor-like ($V_{Ga} + Ga_{As}$) configuration for $E_F < E_v + 0.8$ eV [27]. It is worth noting that these theoretical values of E_{FS} are very close to experimentally determined ranging from $E_v + 0.5$ eV to $E_v + 0.7$ eV [41].

Most recent theoretical calculations have shown that the amphoteric behavior of native defects is a feature common to many different compound semiconductor systems, including II–VI and III–V semiconductors and the group III-Nitrides [51]. The calculations have confirmed that the reaction (1) is responsible for the amphoteric behavior of V_{Ga}. However it has been found that in the case of V_{As} a transformation from a donor like V_{As} to an acceptor like configuration occurs through a dimerization of the three-fold coordinated Ga atoms surrounding the As vacancy rather than reaction (2). Although, a different type of a structural relaxation is predicted in this case it does not change the overall conclusion that both cation and anion site vacancies are amphoteric defects and, when introduced in large concentrations, will lead to a stabilization of the Fermi energy.

Since E_{FS} is associated with highly localized defects, its location has no correlation with the conduction or valence band edges. Thus as is seen in Figure 1, E_{FS} can be located anywhere in the gap or even in the conduction band. In the case of GaAs, E_{FS} is located close to the midgap energy. Therefore high energy particle damage always leads to a high resistivity GaAs [41]. On the other hand, in the unusual case of InAs, E_{FS} is located deep in the conduction band. Consequently any high energy particle damage leads to high n-type conductivity in this material [44]. We will show in this chapter that the location of E_{FS} relative to the band edges is the single most important factor affecting defect-related phenomena in semiconductors.

3. MAXIMUM DOPING LIMITS IN GaAs

Numerous electronic and optoelectronic applications have made GaAs one of the most extensively studied compound semiconductor. It has been realized very early that it is rather easy to dope GaAs with acceptors. Very high concentrations, in excess of 10^{20} cm^{-3}, can be readily obtained by doping with group II atoms [52]. Even higher concentrations close to 10^{21} cm^{-3}

were obtained by doping with carbon [53]. On the other hand n-type doping is much more difficult to achieve. The doping becomes less efficient for donor concentrations larger than about 3×10^{18} cm^{-3} and the maximum electron concentration saturates at a level slightly above 10^{19} cm^{-3} [54–57]. The maximum concentration does not depend on the dopant species or the method by which the dopants are introduced into the crystal therefore it appears to be an intrinsic property of the material rather than a feature attributable to the chemical or electronic characteristics of the dopants.

Over the years numerous attempts were undertaken to understand the nature of this limitation. For example it has been proposed that at high concentrations Se donors form electrically inactive complexes [54]. In the case of group IV dopants an obvious explanation was based on the amphoteric nature of these impurities. It was argued that at high doping levels the dopants begin to occupy both sites forming donors and acceptors that compensate each other [57]. It would be rather surprising if these dopant specific explanations could account for the universal nature of the electron concentration limits.

An obvious absolute limit for free electron concentration is set by the location of the DX centers associated with a specific donor impurity. It is now generally accepted that in its ground state DX center is occupied by two electrons therefore it behaves as a negatively charged acceptor that compensates shallow hydrogenic donors. Consequently the Fermi energy cannot lie higher than the DX center level [58]. This intrinsic instability of the donor sites to the transformation to the DX like configuration is again donor specific and would therefore lead to the free carrier concentration limit that is dependent on the donor species.

A more general explanation of the doping limits in n-GaAs was based on the concept of Fermi level induced formation of compensating native defects. The most obvious choice for the compensating acceptor defect was V_{Ga}. A photoluminescence line at about 1.2 eV whose intensity was found to increase with the doping levels was attributed to a donor-V_{Ga} complex [59]. The complex appears to be responsible for the additional local modes found in GaAs heavily doped with Si [60].

In considering compensation of donors by native defects one needs to know the temperature at which the defects are frozen. It can be argued that this temperature should be somewhat lower than that required to remove the radiation damage in and activate dopants in ion implanted material. In GaAs this temperature appears to be in the range 900 K to 1100 K [61]. Under thermal equilibrium conditions the concentration of gallium vacancies is given by [50],

$$[V_{Ga}] = N_{site}C_{ent} \exp[-E_f/kT] \tag{3}$$

where N_{site} is the concentration of Ga site in GaAs crystal, C_{ent} is the temperature independent term associated with entropy contributions and E_f

is the formation energy of V_{Ga}. Theoretical calculations predict that in GaAs with the Fermi energy above E_{FS}, V_{Ga} is a triply negatively charged center [49, 50]. Therefore its formation energy depends on the Fermi energy and can be written as,

$$E_f = E_{f0} - 3(E_F - E_{FS}) \tag{4}$$

The Fermi energy is determined by the concentration of free electrons that is equal to the net-concentration of donors,

$$n = N_c F_{1/2}[(E_F - E_c)/kT] = N_d - 3[V_{Ga}] \tag{5}$$

where N_c is the conduction band density of states, $F_{1/2}(x)$ is the Fermi-Dirac integral and N_d is the concentration of intentionally introduced donors.

It is commonly assumed that electrons in the conduction band can be described by nondegenerate electron gas statistics for which the Fermi-Dirac integral has the form $F_{1/2}(x) = (\pi)^{1/2}[\exp(x)]/2$. It has been found however, that this assumption is not always accurate and can lead to significant errors, especially for high electron concentrations in semiconductors with small effective masses of the conduction band electrons [62]. Substituting (3) and (4) into Eq. (5) one obtains an equation for E_F or an equivalent equation for free electron concentration in the conduction band. To solve this equation one needs to know the location of E_{FS} with respect to E_c at elevated temperatures. In GaAs at 900 K E_c is located at about 0.46 eV above the intrinsic level [63]. At room temperature $E_{FS} = E_c - 0.8$ eV. Using arguments of ref. 24 about the temperature shift of E_c we find $E_{FS} = E_c - 0.68$ eV at 900 K.

The results of the calculations are shown in Figure 3. They were obtained assuming $C_{ent} = 1$. A good fit to experimental data was obtained adopting the value $E_f = 2.4$ eV for the Fermi energy located at the intrinsic level, $E_F = E_i$. As is seen in Figure 3 the results of the calculations quite well reflect the overall dependence of the electron concentration on the doping level, N_d. At low N_d the Fermi energy is located well below the conduction band, E_f is large and the concentration of V_{Ga} small. Under these conditions all donors are electrically active and $n = N_d$. With increasing doping the Fermi energy shifts upwards towards the conduction band resulting in a lower E_f and higher $[V_{Ga}]$. Gallium vacancies compensate the donors and n becomes a sublinear function of N_d. In fact, it can be shown that in a limited concentration range, n is proportional to $(N_d)^{1/3}$. The 1/3 power reflects the fact that V_{Ga} is a triply charged acceptor. Such dependence is expected when electrons can still be described by nondegenerate statistics. At even higher doping levels the Fermi energy enters the conduction band and becomes strongly dependent on electron concentration. This leads to a rapid reduction of E_f, an increase of V_{Ga} and as a consequence saturation of n.

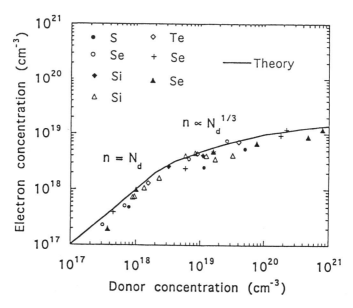

Fig. 3. Electron concentration as a function of donor doping in GaAs. The data points represent experimental results for several different donor species (● – S [55], ○ – Se [55], ◆ – Si [54], △ – Si [57], ◇ – Te [55], + – Se [54] and ▲ – Se [56]).

It is important to note that the value of $E_f = 2.4$ eV appears to be consistent with other determinations of the formation energy of V_{Ga} in intrinsic GaAs. Detailed studies of Ga self-diffusion in undoped GaAs provided the value of the diffusion activation energy, that is a sum of the formation and migration energies of V_{Ga}, $E_{f+m} = E_f + E_m = 3.7$ eV [64]. The entropy of $S = 3.5$ k has also been determined in this study. In addition extensive investigations of V_{Ga} facilitated diffusion of As_{Ga} defects in non-stoichiometric, low temperature grown GaAs have provided the values of V_{Ga} migration energies ranging from 1.4 to 1.7 eV [65, 66]. This leads to E_f ranging from 2.0 to 2.3 eV that is somewhat lower than the value of $E_f = 2.4$ eV needed to explain the free electron concentration limits. The difference can easily be accounted for by the entropy contribution that has been neglected in the present considerations. At 900 K the entropy of 3.5 k leads to an effective formation energy difference of about 0.27 eV.

The success in explaining the doping limitations in n-type GaAs raises the question whether a similar mechanism is responsible for doping limits in p-type GaAs. As is shown in Figure 1 V_{Ga} is an unstable defect for $E_F < E_{FS}$. It relaxes to the $V_{As} + As_{Ga}$ donor like configuration with the formation energy $E_f = E_{f0} + 3(E_F - E_{FS})$. With $E_{f0} = E_f(E_{FS}) = 3.1$ eV one finds that at a temperature $T = 900$ K for E_F located at the valence band edge E_v, the formation energy, $E_f = 1.8$ eV. This formation energy gives a very low value of less than 10^{13} cm^{-3} for the concentration of the defect donors. Since for $E_F = E_v$ the concentration of free holes is equal to about 4×10^{19} cm^{-3}, it is evident that $(V_{As} + As_{Ga})$ donors are not expected to play any role in compensation of intentionally introduced acceptors. This is consistent with experiments that indicate that rather high hole concentrations can be relatively easily achieved in p-type GaAs.

However, it has also been shown that in GaAs doped with column II acceptors the hole concentration saturates at the doping levels slightly above 10^{20} cm^{-3} [67, 68]. This saturation has been attributed to the fact that column II atoms can act either as acceptors, when they substitute Ga atom sites or as donors when they occupy interstitial sites. The concentration ratio of substitutional to interstitial atoms depends on the location of the Fermi energy. At low concentrations all dopant atoms substitute Ga sites acting as acceptors. With increasing doping level the Fermi energy shifts down towards the valence band and more and more dopants occupy interstitial sites acting as donors. As has been shown before [25, 62] this mechanism leads to a saturation of the position of the Fermi energy level and thus also of the concentration of free holes in the valence band. In the case of GaAs, with the maximum hole concentration of 10^{20} cm^{-3} the Fermi energy saturates at about $E_v - 0.2$ eV or at $E_{FS} - 0.67$ eV when measured with respect to E_{FS} as a common energy reference. Vacancy-like defects do not appear to play a significant role in limiting the hole concentration in GaAs doped with group II acceptors. It has been demonstrated however, that under some conditions arsenic vacancies play a critical role in reducing of the electrical activity of C acceptors in GaAs/AlGaAs heterojunction bipolar transistors [69].

4.   OTHER GROUP III–V SEMICONDUCTORS

The above results on the saturation of electron and hole concentrations in GaAs can be interpreted in terms of limitations on the maximum and minimum Fermi energy that can be achieved by doping. In GaAs at room temperature the allowed Fermi energies range approximately from $E_{FS} - 0.8$ eV to $E_{FS} + 1$ eV. It has been shown that similar limits apply as well to other III–V compounds providing a simple and convenient way to estimate maximum doping limits not only in other binary compounds but also in their alloys.

The limiting Fermi energies $E_{F\,min}$ and $E_{F\,max}$ for various III–V compounds are shown in Figure 1. It becomes readily evident from this figure which one of the semiconductors will be easily doped p- or n-type. Thus in InP $E_{F\,max}$ is located deep in the conduction band whereas $E_{F\,min}$ lies at about

0.16 eV above the valence band edge. This suggests that, in contrast to GaAs, it should be much easier to dope InP n-type rather than p-type. This indeed is observed experimentally. Electron concentrations approaching 10^{20} cm^{-3} were reported in Si [70] and Sn [70, 71] doped InP. On the other hand the hole concentration is limited to mid 10^{18} cm^{-3} in this material [68, 72]. It has been shown that a similarly strong correlation exists between the location of E_{FS} and the maximum doping level is found for all III–V semiconductors [73].

An important feature of the present approach is that it provides an easy way to determine the doping limits in alloys. For example a study of Zn doping of In$_{0.5}$(Ga$_{1-x}$Al$_x$)$_{0.5}$P lattice matched to GaAs has shown that the maximum hole concentration decreases with increasing x [74]. For x ranging from 0 to 0.7 the maximum hole concentration decreases from about 5×10^{17} to 1×10^{17} cm^{-3}. Figure 4 shows the location of the Fermi energy calculated for the maximum hole concentration as a function of the Al content for the growth temperature of about 1000 K [75]. As expected the maximum hole concentration corresponds to constant E_F relative to E_{FS}. Similar behavior of decreasing maximum hole concentration with decreasing x has been also observed for Mg doping of these alloys.

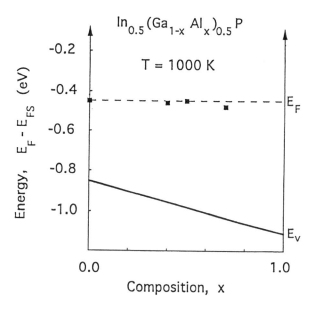

Fig. 4. Composition dependent location of the Fermi energy relative to E_{FS} for the maximum free hole concentration in Zn doped In$_{0.5}$(Ga$_{1-x}$Al$_x$)$_{0.5}$P [73, 74]. Also shown is the position of the valence band edge, E_v.

Since the incorporation of dopants at the levels close to the maximum doping limits leads to the formation of defects it is expected that it may also lead to an enhanced diffusion of the dopants and the host lattice atoms. This issue is especially important for superlattices were the structural integrity of the superlattice and/or distribution of dopants can be affected by such diffusion. As has been shown above, when the concentration of column II acceptors reaches the saturation level an increasing number of them transfer to energetically more favorable interstitial sites. It has been shown that in the uniform case of a single material the interstitials lower their energy by forming precipitates or reacting with host lattice atoms to form new compounds [68].

In the case of heterostructures consisting of layers with different free hole saturation levels one can expect that the interstitials can also lower their energy diffusing into the layers with higher free hole saturation levels. This is exactly what is observed experimentally. A pronounced redistribution of column II acceptors has been found in lattice matched InGaAs/InP/InGaAsP/InP [76], InP/InGaAs [77] and AlGaAs/GaAs [78] heterostructures. The acceptors segregate to the layers with larger maximum hole concentration limits. As an example, Figure 5 shows the profile of free holes in the InGaAs/InP/InGaAsP/InP structure after Zn diffusion [78].

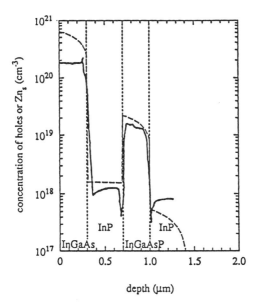

Fig. 5. Hole concentration profile (solid line) in Zn diffused InGaAs/InP/InGaAsP heterostructure lattice matched to InP [78]. The dashed line represents a calculated Zn concentration profile for the substitutional-interstitial Zn diffusion mechanism [79].

Calculations show [79] that the dramatic segregation of Zn atoms can be explained considering the difference in the location of the Fermi level stabilization energy or, equivalently, the valence band offsets in different layers. The well-known band offsets of 0.36 eV for InGaAs/InP and 0.19 eV for InGaAsP/InP heterointerfaces have been used in these calculations [80]. The deviation between the calculated Zn concentration and the free hole concentration in the top InGaAs layer can be attributed to an incomplete activation of Zn acceptors for the concentrations exceeding the free hole saturation level of 2×10^{20} cm^{-3}.

5. GROUP III-NITRIDES

Recent years have witnessed an unprecedented growth of interest in the Group III-Nitrides as a new distinct class of III–V compounds with strongly ionic bonds, smaller lattice constants and large band gaps. These materials form the foundation of a new technology for short wavelength optoelectronic [9] and high power, high-speed electronic devices [81, 82]. Group III-Nitrides have by now been studied for many years. As with many other wide gap materials the main impediment in practical applications was their propensity to exhibit only one type of conductivity. Typically, as grown GaN or InN are found to be highly n-type conducting and for a long time it was impossible to dope them with acceptors. In the case of AlN neither type of doping was possible. These trends in the doping behavior of the nitrides can be again understood in terms of the amphoteric defect model. Figure 6 shows the band offsets of group III-Nitrides. The band edges were located relative to E_{FS} using the fact that there is a negligibly small conduction band offset between GaAs and GaN. It is seen in Figure 6 that E_{FS} is located in the upper part of the bandgap in GaN and slightly above the conduction band edge in InN. Therefore it is expected that it should be relatively easy to dope those materials with donors and much more difficult to dope with acceptors. Free electron concentrations exceeding 10^{20} cm^{-3} [83, 84] and 10^{21} cm^{-3} [85] have been reported in unintentionally doped GaN and InN, respectively. These concentrations correspond to approximately the same Fermi energy at $E_{FS} + 1.3$ eV in both materials. The large difference in the maximum electron concentration is consistent with the large conduction band offset between GaN and InN. Also the very large conduction band offsets of 2 eV between GaN and AlN explains why the n-type doping efficiency decreases with increasing Al content [83].

A saturation of free electron concentration has been recently observed in GaN intentionally doped with Se [86]. At low doping levels all Se atoms were electrically active donors. With increasing doping level the

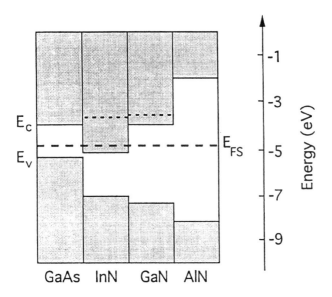

Fig. 6. Band offsets for group III-Nitrides. The dashed lines represent Fermi energy for the maximum achievable free electron concentration in GaN and InN.

electron concentration tends to saturate, showing the characteristic $(N_d)^{1/3}$ dependence on the donor concentration. Similarly, as in GaAs, the saturation of the electron concentration has been explained by incorporation of triply ionized V_{Ga} acceptors. This result confirms the universal nature of the group III vacancies as compensating centers in n-type III–V compounds.

6. GROUP II–VI SEMICONDUCTORS

Wide gap group II–VI semiconductors are the group of materials that exhibit the most severe limitations on the doping efficiency. It is this family of materials for which the problem of doping has been recognized first [3, 4]. Early studies have shown that all wide gap II–VI compounds show a propensity for either n- or p-type conductivity. As grown ZnO, ZnS, HgSe, CdSe and CdS show n-type conductivity and p-type doping is very difficult if not impossible to achieve in these compounds. On the other hand ZnTe typically exhibits p-type conductivity only. It was recognized at that time that the doping limits could originate from compensating native defects that are formed when the Fermi energy shifts towards the band edges [2–5]. It was not clear, however, how within this picture one could explain differences between apparently similar materials exhibiting completely different doping behavior.

Recent advances in the utilization of II–VI compounds for short wavelength light emitting devices have brought the issue of the doping limitations to the forefront and led to intensive efforts aimed at understanding the mechanisms responsible for the limited dopability of these materials [74, 87–91]. Because of its importance for the blue-green light emitters, ZnSe has been considered a prototypical material to study the doping limitations. It can be relatively easily doped n-type but p-type doping is very difficult to accomplish and only recently doping with reactive nitrogen was successful in achieving p-type conductivity. However, even in this case the free hole concentration is limited to 10^{18} cm^{-3} [87, 89].

One explanation of the effect is based on the argument that it is energetically favorable for the dopant species to form new compounds with the host crystal atoms rather than substitute lattice sites and act as donors or acceptors [16]. In the case of N doped ZnSe the calculations suggested that Zn_3N_2 should be easily formed preventing N from acting as a substitutional acceptor [16]. Also these first principle calculations seemed to indicate that the formation energies of native defects are too large and the concentrations are too small to explain low electrical activity of N atoms in ZnSe with compensation by native defects [16]. Later, improved calculations have shown that incorporation of lattice relaxation lowers the formation energy of native defects so that they are likely to play a role in compensation of N acceptors in ZnSe [91]. The concept that the doping with N is limited by formation of new phases has been recently used to explain differences in the doping behavior of a number of II–VI tellurides [92].

Native defects were frequently invoked as the centers compensating electrical activity of intentionally introduced dopants. It was very difficult, however to identify the defects responsible for the compensation or to account for the trends in the doping behavior observed in different II–VI compounds and their alloys. There is evidence that, in the specific case of ZnSe:N, V_{Se} or V_{Se}-N defect complexes are responsible for the compensation of p-type conduction [88, 93]. This finding however, does not provide any guidance on how to identify the compensating defects in other II–VI compounds.

It has been shown again that the trends in the doping behavior of different group II–VI compounds can be understood within the amphoteric defect model without any need to know the specific identity of the compensating defects [94, 95]. The conduction and valence bands for various II–VI semiconductors are shown in Figure 7 [94]. The Fermi level stabilization energy is again located at about 4.9 eV below the vacuum level. As in the case of III–V compounds, it is assumed that there is a band of allowed Fermi energies $\Delta E_F = E_{F\,max} - E_{F\,min}$ determining the maximum electron and hole concentration that can be achieved in a given material.

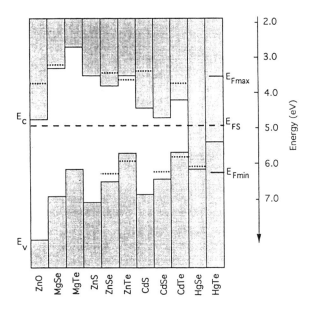

Fig. 7. Band offsets and the Fermi level stabilization energy, E_{FS}, in II–VI compounds. The dashed lines represent positions of the Fermi energy corresponding to the highest hole and electron concentrations reported for the given material.

In the case of ZnSe the highest electron concentration of about 2×10^{19} cm^{-3} [96] defines $E_{F\,max} = E_{FS} + 1.3$ eV as the upper limit of allowed Fermi energies. The lower limit at $E_{FS} - 1.3$ eV corresponds to the maximum free hole concentration of 10^{18} cm^{-3} [87, 89]. Transferring the same limits to other compounds we find that in ZnTe $E_{F\,min}$ is located deep in the valence band confirming the experimental observation that it is very easy to dope this material with acceptors. Indeed free hole concentrations as high as 10^{20} cm^{-3} were reported in ZnTe [97]. On the other hand, since $E_{F\,max}$ is located below the conduction band edge it is expected that n-type conductivity will be much more difficult to achieve. In fact, it was only recently that n-type conduction with a low electron concentration of 4×10^{17} cm^{-3} was reported in ZnTe [98].

As can be seen in Figure 7 for both CdSe and CdS, the upper Fermi energy limit is located in the conduction band in agreement with the observation that both materials are very good n-type conductors. As expected p-type conductivity is much more difficult to realize in this materials. The highest hole concentration of only 10^{17} cm^{-3} was reported in CdSe [99]. It is not surprising that in CdS with its very low position of the valence band no p-type doping was ever achieved.

ZnO represents a case of material with the band edges shifted to very low energies. The conduction band edge is located very close to E_{FS} at $E_{FS} + 0.2$ eV and the valence band edges lies at the very low energy of $E_{FS} - 3.1$ eV. Such an alignment strongly favors n-type conductivity. Existing experimental data indicate that undoped ZnO can exhibit free electron concentrations as large as 1.1×10^{21} cm^{-3} [100]. However the extremely low position of the valence band edge indicates that it will be very difficult, if not impossible, to achieve any p-type doping of this material.

Although the limitations in the doping levels are most apparent in wide gap materials where, either the conduction or the valence band edge can easily fall outside the band of the allowed Fermi energies, it may also be important in lower or even zero gap semiconductors. Mercury Selenide is an example of a zero-gap semiconductor with a distinct asymmetry in the doping behavior. As is seen in Figure 7 in HgSe the energy of the degenerate conduction and valence band edge is located at the very low energy of $E_{FS} - 1.2$ eV. This band alignment indicates that for any Fermi energy position incorporation of donors is energetically favored over incorporation of acceptor like defects. Indeed HgSe is always n-type with the lowest reported electron concentration of about 10^{16} cm^{-3} [101]. This electron concentration corresponds to a Fermi energy $E_{FS} - 1.2$ eV, very close to the lower boundary of the allowed Fermi energies. It also explains why the attempts to dope HgSe with acceptors have never been successful.

This doping behavior can be contrasted with that of another zero-gap semiconductor HgTe where, as is shown in Figure 7, the conduction/valence band edge is located at $E_{FS} - 0.5$ eV, i.e., well within the allowed Fermi energy band. This is consistent with the fact that HgTe can be relatively easily doped with both donors and acceptors.

One of the principal advantages of the amphoteric defect model is that it provides a simple rule to predict doping limits in semiconductor alloys. This is especially important for optoelectronic devices that increasingly require ternary or quaternary alloys for the independent control of the lattice parameters, band gaps and band edge offsets. Thus significant progress for group II–VI based devices has been achieved through the introduction of ZnMgSSe quaternary alloys [102]. A judicious choice of composition allowed the control of the band gap with a lattice parameter matched to a GaAs substrate. It has been found however that increasing the band gap by adding Mg and S to ZnSe leads to quite a dramatic, undesired reduction in the activation efficiency of N acceptors [89, 103]. The drop in the free hole concentration is understandable if one realizes that addition of both Mg and S results in a downward shift of the valence band edge. Figure 8 shows the concentration of free holes measured in ZnMgSSe lattice matched to GaAs. The solid line represents the predicted free hole limit for the valence band offset $\Delta E_v = 0.4 \Delta E_g$, where ΔE_g is the change in the band gap energy [94].

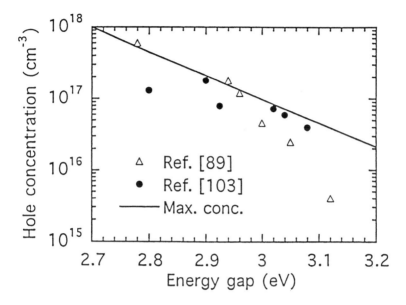

Fig. 8. Free hole concentration as a function of the energy gap, E_g, for ZnMgSSe alloys lattice matched to GaAs. The solid line represents calculated maximum free hole concentration. The data points show the experimental results of Refs. [89] and [103].

Recent studies have shown that the ADM can be also used to predict saturation of the electron concentration in different group II–VI based alloy systems [94, 96, 104]. It has been reported that the doping efficiency of Cl donors is decreasing with increasing Mg content in ZnMgSe [96]. The experimentally observed electron concentration as a function of the composition is shown in Figure 9 [94]. Good agreement between experiment and the calculated concentration limits is obtained assuming the MgSe/ZnSe conduction band offset lies in the range $0.5 < \Delta E_c < 0.7$ eV. Also the experimentally observed rapid decrease in the doping efficiency of Br donors with increasing Mg content in CdMgTe [106] is consistent with the measured [105] very large conduction band offset of 1.55 eV between CdTe and MgTe.

The doping limitations predicted by ADM mean that under thermodynamic equilibrium conditions the free carrier concentration cannot be higher than the limits set by compensating defects. This does not mean however that these limits can always be reached. For example in the case of ZnTe, where the ADM sets the free hole limit at about 4×10^{20} cm^{-3} [94], the highest measured hole concentration has been limited to a value slightly higher than 10^{20} cm^{-3}. It is important to realize that at these high concentrations of acceptors other mechanism, especially those related to limited

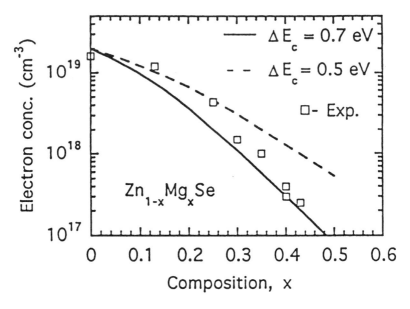

Fig. 9. Electron concentration as a function of the $Zn_{1-x}Mg_xSe$ alloy composition. Experimental points represent the data of Ref. [96]. The curves show calculated maximum electron concentrations for two different conduction band offsets between ZnSe and MgSe.

solubility, acceptor-acceptor interaction or formation of new solid phases could be more restrictive in limiting the incorporation of electrically active substitutional acceptors.

Such restrictions appear to be responsible for a relatively low, about 10^{17} cm^{-3}, concentration of holes found in N doped CdTe [92]. This concentration is much lower than the concentration limit predicted by the ADM [94]. It has been argued that the formation of new solid phase Cd_3N_2 might limit the solubility of substitutional atomic N in the lattice [92]. It should be emphasized however that much higher p-type doping has been attained using other group V elements in CdTe. A hole concentration of 5×10^{19} cm^{-3} obtained through P implantation is close to the maximum concentration in ZnTe and in a reasonable agreement with the limit predicted by ADM [107].

7. GROUP I–III–VI$_2$ TERNARIES

Recently a significant effort has been directed towards the synthesis and studies of group I–III–VI$_2$ ternaries. The effort has been prompted by the

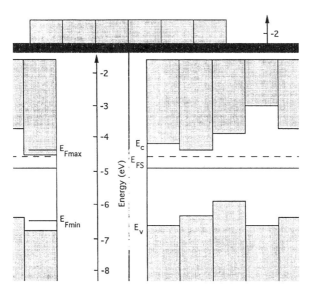

Fig. 10. Band offsets for group I–III–VI$_2$ semiconductors. $E_{F\,max}$ and $E_{F\,min}$ roughly correspond to the experimentally observed maximum and minimum position of the Fermi energy that can be achieved by doping [95].

discovery that some of these materials could be potential candidates for inexpensive solar cells [108, 109]. To fully realize this potential there is a need for a material with well-controlled p- and n-type doping. However, it was found that similarly to binary compounds, many of the ternaries exhibit a very distinct tendency for either n-type or p-type conductivity. An analysis of existing data strongly suggests that the amphoteric defect model applies also to this group of much more complex materials [95]. Band offsets for a number of I–III–VI$_2$ ternaries are shown in Figure 10. The position of the Fermi level stabilization energy for this material system can be determined from the band offset between any of the ternaries and a group II–VI semiconductor. Unfortunately there is still a controversy about the actual value of the conduction band offset even for the most extensively studied CdS/CuInSe$_2$ heterostructure [110, 111]. In Figure 10 the broken line shows E_{FS}, assuming the theoretical value of $\Delta E_c(\text{CdS/CuInSe}_2) = 0.3$ eV [110] while the solid line represents E_{FS} for the experimental result with $\Delta E_c(\text{CdSe/CuInSe}_2) = -0.08$ [111]. The approximate positions of $E_{F\,max} = E_{pin}(n)$ and $E_{F\,min} = E_{pin}(p)$ Fermi energy as determined from the experimental data, are also shown in Figure 10. It is found that for the small value of $\Delta E_c = -0.08$ eV E_{FS} is located closer to the middle of the band of allowed Fermi energies ΔE_F.

While the band gaps of the ternaries are in general smaller than the gaps of II–VI compounds, there exists a large variation of the band offsets implying a possible variation in the doping behavior. Materials like $CuInS_2$ or $CuInSe_2$ show both n- and p-type conductivity. This allows a determination of the width of allowed Fermi energy band ΔE_F in these materials [95]. As is shown in Figure 10, ΔE_F spans an energy of about 1.2 eV. Based on these limits one finds that $CuGaSe_2$, $CuAlSe_2$ and especially $CuInTe_2$ should be easily doped p-type whereas n-type conductivity could be difficult to achieve in these materials. This is especially true for $CuAlSe_2$ with its very high location of the conduction band edge.

8.   OTHER SEMICONDUCTORS

As has been shown above, the ADM works well in explaining the doping limitations in a number of compound semiconductor systems. The obvious question arises whether it can provide any guidance in evaluating doping limits in elemental semiconductors. Figure 11 shows the band offsets for group IV materials with energy gaps ranging from zero-gap in grey tin (α-Sn) to 5.5 eV in diamond. Since both Si and Ge have relatively small energy gaps it is not surprising that there are no serious limitations for the doping of these materials. An implantation study has shown that it is relatively easy to activate acceptors in Ge [112]. It has been found that activation of B acceptors does not require any thermal annealing. Such a behavior is understandable in view of the fact that in Ge E_{FS} is located very close to the valence band edge.

Obviously the most intriguing case is represented by diamond. Although there is some uncertainty regarding the actual position of the diamond conduction band it is now well established that it is located close to the vacuum level [113, 114]. Consequently, as is shown in Figure 11, the conduction band is separated from E_{FS} by almost 5 eV and the valence band is located rather close to E_{FS} at $E_{FS} - 0.6$ eV. This vast asymmetry in the location of the band edges with respect to E_{FS} has significant consequences for the behavior of dopants in diamond. It explains why, despite years of intense efforts, there is still no convincing evidence of n-type conductivity in diamond [115]. It further indicates that there is a fundamental reason why it will not be possible to have a thermally stable, well conducting n-type diamond. On the other hand the close location of the valence band to E_{FS} accounts very well for the relatively high p-type conductivity that can be achieved by doping diamond with boron. It has been reported that the concentrations of electrically active boron in excess of 10^{20} cm^{-3} are possible [116].

Interestingly enough the ADM doping rule applies also to more exotic materials with extremely wide-gaps. Although as is shown in Figure 11,

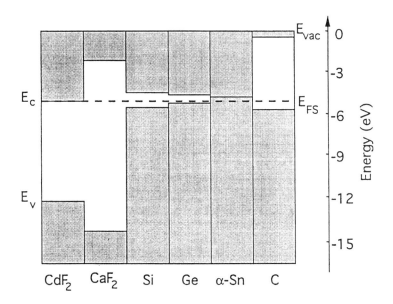

Fig. 11. Band offsets of the group IV and group II–F_2 materials. The distinct asymmetry of the location of the band edges of diamond and CdF_2 with respect to E_{FS} determines the doping characteristics of these materials.

CdF_2 has a very large energy gap of about 8 eV [117], it can be doped quite efficiently with donors. Doping with trivalent impurities followed by annealing converts CdF_2 into conducting material [118]. Free electron concentrations as high as 4×10^{18} cm^{-3} have been reported in this material [119]. Again, the reason for this unusual behavior is the extreme asymmetry in the location of the band edges with respect to E_{FS}. As is shown in Figure 11 E_c of CdF_2 is located almost exactly at E_{FS} [120] which, not only explains a strong propensity towards n-type conductivity but also indicates that p-type doping will not be possible in this material. In contrast CaF_2 with both conduction and the valence band edges far removed from E_{FS}, cannot be doped at all and always acts as a good insulator.

9. UNINTENTIONAL DOPING

The above considerations show that there is a simple phenomenological rule to predict the doping behavior of different semiconductors. A semiconductor can be doped both p- and n-type only when the band of allowed Fermi energies, $\Delta E_F = E_{F\,max} - E_{F\,min}$ spans the whole semiconductor band gap, E_g. Any shift of ΔE_F with respect to E_g will result in an increasing asymmetry in the doping behavior and loss of one type of conductivity. An interesting effect is observed for few extreme cases of semiconductors with a very large displacement of ΔE_F relative to E_g. It has been found that materials with E_{FS} located very close to the conduction or the valence band exhibit n- or p-type conductivity even if they are not intentionally doped.

Thus, as-grown, undoped GaSb with $E_{FS} = E_v + 0.1$ eV is typically p-type with a hole concentration of about 10^{17} cm^{-3} [121]. On the other hand the materials with conduction bands located close to E_{FS} show a very distinct inclination to n-type conductivity. In the case of HgSe discussed earlier, the n-type conduction that is always observed in as grown material is attributable to the location of E_{FS} deep in the conduction band [101]. Also as grown, undoped CdSe, ZnO, ZnS, GaN and InN very often exhibit high electron concentrations. Very high electron concentrations close to 10^{21} cm^{-3} were found in two extreme cases of ZnO [100] and InN [85] in which the Fermi level stabilization energies are located close to or even above the conduction band edge (see Figures 6 and 7).

It is commonly assumed that such unintended conductivity originates from a non-stoichiometric native defects and/or impurity contamination. It has been proposed that the p-type conductivity observed in as-grown GaSb results from Sb deficiency leading to incorporation of antimony vacancies. The vacancies relax to form, $(Ga_{Sb} + V_{Ga})$ acceptors [121]. It should be noted that such acceptor is a defect analogous to the relaxed As vacancy that according to Eq. 1. plays a role of an amphoteric defect in GaAs. On the other hand group III-Nitrides are believed to be N deficient and therefore it was commonly assumed that V_N [122] are responsible for the n-type conductivity in GaN and especially in InN. Recent results indicate that O contamination could be responsible for n-type conductivity in as grown GaN [123].

The effects of non-stoichiometry on the unintended conductivity can be qualitatively understood within the amphoteric defect model. When compound semiconductor crystals are grown at elevated temperature any deviation from stoichiometry is accommodated by native defects in equilibrium with the crystal lattice. The charge balance requires that,

$$rN_A + n = sN_D + p \tag{6}$$

where N_A and r (N_D and s) are the concentration and charge multiplicity of the acceptor-like (donor-like) amphoteric defects and n and p are the concentrations of free electrons and holes, respectively. Because of the amphoteric nature of the native defects the Fermi energy is pinned at E_{FS}. In a material with E_{FS} located close to the mid-gap the concentrations of electrons and holes are relatively low and not very different ($n \simeq p$). This means that

$r N_A \simeq s N_D$. Upon cooling, these two types of oppositely charged defects will tend to recombine forming neutral defect complexes. It has been shown that in non-stoichiometric GaAs the defect complexes consisting of donor- and acceptor-like forms of one type of vacancies are indeed neutral and do not contribute any charge although they can accommodate a significant deviation from the stoichiometry [124]. Therefore, in semiconductors such as GaAs where E_{FS} is located in the mid-gap a deviation from stoichiometry may not be very important as it can be easily accommodated by the neutral defect complexes [124]. Such complexes do not affect the electrical properties of these materials.

A completely different situation is encountered in semiconductors with E_{FS} located close to either the conduction or the valence band. For example, in a material with E_{FS} and thus also E_F situated close to the conduction band edge, n is large, $n \gg p$ which requires, from Eq. 6, that $N_D \gg N_A$. In such a case the entire deviation from stoichiometry is accommodated by donor-like defects only. Since all of them are positively charged they cannot recombine and are frozen during cooling resulting in unintentional n-type conductivity. Analogously, p-type conductivity will be stabilized in a non-stoichiometric material with E_{FS} close to the valence band. In these cases the unintentional conductivity can be eliminated or reduced only by an improved control of crystal stoichiometry.

10. AMPHOTERIC DOPANTS

The asymmetry in the location of E_{FS} has also interesting consequences for the understanding of the behavior of group IV amphoteric dopants in III–V compounds. It can be argued that the group IV elements preferentially substitute the crystal atoms that are most closely matched in terms of size and/or electronegativity. Although such rules appear to provide some guidance, they cannot explain all the aspects of the observed trends. For example, all group IV elements act as donors in InP. It is understandable why the metallic, large size Sn atom substitutes In and acts as a donor. It is surprising, however, that carbon with its small size and high electronegativity is a donor substituting the large size, metallic In site rather than smaller size P site [125]. In stark contrast the same group IV elements are acceptors in GaSb. Even metallic Sn that is such a good donor in so many other III–V's is an acceptor in GaSb [126]. This preferential substitution of amphoteric defects on either group III or group V sites is again understandable in view of the fact that E_{FS} is located close to the conduction band edge in InP and close to the valence band in GaSb.

Interesting trends that confirm the essential role of location of E_{FS} in controlling the behavior of amphoteric impurities, are observed in C doping of semiconductor alloys. In $In_xGa_{1-x}As$ Carbon is a good acceptor for small x values and becomes a donor for larger x. The composition x of the inversion point depends on growth conditions and varies from 0.5 [127, 128] to 0.8 [129]. The dependence on the growth conditions is related to the crystal stoichiometry and presence of hydrogen during the growth. Nonetheless, the factor that eventually controls the substitution of C atoms is the shift of the conduction band towards E_{FS} with increasing x whereas the valence band separation from E_{FS} is practically independent of x [80]. Consequently, for x close to 1 Carbon will predominantly substitute group III sites independently of the growth conditions.

A similar behavior has been observed for $In_xGa_{1-x}As_yP_{1-y}$ lattice matched to InP [130]. For x increasing from 0.53 to 1 and y decreasing correspondingly from 1 to 0 the energy separation $E_{FS} - E_v$ increases by 0.5 eV from 0.5 to 1 eV whereas $E_c - E_{FS}$ increases only by about 0.15 eV. This shift of the valence band edge away from E_{FS} leads to an increased incorporation of C dopants on the group III sites. Experiments show that the transition from p- to n-type occurs at the energy gap of 0.92 eV [130] corresponding to $E_{FS} - E_v = 0.64$ eV and $E_c - E_{FS} = 0.28$ eV.

In all the above considerations it has been assumed that the incorporation of dopants and the saturation of the free carrier concentration occur at high enough temperature so that the semiconductor is under local, thermodynamic equilibrium. Obviously the question arises if these limitations could be overcome and the concentration of the compensating defects reduced by lowering the crystal growth temperature and approaching thermodynamic non-equilibrium conditions. Recent extensive studies of doping of GaAs have shown that C can be a very efficient acceptor. For the growth temperatures below 450°C very high free hole concentrations approaching 10^{21} cm^{-3} were achieved in MOMBE grown GaAs [53]. This concentration is much higher than free hole concentration of about 10^{20} cm^{-3} that has been achieved by doping of GaAs with column II acceptors. However annealing studies of very heavily C doped GaAs have shown that the concentration of free holes is abruptly reduced to mid 10^{19} cm^{-3} [131, 132] for annealing temperatures higher than 600°C. The maximum free concentration is close to that obtained in GaAs implanted with carbon [133]. This indicates that although, in some instances, a significant improvement of the dopant activation can be realized under non-equilibrium conditions, such doping is intrinsically unstable and vulnerable to high temperature processing.

Another example where non-equilibrium doping can play an important role is p-type doping of group III-Nitrides. It has been shown that GaN can be doped with Mg. Room temperature hole concentrations approaching 10^{18} cm^{-3} were observed in Mg doped GaN. Because of the large Mg acceptor binding energy this concentration corresponds to about 10^{20} cm^{-3} of electrically active acceptors. It has been suggested that such high electrical activity is possible because during the MOCVD growth at high temperature of 1100°C the Mg acceptors are almost completely passivated by hydrogen [134]. Under such circumstances the Fermi energy is located at the intrinsic level and no compensating defects are incorporated. The Mg acceptors are activated at much lower temperature of 700°C when the passivating H is removed from the samples.

11. CONCLUSIONS

It has been shown that native defects in a semiconductor crystal lattice exhibit amphoteric behavior. Depending on the location of the Fermi energy they can act either as acceptors or donors. The demarcation energy separating donor- from acceptor-like behavior plays an important role of the energy at which the Fermi level is stabilized in the presence of large concentrations of native defects. It also serves as a convenient energy reference to evaluate the Fermi energy dependent part of the defect formation energy. Based on these observations a model has been developed that addresses the issue of the relationship between the native defects and intentionally introduced dopants. It is shown that the maximum free electron or hole concentration that can be achieved by doping is an intrinsic property of a given semiconductor and is fully determined by the location of the semiconductor band edges with respect to the Fermi level stabilization energy. The Amphoteric Defect Model provides a simple phenomenological rule that explains experimentally observed trends in free carrier saturation in semiconductors. It correctly predicts the maximum concentrations of electrons and holes in a variety of semiconductor materials systems. It has been also used to address other issues including impurity segregation and interdiffusion in semiconductor heterostructures and doping induced suppression of dislocation formation.

Use of complex layered structures of different semiconductor materials plays an increasingly important role in the design of modern optoelectronic devices as they allow not only to tune the emitted light energy but also to control the confinement and separation of free electron and hole systems. This is achieved by the proper tuning of the conduction and the valence band offsets between different component layers of the devices. The problems of the maximum doping and impurity redistribution within the device structure were always treated as entirely separate issues. The Amphoteric Defect Model unifies those two apparently unrelated aspects of optoelectronic devices by providing a simple rule relating the maximum doping levels and dopant diffusion and redistribution to the same conduction and the valence band offsets that control the distribution of free electrons and holes in optoelectronic devices.

Acknowledgements

The author would like to thank Eugene Haller for critical reading of the manuscript. This work was supported by the Director, Office of Energy Research, Office of Basic Energy Sciences, Division of Materials Sciences, of the U.S. Department of Energy under Contract No. DE-AC03-76SF00098.

12. REFERENCES

1. R.L. Longini and R.F. Greene, *Phys. Rev.*, **102**, 992 (1956).
2. F.A. Kroger and H.J. Vink, in *Solid State Physics* vol. III, 307 (1956).
3. G. Mandel, *Phys. Rev.*, **134**, A1073 (1964).
4. R.S. Title, G. Mandel and F.F. Morehead, *Phys. Rev.*, **136**, A300, (1964).
5. G. Mandel, F.F. Morehead and P.R. Wagner, *Phys. Rev.*, **136**, A826 (1964).
6. E.F. Schubert "Doping in III–V Semiconductors" Cambridge University Press (1993).
7. S. Nakamura, *J. Vac. Sci. Technol. A* **13**, 705 (1995).
8. I. Akasaki, S. Sota, H. Sakai, T. Tanaka, M. Koike and H. Amano, *Electron. Lett.*, **32**, 1105 (1996).
9. S. Nakamura and G. Fasol, "The Blue Laser Diode" (Springer, Berlin, 1997).
10. P. Perlin, I Gorczyca, N.E. Christansen, I. Grzegory, H. Teisseyre and T. Suski, *Phys. Rev.*, **B45**, 13 307 (1992).
11. R.M. Park, M.B. Troffer, C.M. Rouleau, J.M. DePuydt and M.A. Haase, *Appl. Phys. Lett.*, **57**, 2127 (1990).
12. K. Ohkawa, T. Karasawa and T. Mitsuyu, *Jpn. J. Appl. Phys.*, **30**, L152 (1991).
13. A. Ishibashi, *Proc. 7th Intern. Conf. on II–VI Comp. and Devices*, Edinburgh, Scotland, UK, Aug. 1995 North Holland, p.555.
14. S. Ito, N. Nakayama, T. Ohata, M. Ozawa, H. Okuyama, K. Nakano, M. Ikeda, A. Ishibashi and Y. Mori, *Jpn. J. Appl. Phys.*, **33**, L639 (1994).
15. E.F. Schubert, G.H. Gilmer, D.P. Monroe, R.F. Kopf and H.S. Luftman, *Phys. Rev.*, **B46**, 15 078 (1992).
16. D.B. Laks, C.G. Van de Walle, G.F. Neumark, P.E. Blochl and S.T. Pantelides, *Phys. Rev.*, **B45**, 10 965 (1992).
17. C.G. Van de Walle and D.B. Laks, *Solid State Comm.*, **93**, 447 (1995).
18. E.E. Haller, *Semicon. Sci. Technol.*, **6**, 73 (1991).
19. S.J. Pearton, J.W. Corbett and T.S. Shi, *Appl. Phys.*, **A43**, 153 (1987).
20. I. Akasaki and H. Amano, *Jpn. J. Appl. Phys.*, **36**, 5393 (1997).
21. S. Nakamura, M. Senoh and T. Mukai, *Japan. J. Appl. Phys.*, **30**, L1708 (1991).
22. J.A. Wolk, J.W Ager III, K.J. Duxstad, W. Walukiewicz, E.E Haller, N.R. Taskar, D.R. Dorman and D.J. Olego, *J. Cryst. Growth*, **138**, 1071 (1994).
23. H. Kukimoto, *J. Cryst. Growth*, **107**, 637 (1991).
24. W. Walukiewicz, *Phys. Rev.*, **B39**, 8776 (1989).
25. W. Walukiewicz, *Mat. Res. Soc. Symp. Proc.*, **300**, 421 (1993).
26. W. Walukiewicz, Inst. Phys. Conf. Ser. No 141, p.259 (1994).
27. W. Walukiewicz, *Appl. Phys. Lett.*, **54**, 2094 (1989).
28. A. Zunger, *Ann. Rev Mat. Sci.*, **15**, 411 (1985).
29. J.M. Langer and H. Heinrich, *Phys. Rev. Lett.*, **55**, 1414 (1985).
30. D.D. Nolte, W. Walukiewicz and E.E. Haller, *Phys. Rev. Lett.*, **59**, 501 (1987).
31. R.J. Nelson, *Appl. Phys. Lett.*, **31**, 351 (1977).
32. D.V. Lang, R.A. Logan and M. Jaros, *Phys. Rev.*, **B19**, 1015 (1979).
33. P.M. Mooney, *J. Appl. Phys.*, **67**, R1 (1990).
34. D.J. Chadi and K.J. Chang, *Phys. Rev. Lett.*, **61**, 873 (1988).
35. J.A. Wolk, M.B. Kruger, J.N. Heyman, W. Walukiewicz, R. Jeanloz and E.E. Haller, *Phys. Rev. Lett.*, **66**, 774 (1991).

36. P. Stallinga, W. Walukiewicz, E.R. Weber, P. Becla and J. Lagowski, *Phys. Rev.*, **B52**, R8609 (1995).
37. J.A. Wolk, W. Walukiewicz, M.L.W. Thewalt and E.E. Haller, *Phys Rev. Lett.*, **68**, 3619 (1992).
38. P. Dreszer and M. Baj, *Acta Phys. Pol.*, **A73**, 219 (1987).
39. P. Dreszer, W.M. Chen, K. Seendripu, J.A. Wolk, W. Walukiewicz, B.W. Ling, C.W. Tu and E.R. Weber, *Phys. Rev.*, **B47**, 4111 (1993).
40. V.N. Brudnyi and V.A. Novikov, *Fiz. Tekh. Poluprovodn.*, **19**, 747 (1985). [*Sov. Phys. Semicon.*, **19**, 460 (1985)].
41. V.N. Brudnyi, M.A. Krivov, A.I. Potapov and V.I. Shakhovostov, *Fiz. Tekh. Poluprovodn*, **16**, 39 (1982) [*Sov. Phys. Semicon.*, **16**, 21 (1982)].
42. J.W Cleland and J.H Crawford, *Phys. Rev.*, **100**, 1614 (1955).
43. V.N. Brudnyi and V.A. Novikov, *Fiz. Tkh. Poluprovodn.*, **16**, 1880 (1982) [*Sov. Phys. Semicon.*, **16**, 1211 (1983)].
44. N.P. Kekelidze and G.P. Kekelidze, in "Radiation Damage and Defects in Semiconductors", Vol. 16, Inst. Phys. Conf. Series (IOP, London, 1972), p.387.
45. T.V. Mashovets and R. Yu. Khansevarov, *Fiz. Tverd. Tela*, **8**, 1690 (1966) [*Sov. Phys-Solid State*, **8**, 1350 (1966)].
46. W. Walukiewicz, *Phys. Rev.*, **B37**, 4760, (1988).
47. W. Walukiewicz, *J. Vac. Sci. Technol.*, **B6**, 1257 (1988).
48. W. Walukiewicz, *J. Vac. Sci. Technol.*, **B5**, 1062, (1987).
49. G.A. Baraff and M.S. Schluter, *Phys. Rev. Lett.*, **55**, 1327 (1985).
50. S.B. Zhang and J.E Northrup, *Phys. Rev. Lett.*, **67**, 2339 (1991).
51. D.J. Chadi, *Mats. Sci. Forum*, **258–263**, p.1321 (1997).
52. E.F. Schubert, J.M. Kuo, R.F. Kopf, H.S. Luftman, L.C. Hopkins and N.J. Sauer, *J. Appl. Phys.*, **67**, 1969 (1990).
53. T. Yamada, E. Tokumitsu, K. Saito, T. Akatsuka, M. Miyauchi, M. Konagai and K. Takahashi, *J. Crystal Growth*, **95**, 145 (1989).
54. L.J. Veiland and I. Kudman, *J. Phys. Chem. Solids*, **24**, 437 (1963).
55. M.G. Milvidskii, V.B. Osvenskii, V.I. Fistul, E.M. Omelyanovskii and S.P. Grishina, *Sov. Phys. Semicond.*, **1**, 813 (1968).
56. O.V. Emelyanenko, T.S. Lagunova and D.B. Nasledov, *Sov. Phys.-Solid State*, **2**, 176 (1960).
57. J.M. Whelan, J.D. Struthers and J.A. Ditzenberger, *Proc. Intern. Conf. On Semicon. Phys.*, Prague (1960) p.943, Academic Press, New York and London (1961).
58. T.N. Theis, P.M. Mooney and S.L. Wright, *Phys. Rev. Lett.*, **60**, 361 (1988).
59. E.W. Williams and H.B. Bebb, in "Semiconductors and Semimetals", edited by R.K. Willardson and A.C. Beer (Academic, New York, 1972) Vol. 8, p.321.
60. J. Maguire, R. Murray, R.C. Newman, R.B. Beal and J.J. Harris, *Appl. Phys. Lett.*, **50**, 516 (1987).
61. S.J. Pearton, J.S. Williams, K.T. Short, S.T. Johnson, D.C. Jacobsen, J.M. Poate, J.M. Gibson and D.O. Boerma, *J. Appl. Phys.*, **65**, 1089 (1989).
62. W. Walukiewicz, *Phys. Rev.*, **B50**, 5221 (1994).
63. J.S. Blakemore, *J. Appl. Phys.*, **53**, R123 (1982).
64. H. Bracht, E.E Haller, K. Eberl and M. Cardona, *Appl. Phys. Lett.*, **74**, 49 (1999).
65. D.E. Bliss, W. Walukiewicz, J.W. Ager III, E.E. Haller, K.T. Chan and S. Tanigawa, *J. Appl. Phys.*, **71**, 1699 (1992).
66. Z. Liliental-Weber, X.W. Lin and J. Washburn, *Appl. Phys. Lett.*, **66**, 2086 (1995).
67. E.F. Schubert, G.H. Gilmer, R.F Kopf and H.S. Luftman, *Phys. Rev.*, **B46**, 15 078 (1992).
68. L.Y. Chan, K.M. Yu, M. Ben-Tzur, E.E Haller, J.M. Jaklevic, W. Walukiewicz, and C.M. Hanson, *J. Appl. Phys.*, **69**, 2998 (1991).
69. K. Wada, H. Fushimi, N. Watanabe and M. Uematsu, Inst. Phys. Conf. Ser. No. 149, p.31, IOP Publishing Ltd. (1996).
70. S.L. Jackson, M.T. Fresina, J.E. Baker and G.E. Stillman, *Appl. Phys. Lett.*, **64**, 2867 (1994).
71. M.G. Astles, F.G.H. Smith and E.W. Williams, *J. Elechtrochem. Soc.*, **120**, 1750 (1973).
72. R. Hirano, T. Kanazawa and T. Inoue, *J. Appl. Phys.*, **71**, 659 (1992).
73. E. Tokumitsu, *Jpn. J. Appl. Phys.*, **29**, L698 (1990).

74. Y. Nishkawa, Y. Tsuburai, C. Nozaki, Y. Ohba, Y. Kokubun and H. Kinoshita, *Appl. Phys. Lett.*, **53**, 2182 (1988).
75. W. Walukiewicz, Inst. Phys. Conf. Series, No. 141, 259 (1995).
76. R. Weber, A. Paraskevopoulos, H. Schroeter-Janssen and A.G. Bach, *J. Elechtrochem. Soc.*, **138**, 2812 (1991).
77. F. Dildey, M.C. Amann and R. Treichler, *Jpn. J. Appl. Phys.*, **29**, 809 (1990).
78. T. Humer-Hager, R. Treichler, P. Wurzinger, H. Tews and P. Zwicknagl, *J. Appl. Phys.*, **66**, 181 (1989).
79. H. Bracht, W. Walukiewicz and E.E. Haller, *Mat. Res. Soc. Symp. Proc.*, **490**, 93 (1998).
80. S. Tiwari and D.J. Frank, *Appl. Phys. Lett.*, **60**, 630 (1992).
81. Y.-F. Wu, B.P. Keller, S. Keller, D. Kapolnek, P. Kozodoy, S.P. Denbaaars and U.K. Mishra, *Solid-State Electronics*, **41**, 1569 (1997).
82. M. Asif Khan, Q. Chen, M.S. Shur, B.T. Dermott, J.A. Higgins, J. Burm, W.J. Schaff and L.F. Eastman, *Solid-State Electronics*, **41**, 1555 (1997).
83. S. Yoshida, S. Misawa and S. Gonda, *J. Appl. Phys.*, **53**, 6844 (1982).
84. Y. Koide, H. Itoh, N. Sawaki, I Akasaki, *et al.*, *J. Elechtrochem. Soc.*, **133**, 1956 (1986).
85. Y. Sato and S. Sato, *J. Cryst. Growth*, **144**, 15 (1994).
86. Gyu-Chul Yi and B.W. Wessels, *Appl. Phys. Lett.*, **69**, 3028, (1996).
87. J. Qiu, J.M. DePuydt, H. Cheng and M.A. Haase, *Appl. Phys. Lett.*, **59**, 2992 (1989).
88. I.S. Haukson, J. Simpson, S.Y. Wang, K. Prior and B.C. Cavennett, *Appl. Phys. Lett.*, **61**, 2208 (1992).
89. H. Okuyama, Y. Kishita, T. Miyajima and A. Ishibashi, *Appl. Phys Lett.*, **64**, 904 (1994).
90. D.J. Chadi, *Phys. Rev. Lett.*, **72**, 534 (1994).
91. A. Garcia and J.E. Northrup, *Phys. Rev. Lett.*, **74**, 1131 (1995).
92. T. Baron, K. Saminadayar and N. Magnea, *Appl. Phys. Lett.*, **67**, 2972 (1995).
93. A.L. Chen, W. Walukiewicz, K. Duxstad and E.E. Haller, *Appl. Phys. Lett.*, **68**, 1522 (1996).
94. W. Walukiewicz, *J. Cryst. Growth*, **159**, 244 (1996).
95. S.B. Zhang, Su-Huai Wei and A. Zunger, *J. Appl. Phys.*, **83**, 3192 (1998).
96. S.O. Ferreira, H. Sitter and W. Faschinger, *Appl. Phys. Lett.*, **66**, 1518 (1995).
97. S.O. Ferreira, H. Sitter, W. Faschinger and G. Brunthaler, *J. Cryst. Growth*, **140**, 282 (1994).
98. H. Ogawa, G.S. Ifran, H. Nakayama, M. Nishio and A. Yoshida, *Jpn. J. Appl. Phys.*, **33**, L980 (1994).
99. T. Ohtsuka, J. Kawamata, Z. Zhu, T. Yao, *Appl. Phys. Lett.*, **65**, 466 (1994).
100. M. Sondergeld, *Phys. Stat. Solidi*, **B81**, 253 (1977).
101. S.L. Lehoczky, J.G. Broerman, D.A. Nelson, C. Whitsett, *Phys. Rev.*, **B9**, 1598 (1974).
102. H. Okuyama, K. Nakano, T. Miyajima and K. Akimoto, *Jpn. J. Appl. Phys.*, **30**, L1320 (1991).
103. B.J. Wu, J.M. DePuydt, G.M. Haugen, G.F. Hofler, M.A. Haase, H. Cheng, S. Guha, H.L. Kuo and L. Salamanca-Riba, *Appl. Phys. Lett.*, **64**, 3462 (1995).
104. W. Faschinger, *J. Cryst. Growth*, **159**, 221 (1996).
105. M.W. Wang, J.F. Swenberg, M.C. Philips, E.T. Yu, J.O. McCaldin, R.W. Grant and T.C. McGill, *Appl. Phys. Lett.*, **64**, 3455, (1994).
106. A. Waag, Th. Litz, F. Fischer, H. Heinke, S. Scholl, D. Hommel, G. Landwehr and G. Biliger, *J. Cryst. Growth*, **138**, 437 (1994).
107. H.L. Hwang, K.Y.J. Hsu and H.Y. Ueng, *J. Cryst. Growth*, **161**, 73 (1996).
108. K. Mitchell, C. Eberspacher, J. Ermer, K. Pauls, D. Pier and D. Tanner, *Solar Cells*, **27**, 69 (1989).
109. L. Stolt, J. Hedstrom, J. Kessler, M. Ruckh, K.O. Velthaus and K.W. Schock, *Appl. Phys. Lett.*, **62**, 597 (1993).
110. Su-Huai Wei, A. Zunger, *Appl. Phys. Lett.*, **63**, 2549 (1993).
111. M. Turowski, G. Margaritondo, M.K. Kelly and R.D. Thomlinson, *Phys. Rev.*, **B31**, 1022 (1985).
112. K.S. Jones and E.E. Haller, *J. Appl. Phys.*, **61**, 2469 (1987).
113. F.J. Himpsel, J.A. Knapp, J.A. Van Vechten and D.E. Eastman, *Phys. Rev.*, **B20**, 624 (1979).

114. J. van der Weide, Z. Zhang, P.K. Baumann, M.G. Wensell, J. Bernholc and R.J. Nemanich, *Phys. Rev.*, **B50**, 5803 (1994).
115. see e.g. A.T. Collins, *Properties and Growth of Diamond*, emis, Datareview series No. 9, p.284 (1994).
116. K. Nishimura, K. Das and J.T. Glass, *J. Appl. Phys.*, **69**, 3142 (1991).
117. B.A. Orlowski and J.M. Langer, *Phys. Status Solidi*, **B91**, K53 (1979).
118. J.D. Kingsley and J.S. Prener, *Phys. Rev. Lett.*, **8**, 315 (1962).
119. R.P. Khosla, *Phys Rev.*, **183**, 695 (1969).
120. A. Izumi, Y. Hirai, K. Tsutsui and N.S. Sokolov, *Appl. Phys. Lett.*, **67**, 2792 (1995).
121. Y.J. Van Der Meulen, *J. Phys. Chem. Solids*, **28**, 25 (1967).
122. P. Perlin, T. Suski, H. Teisseyre, M. Leszczynski, I. Grzegory, J. Jun, S. Porowski, P. Boguslawski, J. Bernholc, J.C. Chervin, A. Plian and T.D. Moustakas, *Phys. Rev. Lett.*, **75**, 296 (1995).
123. M.D. McCluskey, N.M. Johnson, C.G. Vande Walle, D.P. Bour, M. Kneissl and W. Walukiewicz, *Phys. Rev. Lett.*, **80**, 4008 (1998).
124. D.J. Chadi and S.B. Zhang, *Phys. Rev.*, **B41**, 5444 (1990).
125. Je-Hwan Oh, Jun-ichi Shirakashi, F. Fukuchi and M. Konagai, *Appl. Phys. Lett.*, **66**, 2891 (1995).
126. K.L. Longenbach, S. Xin and W.I. Wang, *J. Appl. Phys.*, **69**, 3393 (1991).
127. C.R. Abernathy, S.J. Pearton, F. Ren, W.S. Hobson, T.R. Fullowan, A. Katz, A.S. Jordan and J. Kovalchik, *J. Cryst. Growth*, **105**, 375 (1990).
128. C.-S. Son, S.-I. Kim, Y. Kim, M.-S. Lee, M.-S. Kim, S.-K. Min and I.-H. Choi, *J. Cryst. Growth*, **165**, 222 (1996).
129. J. Shirakashi, T. Yamada, M. Qi, S. Nozaki, K. Takahashi, E. Tokumitsu and M. Konagai, *Jpn. J. Appl. Phys.*, **30**, L1609 (1991).
130. G.M. Cohen, J.L. Benchimol, G. Le Roux, P. Legay and J. Sapriel, *Appl. Phys. Lett.*, **68**, 3793 (1996).
131. G.E. Hofler, H.J. Hofler, N. Holonyak, Jr. and K.C. Hsieh, *J. Appl. Phys.*, **72**, 5318 (1992).
132. K. Watanabe and H. Yamazaki, *Appl. Phys. Lett.*, **59**, 434 (1991).
133. A.J. Moll, E.E. Haller, J.W. Ager III and W. Walukiewicz, *Appl. Phys. Lett.*, **65**, 1145 (1994).
134. J. Neugebauer and C.G. Van de Walle, *Appl. Phys. Lett.*, **68**, 1829 (1996).

CHAPTER 2

Point Defect Formation Near Surfaces

KAZUMI WADA

Department of Materials Science and Engineering
Massachusetts Institute of Technology
77 Massachusetts Avenue, Cambridge, MA 02139

1. INTRODUCTION

Intrinsic point defects are induced by various semiconductor processing such as epitaxial growth, diffusion, etching, and deposition, and play a significant role in determining electrical and optical properties of semiconductors. Thus, it is quite important to understand and control defect incorporation during these processing. In recent years, considerable progress has been made in understanding point defect equilibria, especially Ga vacancies in GaAs. Anomalies in carrier saturation in various compound semiconductors have been explained by "Amphoteric Defect Model (ADM)" [1], as shown in the chapter 1.1.1. The model is based on the *ab-initio* calculation of various point defects in GaAs [2, 3].

Since semiconductor devices are mainly fabricated just below the surfaces, it is quite important to understand point defect equilibria near surfaces. This chapter, thus, extends the ADM to sub-surface layers. The extension is semi-quantitative and is justified by using near-surface defect related phenomena. A surface "bottleneck effect" is naturally introduced. Based on this effect, the formation mechanism of point defects is discussed. A bulk "bottle neck effect" due to dislocations is also addressed, which is

contradictory to the previous understanding that dislocations act as perfect
sinks and sources of point defects.

2. POINT DEFECT EQUILIBRIA NEAR THE SEMICONDUCTOR SURFACES

First, how the point defect equilibrium concentration is estimated is briefly
shown according to the ADM, employing Ga vacancies. The formation
energy of Ga vacancies $E_f(V_{Ga}^{3-})$ and the equilibrium concentration $[V_{Ga}^{3-}]_b^{eq}$
in n-type GaAs bulk can be expressed as;

$$E_{f V_{Ga}^{3-}} = E_{f0} - 3E_{FS} \tag{1}$$

$$\left[V_{Ga}^{3-}\right]_b^{eq} = C \exp\left\{-E_f(V_{Ga}^{3-})/kT\right\} = C' \exp(3E_F/kT), \tag{2}$$

where E_{f0} is the formation energy at which the Fermi-level is located at
the Fermi-level stabilization energy, and E_F is the Fermi-level measured
with respect to the Fermi-level stabilization energy E_{FS}. The Ga vacancy
equilibrium concentration is found to be higher in heavily donor-doped GaAs
than in intrinsic GaAs. This model as being an effect of Ga vacancies [1–4]
successfully explains the poorly understood phenomena of carrier saturation
in n-type GaAs.

The Fermi-level of GaAs surfaces is known to be pinned at the Fermi-level
stabilization energy [5]. In terms of surface pinning, bands should be bent as
shown in Figure 1, resulting in the formation of a surface depletion layer.
This forms an equilibrium concentration gradient in the layer, since the
Fermi-level is now a function of the depth from the surface. This induces
the depth dependence of the formation energy of Ga vacancies.

$$E_{V_{Ga}^{3-}}(x) = E_{f0} - 3E_{FS}(x) \tag{1'}$$

Under depletion layer approximation [6], the depth profile of Ga vacancies
can be expressed by

$$\left[V_{Ga}^{3-}(x)\right] = C \exp\left(\frac{3qN_d\varepsilon_S(Wx - \frac{1}{2}x^2)}{kT}\right) \tag{3}$$

Based on Eq. (3), the depth profile of the equilibrium concentration of
Ga vacancies is obtained, as shown in Figure 2. It is understood that the
equilibrium concentration of Ga vacancies near the surface has a steep
gradient in the surface depletion layer: It is intrinsic at the surface and

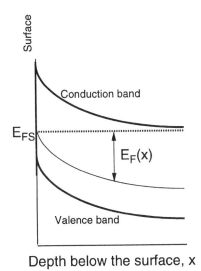

Surface

E_FS .... E_F(x)

Depth below the surface, x

E_{FS} : Fermi level stabilization energy

Fig. 1. Fermi level pinning at a surface in n-GaAs.

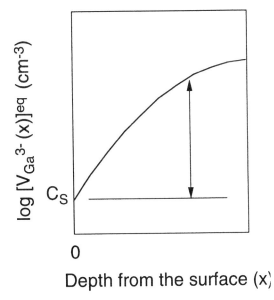

Fig. 2. Depth profile of Ga vacancies in n-GaAs.

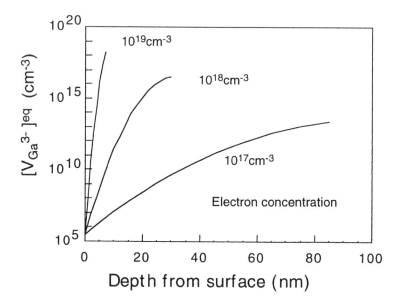

Fig. 3. Semi-quantitative calculation of depth profile of Ga vacancies. The parameter is electron concentration.

increases by several orders of magnitude below the surface to approach the equilibrium concentration in the bulk.

Figure 3 shows a calculation of equilibrium concentration of Ga vacancies below the surface, based on Eq. (3). Here, the equilibrium concentration at intrinsic condition is assumed to be 1×10^5 cm^{-3}. It is clearly shown that the depth profiles of Ga vacancies in equilibrium concentration are strongly electron density dependent.

Quite recently, Ogawa and Kawabe [7] have reported that Si-doped GaAs/AlAs superlattice disordering is delayed near the surface, as shown in Figure 4. It is well accepted that V_{Ga}^{3-} diffusion disorders n-type GaAs/AlAs superlattices. Their conclusion is in accordance with our surface extension of the ADM noted above. In other words, the equilibrium concentration of V_{Ga}^{3-} is lower near the surface, as shown in Figure 3.

3. POINT DEFECT FORMATION KINETICS IN THE SUB-SURFACE LAYER — BOTTLENECK EFFECT

How point defects attain their equilibrium concentration profile near the surface, i.e., point defect formation kinetics is discussed by using Ga

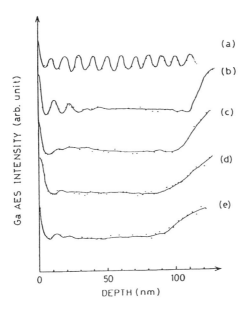

Fig. 4. Delay in superlattice disordering near the surface [Ref. 7]. (a) as grown, (b) 800°C for 2 min, (c) 800°C for 5 min, (d) 800°C for 20 min, and (e) 800°C for 60 min. Delay in disordering is observed in the surface layer 30 nm thick.

vacancies as an example. There are two mechanisms with which to form Ga vacancies. The first one is Schottky-type formation and the second is the Frenkel-type formation of Ga vacancies accompanying Ga interstitials. At the surface the Schottky-type defect formation should occur to attain the equilibrium concentration. This is consistent with the previous assumption that the surface acts as perfect sink and source of intrinsic point defects to attain point defect equilibria. The time taken to attain the equilibrium concentration in the bulk, t_S, can be expressed by

$$t_S = \frac{\int_0 \left[V_{Ga}^{3-}(x) \right]^{eq} dx}{D_{V_{Ga}^{3-}} \mathrm{grad} \left(\dfrac{C_S}{\left[V_{Ga}^{3-}(x) \right]^{eq}} \right)}. \tag{4}$$

Here, C_S is the Ga vacancy equilibrium concentration at the surface. In this context, it is assumed that formation of the Ga vacancy equilibrium concentration profile is diffusion-limited and not limited by the Ga vacancy formation rate at the surface. Since Ga vacancies undergo typical uphill diffusion from the surface to the bulk in this case, the diffusion flux is expressed in the denominator. On the other hand, the time taken to attain the equilibrium concentration in terms of the Frenkel-type defect formation t_F can be expressed by

$$t_F = \frac{\int_0 \left[V_{Ga}^{3-}(x) \right]^{eq} dx}{G}. \tag{5}$$

Here, G is formation rate of the Frenkel-type defects. Therefore, the Frenkel-type Ga vacancy formation dominates when $t_F < t_S$. It is to be noted that, in some cases, intrinsic point defects cannot be provided from surfaces due to kinetic limit to equilibrium, even though surfaces act as perfect sinks and sources of point defects in concentration of their intrinsic equilibria. This can be called the bottleneck effect [8].

When considering point defect equilibrium concentration, it is generally assumed that dislocations act as point defect sources and/or sinks to attain their equilibrium concentration. It is also assumed that dislocations act as Fermi-level pinning centers, as shown in Figure 1. Accordingly, it is likely that the action of dislocations as point defect sources and sinks is kinetically affected by the bottleneck effect shown above. In other words, as far as n-type GaAs is concerned, Ga vacancies equilibrium concentration should be low near the dislocation cores and gradually high away from dislocations by the depletion layer width, as shown in Figure 2.

4. BOTTLENECK RELATED PHENOMENA

The kinetics is considered for attaining the equilibrium concentration of the Ga vacancy during simple thermal annealing. The surface and bulk bottleneck effect presented here should be working behind poorly understood phenomena as follows. Here, it should be noted that the results of Ogawa and Kawabe [7] shown in Figure 4 can be regarded as one typical example of how the surface bottle neck effect works. In other words, if the Schottky-type formation of V_{Ga}^{3-} is dominant in the superlattices, the total concentration of V_{Ga}^{3-} required to attain the superlattice layer in the bulk should be supplied from the surface. This means that disordering should start from the surface, since all the Ga vacancies pass through the sub-surface layer. However, their report shows a delay in disordering. This leads us to conclude that the V_{Ga}^{3-} is generated via the Frenkel-type defect formation in the bulk in their experiment. The Ga interstitial, which is simultaneously formed by the Frenkel-type defect formation, could diffuse to the sample because of its high diffusivity and disappears. This is indeed observed when column II acceptors on the Ga sub-lattices are present nearby the n-type layer, as shown next.

Such severe undersaturation of Ga vacancies in the bulk especially occurs during epitaxial growth. During epitaxial growth, the growth surface is assumed to be pinned. This means that the equilibrium concentration of Ga vacancies is lower at the surface than in the bulk, as shown in Figure 3. This condition favors the Frenkel type defect formation. Accordingly, Ga vacancies incorporated at the surface in equilibrium become undersaturated in course of the continued growth. Ga vacancies could be formed via the Schottky-type defect formation at the surface and provided to the bulk via uphill diffusion. However, when t_S is shorter than t_F, the Frenkel-type defect formation dominates Ga vacancy formation. This has been already observed in device epilayer growth as follows.

Enquist *et al.* reported an anomalous Zn redistribution during epitaxial growth of a hetero-junction bipolar transistor (HBT) made from an Al-GaAs/GaAs system [9]. They found that the n^+ GaAs emitter contact layer growth induces redistribution of Zn acceptor impurities which were originally doped in the GaAs base underneath the AlGaAs emitter. Their growth temperature was 600°C. Zn redistribution occurs not only in the AlGaAs emitter layer but also in the GaAs collector layer. They concluded that it is not due to Zn segregation at the growth surface but due to anomalous diffusion of Zn induced by grown n-type emitter contact layers. In other words, the redistribution strongly depends on the donor impurity concentration in the GaAs emitter contact layers: The redistribution clearly occurs only when the donor concentration is high 10^{18} cm^{-3}. A similar phenomenon has been reported in a Be doped base layer in HBT [10], as shown in Figure 5. It should be noted that Zn and Be are column II acceptors on the Ga sublattice where the anomalous formation of Ga interstitials is responsible for the impurity diffusion.

Based on these findings, Deppe has proposed a conceptional idea that the anomalous formation of Ga interstitials is due to the Frenkel-type defect formation by means of the growth surface pinning [10]. Our calculation indeed supports his idea. According to Figure 3, the equilibrium concentration of Ga vacancies in the bulk is about three orders of magnitude higher when the electron concentration is in the high 10^{18} cm^{-3} than when it is in the low 10^{18} cm^{-3}. Since these vacancies are formed via the Frenkel-type defect formation during n^+ contact layer growth, the same concentration of Ga interstitials should be simultaneously formed. They diffuse into the base layer to diffuse out Zn or Be.

Recently, Kurishima *et al.* reported anomalous Zn redistribution in the InGaAs/InP system [12], as also shown in Figure 5. Basically, their findings are the same as those reported by Enquist *et al.* [9]. Although no calculation of point defect formation energy is available for InGaAs, the bottle neck effect should be valid in InGaAs. In their case, Zn redistribution is induced by growth of the n^+ collector contact layer which is underneath the Zn doped

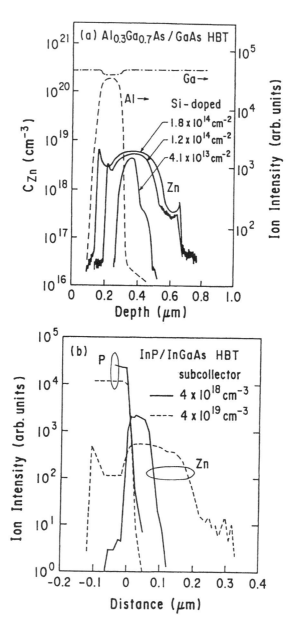

Fig. 5. Zn redistribution during epitaxial growth. (a) GaAs Based HBT [10] and (b) InP based HBT [12].

Base layer. The difference from the GaAs/AlGaAs system is possibly due to the time required for establishing the Ga vacancy equilibrium concentration. Their InP system is grown at 450 Å, which is a much lower temperature than used for the GaAs system. Kobayashi *et al.*, have succeeded in minimizing the Zn redistribution by utilizing growth interruption by which self-interstitials on column III sub-lattice formed in the n^+ collector contact layer are eliminated at the growth interrupted surface [13]. The time for the growth interruption was 30 min. It is open to question whether this time is necessary for attainment of the equilibrium concentration of the vacancy on the column III sub-lattice via Frenkel-type defect formation.

Marioton *et al.*, have recently shown that dislocations are active with a limited efficiency as point defect sources and sinks [14]. An introduction of their results is beyond of this work, but the phenomenon should be related to the bottleneck effect in the bulk, called this "bulk" bottleneck effect.

5. CONCLUSION

Current understanding on point defect equilibria and kinetics near the surface is reviewed. The equilibrium concentration of intrinsic point defects near the surface of GaAs is analyzed by extending the Amphoteric Defect Model. The equilibrium concentration of Ga vacancies at surface is intrinsic equilibrium concentration which should be several orders of magnitude lower than in the bulk. The surface bottle neck effect is, thus, presented, in which point defects can not be provided to the bulk due to kinetic limit, since the point defect diffusion flux is too small to attain equilibrium concentration in the bulk from the surfaces. Based on this effect, a model is proposed in which the Frenkel-type defect formation would be dominantly working to attain an equilibrium concentration in some cases. The model is consistent with delays in disordering of GaAs/AlAs superlattices near the surface and the phenomenon of redistribution of Zn in the hetero-junction bipolar transistor structures based on the GaAs/AlGaAs and InGaAs/InP systems. It is proposed that dislocations which pin the Fermi level induce the bottleneck effect in the bulk. This bottleneck effect might be a clue to understand unsolved problems in GaAs and related compound semiconductors.

References

1. See 1.1.1 in this book, W. Walukiewicz, *Appl. Phys. Lett.*, **54**, 2094 (1989).
2. G.A. Baraff and M. Schulter, *Phys. Rev. Lett.*, **55**, 1327 (1985), and G.A. Baraff and M. Schulter, *Phys. Rev.*, **B33**, 7346 (1986).
3. R.W. Jansen and O.F. Sankey, *Phys. Rev.*, **B39**, 3192 (1989).
4. W. Walukiewicz, *Phys. Rev.*, **B14**, 10218 (1990).

5. W. Walukiewicz, *Phys. Rev.*, **B37**, 4760 (1988).
6. E.H. Nicollian and J.R. Brews, *MOS Physics and Technology* (John Wiley & Sons, New York, 1982) p.54.
7. K. Ogawa and M. Kawabe, *Jpn. J. Appl. Phys.*, **29**, 1240 (1990).
8. K. Wada, *Appl. Surf. Sci.*, **85**, 246 (1995).
9. P. Enquist, J.A. Hutchby and T.J. de Lyon, *J. Appl. Phys.*, **63**, 4485 (1988).
10. P. Enquist, G.W. Wicks, L.F. Eastman and C. Hitzman, *J. Appl. Phys.*, **58**, 4130 (1985).
11. D.G. Deppe, *Appl. Phys. Lett.*, **56**, 370 (1990).
12. K. Kurishima, T. Kobayashi and U. Goesele, *Appl. Phys. Lett.*, **60**, 2496 (1992).
13. T. Kobayashi and K. Kurishima, *Appl. Phys. Lett.*, **62**, 284 (1993).
14. B.P.R. Marioton, T.Y. Tan and U. Goesele, *Appl. Phys. Lett.*, **54**, 849 (1989).

CHAPTER 3

Optical Characterization of Plasma Etching Induced Damage

EVELYN L. HU AND CHING-HUI CHEN

Department of Electrical and Computer Engineering, University of California, Santa Barbara, CA 93106

1. INTRODUCTION

Ion-enhanced dry etching techniques have proven to be a critical enabling technology for shaping optoelectronic materials into devices. The ability to produce well-controlled and accurate pattern transfer having high spatial resolution has been used to etch waveguides [18], define in-plane and vertical cavity laser structures [28, 35, 47], and to form mirrored facets for lasers and waveguides [15, 31]. The power of such dry-etch technology lies in the combination of chemical reactivity and selectivity afforded by the appropriate

43

choice of etch gases, together with the enhanced control of etch profile and
rate achieved through the directionality and energy of ion beams. (In addition
to this ion-assisted chemistry, electron beam [25], or photon-enhanced [6]
gas phase chemical etching can also be effective. Since ion-assisted etching
technology is the one in most widespread use, we will focus our discussions
in this area). Nevertheless, the reduced dimensions and increased sensitivities
of current devices render them more susceptible to the possibilities of
materials damage introduced by the etch process. This forms the motivation
for the discussions undertaken in this Chapter. Beginning with a very brief
consideration of the ion-assisted etching process and the means of ion
interaction with the substrate (Section 2), we will then select out a particular
means of optical characterization (Section 3) in order to more sensitively
explore the issues of etch-induced damage. Section 4 will systematically
trace a series of studies that identify the most important components in
etch-induced damage, also following the progressively better correspondence
between experiment and simulation. Finally, we will consider in Section 5
the issues of etch-induced damage in a variety of different material structures.
We do this in order to better extract out general trends of damage propagation,
but also to understand how we might engineer material structures to better
withstand etch-induced damage.

2. ION-ASSISTED ETCHING: UNDERSTANDING THE PROBLEM

There are currently a number of different techniques available to achieve ion-
assisted etching, including Reactive Ion Etching (RIE), Electron Cyclotron
Resonant Etching (ECR), and Inductively Coupled Plasmas (ICP). These
techniques differ in the characteristics and the formation of the plasmas that
give rise to the reactive ions, and can be used to obtain high etch rates,
greater etch uniformity, and other desired etch features. Further information
about these specific processes, as well as information about the design and
optimization of ion-assisted etch processes can be found in a number of
references. Our intention here is to extract the basic features of any etch
process that will help us to better understand the origins of etch-induced
damage in optoelectronic materials. Therefore, we will begin with a very
simple schematic of a gas-phase etch process, shown in Figure 1, which
delineates the role that the ions may play in the etch process.

We use the chlorine etching of GaAs as a model, highlighting the steps
corresponding to (1) physisorbtion, then chemisorption of the reactive gas
onto the substrate, (2) reaction with the substrate to form a product molecule
which if sufficiently volatile, will subsequently desorb from the substrate
(3). The kinetic energy of ions can dramatically enhance the set of processes
shown in Figure 1 by increasing the speed of the rate-limiting step in the etch

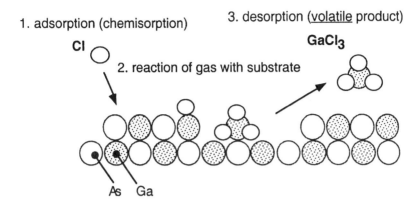

Fig. 1. Simple schematic of the components of a gas-phase etch process.

process. For example, with product molecules that are relatively non-volatile under the etch conditions, ion bombardment may assist in the removal of those products from the surface. Since the direction of the ions can be controlled through the use of electric fields, so then can the directionality of the ion-promoted etch process. Using the simple mechanisms shown in Figure 1, it would seem reasonable that ion enhanced etching would be effective for ion energies on the order of, or a few times larger than typical bond strengths, say tens of electron volts of energy. In fact, most ion-assisted etching processes utilize ion energies of about 100 to a few hundred electron volts. With energies this high, it is reasonable to imagine that the ion interaction with the substrate material may result in the creation of defects in the material. Figure 2 represents schematically the multiplicity of ways in which ions may interact with the substrate to produce material damage.

3. OPTICAL DAMAGE ASSESSMENT TECHNIQUES: CHOOSING A METHOD

Etch-induced damage in materials pragmatically encompasses the range of deleterious modifications introduced to the material through the processing, and may be manifested as changes in conductivity, optical efficiency, or modifications of interfaces with other materials, such a metals.

Although chemical modification of the surface, shown in Figure 3, may have undesirable consequences on device performance (for example, a change in the stoichiometry of the surface material), we will focus in this

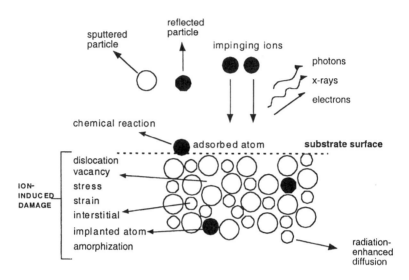

Fig. 2.   A schematic representation of ion-solid interactions that can lead to material degradation, after Malherbe [Malherbe, 1994].

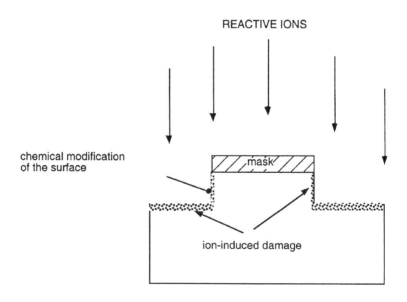

Fig. 3.   Modification of the ion-etched material.

chapter on the better understanding of the physical component of damage introduced by the use of energetic ions in the etch process. And although a variety of sophisticated characterization techniques have been employed in the study of etch-induced damage, we will focus on a particular strategy for assessing the optical efficiency of ion-bombarded materials. Schottky Barrier measurements [51, 52], Deep Level Transient Spectroscopy [20], Raman Spectroscopy [34], Rutherford Backscattering [3], and Transmission Electron Microscopy, in addition to examination of device characteristics, are among the other techniques that have been used to better understand the consequences of ion-assisted etching processes. Full understanding of the complexity of interactions that propagate ion damage into the etched material, that accounts for variations in material composition, etch gas chemistry, etch technique, is as multi-faceted a problem as understanding the detailed reactions and components of the etch process itself. Nevertheless, there have been some clear trends and general progress made in recent years, and we hope that the presentation of this chapter will illuminate some important advances in understanding made over that time period.

Photoluminescence (PL) measurements of optoelectronic materials before and after ion-assisted etching should provide an excellent measure of possible optical damage introduced into the material. A number of early papers did indeed carry out such studies [1, 33, 65]. A major concern for simple assessments of luminescence efficiency of bulk materials is the sensitivity of the technique, and the minimal levels of damage detectable. Initially, it was believed that the ion range, and hence the damage range of these few hundred eV ions would be no deeper than a few tens or hundreds of angstroms. Compared with absorption lengths of ~ 1 μm for the laser lines used for the PL measurements, one might expect to see a modulation of the PL intensity of less than a fraction of a percent. In addition, such measurements do not yield information about the profile of the damage: how the concentration of defects may vary with increasing distance from the surface. It would be extremely useful to somehow be able to place discrete optical 'markers' at varying depths into the substrate material, which would carry that spatial information, as well as information about the optical efficiency of the material at that point.

The capability of epitaxial growth techniques like Molecular Beam Epitaxy (MBE) and Metalorganic Chemical Vapor Deposition (MOCVD) to 'routinely' grow multiple quantum well material, such as that shown in Figure 4(a), produced that ideal marker structure for ion damage studies: quantum wells of differing widths, situated at different depths from the substrate surface, would each give a unique signature in energy position of their luminescence peak. The cathodoluminescence (CL) spectra of Figure 4(b) show the material signature before and after ion bombardment conditions, and allows us to generate profiles in depth of the damaging

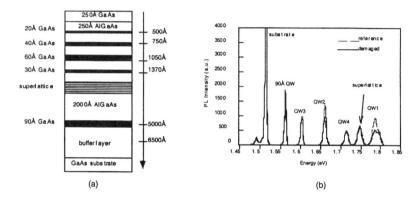

(a)   (b)

Fig. 4.   (a) Schematic of multiple quantum well (MQW) probe material and (b) corresponding photoluminescence spectra for a reference sample and a substrate bombarded by argon ions at 500 eV for 3 minutes at 50 μA/cm^2.

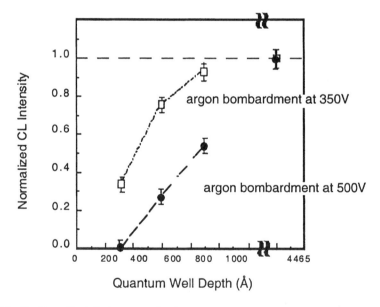

Fig. 5.   Normalized CL damage profiles for incident Ar ions of 350 eV and 500 eV.

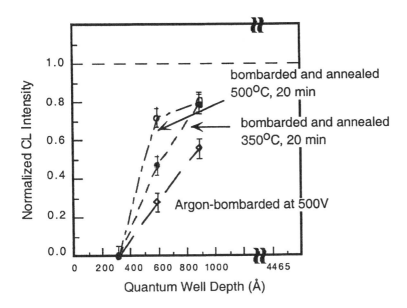

Fig. 6.   Effect of post-bombardment anneals on damage profile.

effects of ion bombardment, such as shown in Figure 5. All the CL and PL measurements that will be shown in this chapter are taken at low temperature (~4K–10K). These data provided an exciting new vision of the propagation of etch-induced defects into the material, and allowed systematic comparisons to be made for changing etch/bombardment parameters, and for different post-etch treatments (Figure 6). The MQW probe technique is an extremely powerful one, with excellent sensitivity, and spatial resolution that is in many senses limited only by the number and proximity of quantum wells placed within the MQW structure.

Not surprisingly, the data of Figure 5 show the increased damage (i.e. reduction in CL luminescence), at a given material depth, for increased ion bombardment energy. Figure 6 shows the effects of various post-bombardment anneals on the material quality. Although the luminescence of the deeper quantum wells (presumably the 'less damaged' material) improves slightly upon annealing at 350°C and 500°C, full recovery of luminescence is not achieved for these conditions, and no recovery of the most heavily damage quantum well is observed for either annealing condition.

The principal strategy that we shall follow in the development of experiments and ideas in the rest of the chapter will be to continue to utilize various multiple quantum well materials, such as that shown in Figure 4(a), subjecting the material to argon ion bombardment to simulate the physical component of ion-assisted etching, and utilizing the comparison of CL or PL spectra before and after the bombardment to trace the propagation of etch induced damage. Aware that a 'dynamic', chemically amplified etch process is far more complex, and that the chemical component may alter the actual damage profiles obtained, we nevertheless seek to lay the groundwork of understanding by working with as simple a representative system as possible. We will briefly consider the effects of etch-removal of the substrate surface, but will primarily work in a regime where the actual removal of material through ion bombardment is negligible. We note that experiments have been carried out that do seek to compare the effects of chemical etching on the damage profile [57, 58] and that Deng and Wilkinson have offered a model of etch damage that differentiates among the ion species in more complex etch gases [17].

A number of research groups have by now made use of such quantum-well substrates in their etching experiments [14, 23, 70]. One surprising, but general result that emerged is the unusually deep penetration of the damage profile into the material: about 1000 Å for incident ion energies of only a few hundred electron volts. The new data on damage profiles therefore immediately stimulated further questions concerning the nature of the damage propagation.

4. THE RANGE OF ION-INDUCED DAMAGE

Estimates of the extent of ion damage introduced into a substrate can be guided by the approach of Lindhard, Scharff and Schiott (LSS), whose 'unified' energy-range calculations formed the basis for the understanding of ion range and straggle in ion implantation processes [41, 42]. Such calculations form the basis of generally available simulation programs like TRIM (The Range of Ions in Matter). Using those calculations, and extrapolating to the much lower ion energies (few hundred eV) characteristic of ion-assisted etching processes, we would estimate that the ion range extends no more than a few hundred angstroms below the surface of the substrate. Although the ion and defect distributions are not necessarily coincident [48], we will operate on that premise, as a first approximation. In a 'dynamic' etching situation, where the substrate surface is being chemically removed at some rate, at the same time that ion damage is being introduced into the substrate, we might expect that the ion-damaged 'zone' would actually be more shallow. This estimate is obviously in contradiction to the experimental results obtained. We might question the accuracy of calculations generally applied to ion implantation energies, of 10's or hundreds of keV: there is no guarantee that the calculations will be accurate for our ion

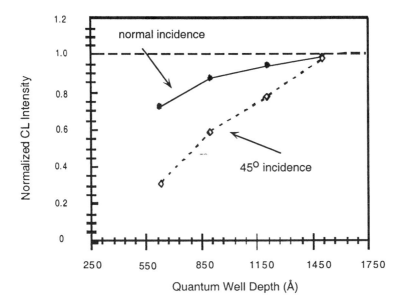

Fig. 7. Difference in MQW damage profiles obtained for 450 eV argon ions at normal and 45° angle of incidence to substrate.

bombardment processes, with ion energies about a 1000 times lower. Secondly, we have not taken into account a phenomenon known to occur in ion implantation: ion channeling through the crystalline substrate.

4.1. The Role of Channeling: A Major Discrepancy Removed

Perhaps one of the first experiments to clearly point out that ion channeling might be the reason for these deep damage profiles, was that carried out by Germann *et al.* [23] , where a single quantum well probe structure was used, and the PL of that well was measured as a function of the angle of the ion gun with respect to the substrate surface. A clear dip in the luminescence was observed for an angle of incidence of 45°, corresponding to the (011) direction in the substrate, in apparent confirmation of the role of channeling in the propagation of ion damage. Our own multiple quantum well data, shown in Figure 7, also shows the greater optical degradation at 45° ion incidence.

Despite this confirmation, puzzles remained: Germann's experiment utilized a focused ion source, where there was a clearly definable angle and direction between the ion source and the substrate. However, this deep

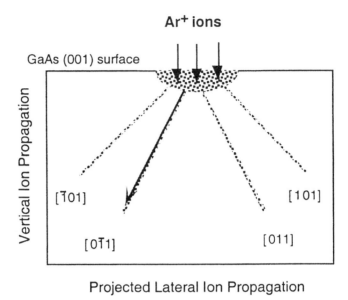

Fig. 8.   Schematic of SCHLEICH simulation, illustrating how ions can channel and penetrate more deeply into the substrate material.

ion-damage range had been observed for a variety of groups, using differing ion sources (parallel plate reactive ion etchers, ion guns) on substrates which may not have been oriented in a particular manner during the etch process. How do we reconcile the importance of the channeling process, which seems so critically dependent on orientation, with the general results obtained by different research groups, with experiments done in a variety of ways, carried out without particular thought being given to substrate orientation?

The answer was provided by N. Stoffel, in his molecular dynamics simulations of low energy ion introduction into a substrate [64]. Stoffel's program, SCHLEICH (SCattering of Heavy, Low Energy Ions into CHannels) showed that some small percentage of ions, initially incident perpendicular to the substrate surface, were fortuitously scattered into a channeling direction. This situation is represented schematically in Figure 8. Although a very small percentage of the incident ion flux, once scattered into the channeling direction, these 'lucky ions' were able to penetrate deeply into the substrate material. Stoffel used Germann's data for program calculations, and found far better agreement between experiment and model than had previously been possible using TRIM calculations. However, the experimentally determined ranges still exceeded the calculated ranges by a factor of two or so.

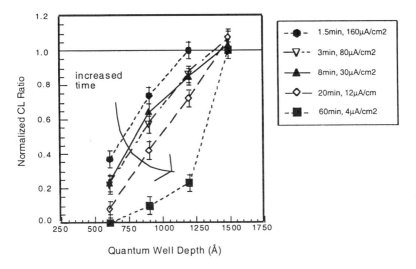

Fig. 9. Normalized cathodoluminescence intensities for substrates bombarded by argon ion beams at 400 eV. Varying times and current densities were used to produce a constant total dose of 9×10^{16} ions/cm^2.

4.2. Diffusion of Defects: Beyond Channeling Alone

In addition to channeling, another mechanism that would introduce defects more deeply into the substrate might be that of defect diffusion. In order to explore this possibility, we designed some experiments that would highlight diffusion effects [26]. MQW structures were used to assess the effect of varying temperature or varied time of irradiation (but with total dose constant) on the resulting profiles of damage. To first order, channeling probabilities should be insensitive to these variations, and therefore there should be no difference observed in damage distributions for these samples. Figure 9 shows the damage profiles obtained for a succession of ion bombardment conditions, having a constant total dose (9×10^{16} cm^{-2}), but with differing values of instantaneous current density and total irradiation time. Quite clearly, for the longest irradiation time, the reduction in the luminescence intensity is the most profound, and the damage profile appears to extend more deeply into the substrate. One could argue that high instantaneous ion impingement rates could either change the nature of the substrate surface, or alter channeling probabilities (i.e., two ions coincident along the same channel). Our bombardment rates are quite low, however; moreover SCHLEICH simulation results indicate that only about 0.22% of the ions directed at normal incidence will channel into the deeper regions of the material, further reducing the actual instantaneous flux of ions.

Superlattice-MQW Probe

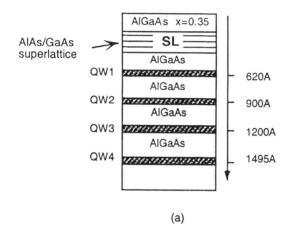

(a)

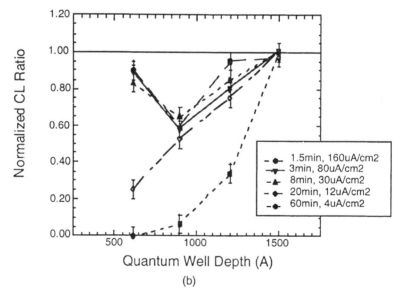

Quantum Well Depth (A)

(b)

Fig. 10.   (a) Use of the 'superlattice-MQW-probe' with the interposition of a lattice-matched AlAs/GaAs superlattice, resulting in the damage curves shown in (b).

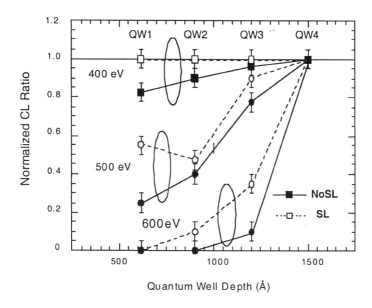

Fig. 11. Damage profiles for samples with (filled) or without (open) SL, following Ar ion bombardment at various energies for 3 minutes at 50 μA/cm^2.

The incorporation of a superlattice into the MQW structure (Figure 10(a)) produced unusual damage profiles. By interposing a AlAs/GaAs (10 periods of 1 nm/1 nm AlAs/GaAs), between the substrate surface and the first quantum well, an improved damage profile resulted (comparing Figure 10(b) to Figure 9) but the improvement was observed primarily for the uppermost quantum well. Furthermore, as either the total irradiation time, or the energy of the incident ions is increased (Figure 11), the effect of the superlattice is diminished.

SCHLEICH simulations of the ion ranges were not able to delineate any differences for the two different structures [27]. This indicates that the introduction of a lattice-matched superlattice does not alter the channeling probabilities for the incident ions. However, superlattices have been shown to be effective in gettering defects, and have been used in the epitaxial growth of high quality optoelectronic materials to prevent propagation of defects from the substrate material into the overlying device layers [53]. Using this information, we can begin to construct a framework of understanding of the 'superlattice data', and more clearly single out the contribution of defect diffusion.

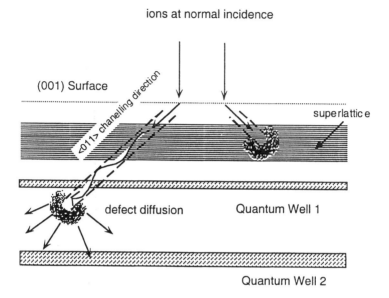

Fig. 12. Schematic of damage propagation from low energy ions. Ions incident normal to the substrate are scattered along a channeling direction, and propagate deeply into the substrate. Defects created by the stopping ions diffuse more deeply into the substrate; their presence is 'probed' by proximal quantum wells. The interposition of a superlattice may retard defect propagation into the adjacent quantum well.

Channeling allows the passage of low energy ions deep into the substrate. Particularly at the end of the range, transfer of ion energy to the material leads to displacement and defect creation. Assuming that there is appreciable diffusion of those defects, the QWs in our structures probe the defects that are spatially generated within a diffusion length of their position. This is indicated schematically in Figure 12. The superlattices used in these experiments have the ability to getter and retard the transport of defects. The effect of the superlattice will only be sensed by the quantum well probe in closest proximity to that superlattice (in our case, QW1). If we compare the SCHLEICH-calculated ranges of ions of 400 eV, 500 eV, and 600 eV with the position of the superlattice (Figure 13), we can provide a first interpretation of the data of Figure 11: defects created within the spatial extent of the superlattice may be gettered, and their further propagation into the material retarded. The range of the 600 eV ions extends beyond the superlattice, hence a relatively small percentage of generated defects are expected to be affected by the presence of the superlattice.

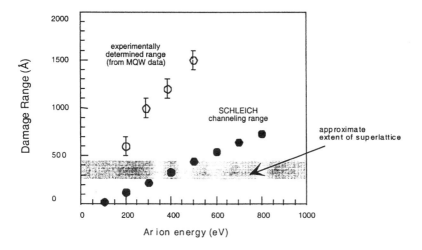

Fig. 13. Comparison of SCHLEICH-calculated channeling ranges with experimentally determined defect depths. Also shown is the approximate extent of the superlattice region.

4.3. Channeling and Diffusion in Ion Damage: A CHANDID Approach

The experiments above provide evidence for the importance of defect diffusion in determining the defect profile; can we, in fact, obtain better quantitative correspondence to a simulation that included both channeling *and* diffusion of defects? To help address this question, we developed such a simulation program, building upon the SCHLEICH calculations, and called CHANDID: CHanneling ANd Diffusion in Ion Damage [8]. This first attempt at a phenomenological modeling of the effects of channeling and diffusion necessarily made a number of simplifying assumptions: (1) although there may be a variety of defects (e.g. vacancies, interstitials) that contribute to the final damage profile, and that have differing statistics of formation, differing characteristic diffusion constants and varying energy levels, we assumed only a single generic defect with a single value of diffusion constant, depending on the region in the substrate through which that defect is propagating. We assumed that there is a single defect diffusion constant in the GaAs and AlGaAs material, D_{eff}. (2) The characteristic diffusion constant in the superlattice region differs from D_{eff}. We simulated the apparent 'slower' transit of the defects through the superlattice region by fitting to a $D_{\text{eff}.SL} < D_{\text{eff}}$. (3) Although we expect that the spatial distribution of the defects will be different from the distribution of the ions themselves, the model assumes that the distributions are identical.

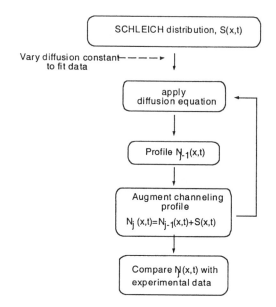

Fig. 14. Flowchart of CHANDID calculations.

The simulation proceeded by an alternating introduction of an ion (defect) distribution calculated by SCHLEICH with the subsequent application of Fick's equation to that distribution to account for defect diffusion for a given time interval. The channeling profile obtained from SCHLEICH, $S(x, t)$ forms the initial condition, $N_1(x, t)$, where x is the direction into the substrate, with $x = 0$ at the substrate surface; t is the time. Fick's equation is solved by the finite difference method:

$$\frac{\partial N_j(x, t)}{\partial t} = D_k \frac{\partial^2 N_j(x, t)}{\partial x^2}$$

where $N_j(x, t) = N_{j-1}(x, t) + S(x, t)$; $N_j(x, t)$ is the defect distribution at the jth time interval during the calculation, and k ($= 1, 2, 3 \ldots$) represents different materials in the layered structure. This was repeated successively until the total ion bombardment time is accounted for. Further details of the calculation can be found in reference [8], and the algorithmic approach is indicated in Figure 14.

We applied CHANDID to our experimental data of defect profiles generated with or without an intervening superlattice structure. Figure 15 shows the CHANDID evaluation of the effect of the superlattice on the defect distribution: the defects 'pile up', in the superlattice region, resulting in a

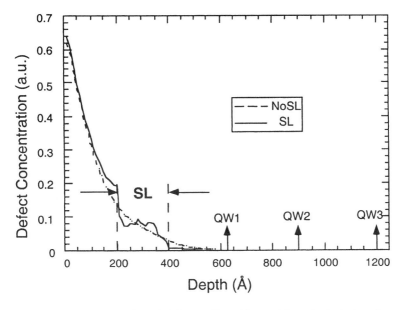

Fig. 15. CHANDID simulation of defect profiles with or without the inclusion of an intervening superlattice.

lower concentration of defects in QW1. The best fits to the data yielded values of D_{eff} in the GaAs material $\sim 3 \times 10^{-15}$ cm^2/s, and $D_{eff,SL} \sim 1 \times 10^{-17}$ cm^2/s. In addition, using the range of total defects calculated by CHANDID, and comparing it to that obtained experimentally, we find a far better agreement between simulation and experiment than had been obtained by either TRIM or SCHLEICH alone.

The comparison between the experimentally obtained range of damage propagation, with the successively closer calculated values given by TRIM, SCHLEICH (with channeling) and CHANDID (channeling and diffusion) are shown in Figure 16.

4.4.   Radiation-Enhanced Defect Diffusion — The Final Link?

The inclusion of defect diffusion into the ion damage model, even at a very simple level, seems to provide good agreement with experimental results. The range of damage propagation here appears to be far better than a simple channeling-only model could provide. However, in characteristic fashion, the increased understanding is accompanied by additional puzzles. The outstanding question relates to the very high value of the diffusion constant: 3×10^{-15} cm^2/sec at room temperature! The comparable temperature for the

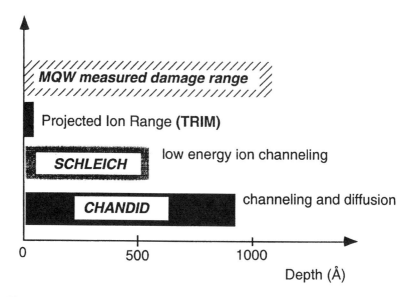

Fig. 16. Schematic representation of the various ion ranges calculated for 500 eV argon ions in GaAs, using TRIM, SCHLEICH and CHANDID. Also shown in the experimentally observed range, obtained from measuring luminescence of a MQW probe structure.

same value of Ga diffusion in GaAs would be $>800°C$, and considerably higher for that value of As diffusion in GaAs [24]. Further, the high value of the diffusion constant can only be in effect during the ion bombardment itself. Otherwise, the damage profiles that we measure would change dramatically over the period of a few days or weeks; our measurements show that this is not the case. It is important to determine whether the value of diffusion constant we have extracted is simply a fortuitous fitting parameter that helps to better rationalize our data, or if it indeed has some basis in an actual physical process. We note that a number of researchers have made calculations relating to defect diffusion in these processes, and have produced a range of rather high values, spanning 10^{-15} cm^2/s to 10^{-13} cm^2/s. Rahman *et al.* have made calculations based on conductance measurements on dry etched wires and epilayers, and obtained values of $D \sim 10^{-13}$ cm^2/s [54, 55]. Frost *et al.* have used a technique that combines a MQW probe together with a beveled sectioning technique, and deduced $D \sim 4 \times 10^{-15}$ cm^2/s for 500V nitrogen ion beam etching. Davis *et al.* have made a calculation of diffusion in a 'moving boundary' configuration, and deduced values of $D > 10^{-15}$ cm^2/s [16].

How then can we account for the dramatically large value of defect diffusion? One possible mechanism may lie in the nature of the semiconductor structures being studied, and additional excitations that can occur during

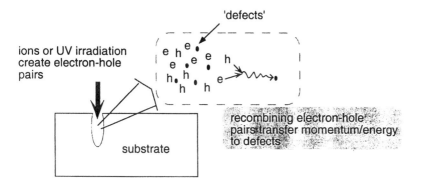

Fig. 17.   Schematic of the radiation-enhanced recombination process in a semiconductor.

the ion-bombardment/etching process. Earlier work had found that the vibrational energy released in non-radiative electron-hole recombination in semiconductors could transfer momentum and energy to defects in the material, resulting in enhanced motion of the defects, or indeed, enhanced annealing of the defects [5, 36, 66]. A low temperature annealing of electron or implantation-induced damage can be achieved by minority carrier injection [36, 63]. These earlier studies were generally carried out in the context of high-energy ion implantation or electron irradiation; very little has been done on semiconductors irradiated with low energy ions for the dry etching process. However, in an ion-enhanced etch process, it is clearly possible that excess electron-hole pairs may be generated through coupling of energy to the substrate from an ambient plasma emission, or from energy transferred from the electronic stopping of the ions. This process of 'radiation-enhanced recombination' is schematically illustrated in Figure 17.

   In 1995, Chen *et al.* conducted a series of studies to investigate the role of excess carriers in the enhanced diffusion of ion defects during reactive ion beam etching of GaAs and InP [9, 10, 11]. The excess carriers were generated through the addition of above-band gap laser illumination, applied during the ion bombardment process. Figure 18 compares the damage profiles of a sample that was bombarded with 500 eV Ar ions for 3 minutes, either with or without 250 mW/cm^2 HeNe laser illumination during the ion exposure. There is a clearly visible change brought about by the addition of the laser illumination; Figure 19 shows that an increase in laser power density increases the relative degradation of each QW peak.

   To provide a calibration, similar studies were undertaken using an InGaAs diode laser, with 980-nm wavelength, providing illumination below the bandgap of the Ga (Al) As sample. In that case, there is only a slight decrease

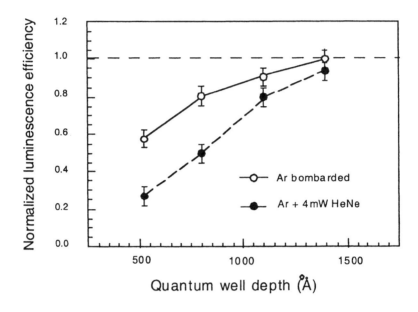

Fig. 18. Damage profile for the GaAs MQW for samples with (dots) and without (circles) 250 mW/cm^2 HeNe laser illuminations during ion exposure (above bandgap).

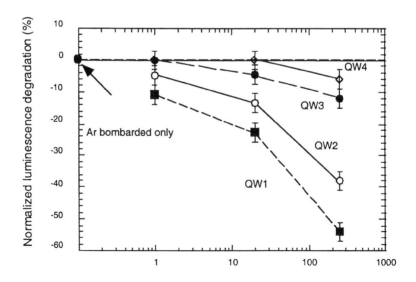

Fig. 19. Degradation of QW luminescence efficiency versus HeNe laser power density.

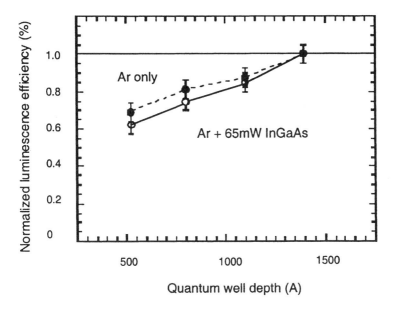

Fig. 20. Damage profile for the GaAs MQW for samples with (dots) and without (circles) InGaAs laser illumination during ion exposure (below bandgap).

in luminescence, with no significant increase in damage as the laser power density is increased (see Figure 20). Below-bandgap illumination results in a minimal increase in damage profile. Rather than radiation-enhanced diffusion, one might question whether the effect of the laser would be to provide local heating of the material, thereby augmenting defect diffusion. In fact, the laser-induced temperature rise is negligible at the power density used in the experiments. The spot size of the laser beam used in the experiments is ∼ 1 mm. or even larger. Lax's theoretical calculations [37] of temperature rise induced by Gaussian laser beam irradiation absorbed in the solid indicate that the maximum temperature rise is ∼ 0.1°C at the MQW surface for the 4mW HeNe laser used in the experiments. This is not expected to be high enough to produce a change in the MQW damage profile.

It is also important to note that laser illumination of samples that were not carried out concurrently with ion bombardment produced *no* change in the PL intensity of the sample. That is, laser-induced changes in the PL spectra could only be brought about during ion bombardment. Ion bombardment creates an environment local to the defects that lowers the effective barrier to defect diffusion. The energy transfer between lattice and incident ions in collision cascades will result in increased lattice vibration and local heating. This can be far more effective on a local scale than changes in defect diffusion and annealing brought about by temperature changes made to the entire substrate.

Our results show clear evidence of radiation-enhanced diffusion of etch induced defects, but it is not yet clear if recombination enhanced diffusion alone can explain the high mobility of ion-induced defects. Further analysis and experiments are clearly required to better understand the enhanced diffusion of etch-induced defects.

4.5.   Etch-Rate Dependence of Damage

Our discussions of damage propagation, and the simulations have up to now not taken into account the chemical component of the ion-assisted etch process, and the possibility of removal of defects during the etch process. Intuitively, shifting the balance of the etch process to a more chemically (rather than physically) driven process should also produce a lower damage process. Experiments carried out by Skidmore *et al.* showed improvements in both surface quality and bulk damage when utilizing a greater chemical component in the etch process [57]. We have used CHANDID to calculate the effect of a finite etch rate and etch-removal of surface damage; in this case, the surface of the substrate moves with a velocity corresponding to the sputter/etch rate, R_e, and the position of the surface is therefore 'reset' at each time interval of the numerical calculation.

Figure 21 shows the results of that calculation on the defect profile and the total accumulated damage in the material. The faster sputter (or etch) rate results in lower total accumulated damage. In our case, the total ion dose is the same in all cases: the simulations are not simply reflecting lower integrated ion dose (that is, a faster etch rate at the same ion dose results in a lower integrated ion dose to produce a give etched depth). The critical process in this case is the rapid removal of the highly defect-ridden surface region, on a time scale that is fast compared to the time required for diffusion of those defects away from the surface. However, although the total integrated damage is reduced in samples with higher argon sputtering (etch) rates, the damage at the surface is higher.

4.6.   Ion Damage to Etched Sidewalls

The discussion up to now has centered on the introduction of ion damage directly into the bulk of the substrate. In many cases, the critical, device-active material will be covered by a mask during the etch process, which should impede the introduction of the defects into the underlying material. However, in spite of the highly anisotropic nature of the process, etch damage can be introduced to the sidewalls of the structures being etched, either through chemical modification of the surface, or through some amount of ion

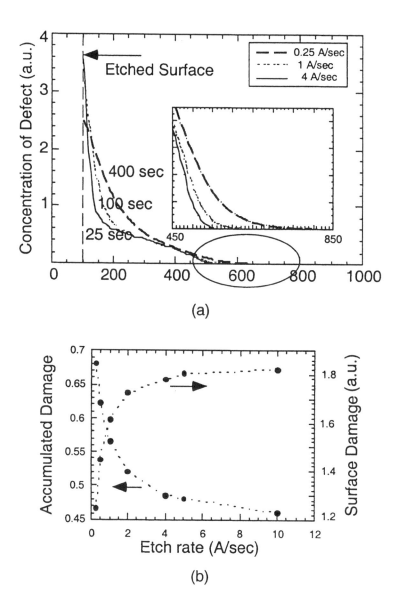

Fig. 21. CHANDID simulations of (a) the influence of the etch-rate on the time dependent damage distribution, and (b) time dependence of accumulated damage in the bulk and at the surface of the etched sample. The simulations correspond to 500 eV Ar ion exposure, at various current densities to achieve a constant etched depth of 100 Å .

scattering that attacks the sidewalls. This is particularly critical for structures that have a large ratio of surface to volume, and has limited the minimum diameter of etched vertical cavity surface emitting lasers (VCSELs) [15]. Utilizing PL data taken from InGaAs/InP and GaAs/AlGaAs wires etched through a various dry etch processes, Maile *et al.* determined that the ion-assisted etching could give rise to optically inactive sidewall layers on the order of ~10's of nms [45, 46]. Wet etch-removal of the optically 'dead' layer, followed by passivation through surface treatment and regrowth has been used to overcome this process-induced materials limitation.

5. ENGINEERING MATERIALS FOR GREATER ROBUSTNESS TO ION DAMAGE

A fairly consistent picture of ion-induced defect generation is emerging, with important implications for the design of low-damage etch processes. The experiments described earlier in this chapter have confirmed the wisdom of designing etch processes with high chemical component, and low ion energies. They have provided some guidelines for carrying out *in situ*, effective annealing out of defects. Simulations and interpretations are still very much at the phenomenological stage: much more work remains to be done to identify particular defects, understand their formation and interactions with the lattice and with other defects. Use of techniques such as Deep Level Transient Spectroscopy (DLTS) has helped to trace the evolution of particular ion bombardment-induced traps during low energy ion exposures, and shown that the highly mobile defects generated during low energy ion exposures are basically associated with the components of primary point defects, such as such as interstitials and vacancies [11, 32, 73].

There is a great deal of work going on in the development of dry-etch schemes that will minimize damage to the substrate. We will proceed in this chapter, not by consideration of defect behavior associated with various etch schemes, but by further examining defect behavior in different materials, having different microstructure. The goal here is twofold: (1) to be able to ultimately extend our simulations and insights from the Ga(Al)As materials to other important optoelectronic materials, and to thereby augment the accuracy of the simulations and begin to test its predictive capabilities. The second goal is to see if we can engineer materials, like the superlattice structures discussed earlier, which may either be themselves more robust with respect to etch-associated damage, or which may serve as barriers to or getters of defects. In the next section, we describe some of our initial studies in this area, examining the damage profiles of InP-based materials, non-stoichiometric GaAs (or low temperature-grown GaAs), strained quantum well material, and finally, quantum dot materials.

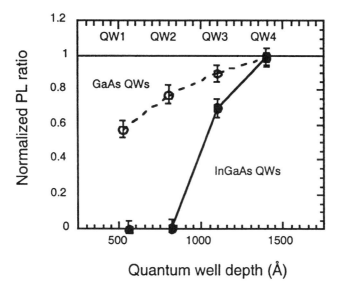

Fig. 22. Damage profiles for GaAs QWs and InGaAs QWs after 500 V Ar ion bombardment for 3 minutes, at 50 μA/cm^2.

5.1. Etch Damage of InP

InP-based materials are critical optoelectronic materials, particularly for the longer wavelength (1.3, 1.55 μm) device applications. Multiple quantum well probe materials can be grown in this material system, utilizing InGaAs quantum wells. We have carried out a number of comparative studies between GaAs and InP-based MQWs, subjecting samples to the same conditions of ion bombardment, and subsequently using PL to analyze the resulting damage profiles. Figure 22 gives a comparison of the damage profiles for GaAs and InP bombarded by 500 V Argon ions for 3 minutes at 50 μA/cm^2. Clearly, there is a more profound effect on the InP sample than on the GaAs material. If we monitor the effects of increased irradiation time (with constant total dose) on the PL intensity, we see in Figure 23 that the InP material degrades more rapidly with increased ion exposure. Finally, laser-illumination during ion bombardment, produces a degradation in the InP MQW damage profile, similar to that observed for the GaAs MQW material.

The most extensive comparisons of ion damage in compound semi-conductors have been carried out for ion energies in the keV and MeV range, characteristic of energies used for ion implantation. Using Rutherford Backscattering Channeling (RBC), and optical absorption measurements, Wesch *et al.* have reported higher defect densities in the phosphides,

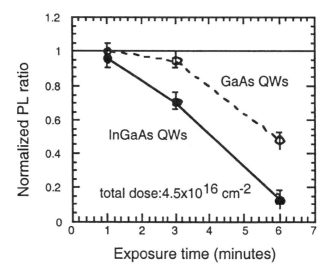

Fig. 23. Time dependence of PL degradation of QW3 in the GaAs and InP-based MQW samples. 500 V Ar ion bombardment was carried out at varied ion currents and exposure times to maintain a constant dose, $4.5 \times 10^{16}/cm^2$.

compared to the arsenides, for incident ion energies >200 keV [68]. The primary defects created in GaAs are believed to be point defects, such as vacancies and interstitials, with sufficient mobilities that a substantial fraction of the defects are annealed out *during* the bombardment process itself, at room temperature. Such self-annealing is believed not to take place for defects and defect clusters in InP. Williams subjected Si, GaAs and InP to argon ion bombardment at energies of 1, 2 and 3 keV [69]. Total doses were on the order of $10^{15} - 10^{16}$ cm^2. InP was found to be the most sensitive to ion bombardment, with differential sputter removal of P, as well as 'bombardment driven' defect diffusion. At 1 keV, and for the same ion dose, the amount of damaged material (as determined by RBC) was twice as great for InP as for GaAs. These results are consistent with the values of displacement energy calculated by Bäuerlein [4] and incorporated by Stoffel into his SCHLEICH program. A partial listing of the derived displacement energies derived is listed in Table I.

The asymmetry in the sizes of the In and P, compared to GaAs, produces the asymmetry and lowered displacement energies. We can follow up these trends in the lower ion energy regime of interest to us by using SCHLEICH calculations to compare the channeling depths and number of

TABLE I
Displacement Energy of Semiconductors [4]

Substance	Si	InP		GaAs		InAs		InSb	
Displaced Atom	Si	In	P	Ga	As	In	As	In	Sb
Displacement Energy (eV)	12.9	6.7	8.7	9.0	9.4	6.7	8.3	5.7	6.6

atomic displacements in GaAs, and InP subject to incident Ar ions of varying energy. The results are plotted in Figure 24(a) and (b). At each ion energy, the channeling depth in InP is slightly larger than for GaAs; more noticeable is the difference in the number of atomic displacements in the two materials, and the different rates of increase of atomic displacements with increasing energy. These differences may manifest themselves in the rate of formation, and ultimate concentration of defects, per incident ion.

Having seen that SCHLEICH itself can indicate the trends of damage to be observed experimentally in the various materials, let us go further and compare experiment and simulations that invoke diffusion as well as channeling in the defect propagation. In a manner similar to the experiments carried out for GaAs MQWs, we designed InGaAs MQW structures, with and without the interposition of a lattice-matched superlattice comprising 4 periods of 10 Å InGaAs QWs separated by 150 Å InP barrier layers. (Another SL structure was designed and used, having twice as many QWs and half the thickness in the barrier layer to maintain the same overall material thickness. Within experimental error, the results obtained for the two superlattice structures were the same). The SL and NSL samples were then subject to the same conditions of Ar ion bombardment: different total irradiation times for a fixed constant dose ($2.25 \times 10^{16}/cm^2$) at 500 eV energy (See Figure 25). As in the case of the GaAs MQWs, the introduction of the SL layer improved the luminescence properties of the material; in the case of the InGaAs MQWs, the superlattice seemed to have a larger, and longer range effect than was true for the GaAs material.

Figure 26 compares the percentage improvement introduced by superlattices in the two materials. These data were then used in CHANDID to extract a value for effective defect diffusion constant. The best fit to these data yielded $D_{InP} = 4 \times 10^{-15}$ cm^2/s $- 1 \times 10^{-14}$ cm^2/s, and $D_{SL} = 1 \times 10^{-16}$ cm^2/s. The defect diffusivity in InP is slightly larger than that fitted to the data for GaAs.

In summary, these experiments have shown that under comparable ion bombardment conditions, the InP-based material seems to degrade more substantially. This is borne out by the simulations and data that give deeper channeling ranges, a higher number of displacements, and yields a larger

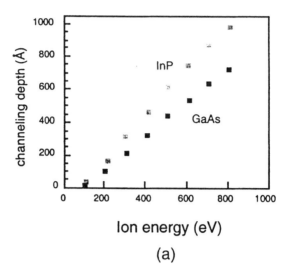

(a)

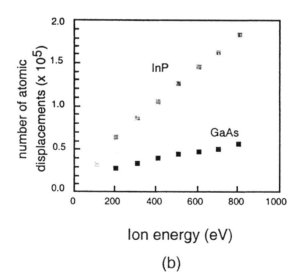

(b)

Fig. 24. SCHLEICH calculations of (a) channeling depth and (b) number of atoms displaced in GaAs (dots) and InP (circles) for 10,000 argon ions incident along the ⟨011⟩ direction, at ion energies of 100–800 eV.

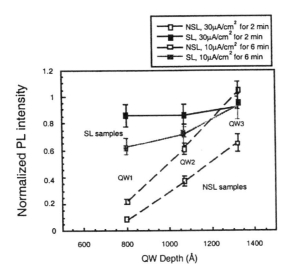

Fig. 25. Comparison of damage profiles in InGaAs MQW structures with (SL) and without (NSL) a superlattice. Argon ion energy was 500 eV, for a total dose of $2.25 \times 10^{16}/\text{cm}^2$.

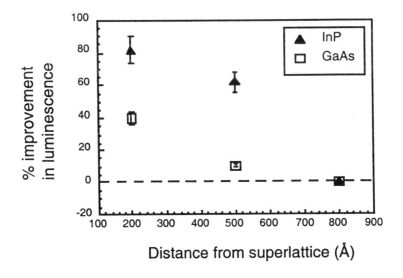

Fig. 26. Comparison of percentage improvement introduced by SL layers, for both GaAs-based and InP-based multiple quantum well materials. The effect is shown for QWs situated at various distances from the SL region.

defect diffusion constant. Measurements by Steffensen *et al.* have shown that InGaAs QW peaks actually blueshift under high enough irradiation conditions [62]. The blueshift may be related to defect-enhanced intermixing of the QW hetero-interfaces, the defects having been generated through the ion bombardment process. No such shifts have been observed in our experience with GaAs MQWs. Further characterization of the nature of the defects formed in InP will be required, but at the moment it appears that defects can more easily form, and more readily propagate deeper into the substrate.

5.2. Non-Stoichiometric GaAs

We have seen that the microstructure and composition of materials may influence the generation and propagation of ion-induced defects. The superlattice structures mentioned earlier in this chapter have had some influence on retarding the propagation of defects further into the material. The effectiveness of the superlattices encourages us to seek out other possibilities for materials that may getter or serve as barriers to defect propagation. One likely candidate material is non-stoichiometric, or low-temperature grown (LT) GaAs. Grown by techniques such as MBE at lower temperatures ($200°C$–$300°C$) than would be thought optimal for the highest quality materials, LT GaAs contains 1–2% excess arsenic [59], which corresponds to 10^{20}–10^{21} cm^{-3} excess As atoms. The excess arsenic exists in the as-grown material as point defects, As interstitials or As_{Ga} antisite defects. The high concentration of As_{Ga} antisite defects acts as a deep donor and allows hopping conductivity [29]. In addition, the large amount of excess As in the LT GaAs films is expected to expand the GaAs lattice, and a high dependence of lattice constant on growth temperature has been found [39, 40, 44].

Following a post thermal annealing at a higher temperature ($>400°C$), the excess arsenic begins to form As precipitates embedded in the GaAs matrix. For GaAs/AlAs or GaAs/AlGaAs heterostructures, As precipitates are preferentially formed in GaAs or AlGaAs layers with lower Al composition. The size and density of precipitates are strongly dependent on the annealing temperature [67]. The lattice constant returns to the normal GaAs value after thermal annealing and the resistivity increases to 10^6–10^7 Ω-cm. The unusual microstructure and electronic properties of the LT GaAs has led to a number of device applications: Incorporating an LT GaAs buffer layer between the substrate and active region can eliminate backgating in GaAs MESFETs [59]. In addition to serving as a buffer layer, LT-GaAs has been also employed as a surface passivation layer in a MISFET type structure to enhance the breakdown voltage or reduce the $1/f$ noise at frequencies below 100 Hz of a GaAs FET [49, 71]. Because of a high concentration of defects situated at midgap, the lifetime of photo-induced carriers typically is of the order

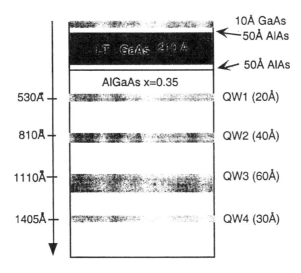

10Å GaAs
← 50Å AlAs

← 50Å AlAs

AlGaAs x=0.35

530Å — QW1 (20Å)

810Å — QW2 (40Å)

1110Å — QW3 (60Å)

1405Å — QW4 (30Å)

Fig. 27. MQW structure with LT GaAs capping layer. A similar structure with a normally grown, rather than LT GaAs cap was used for comparison.

of a few hundred femtoseconds [60] leading to applications as high speed LT-GaAs photodetectors [13].

We have previously established the two main phenomena giving rise to deeply penetrating damage profiles: channeling of ions, and rapid diffusion of ion-created defects. It is possible that the LT GaAs may serve as an effective barrier to the propagation of ion damage, both because of its non-crystalline structure, and because of an ability to getter defects. A number of experiments were undertaken to explore this possibility. The material structures used in this study were standard MQW structures, grown on semi-insulating (100) GaAs substrates by molecular beam epitaxy (MBE). Above the quantum wells, a thin layer of LT-GaAs (210 Å) and 50 Å AlAs were grown at 250°C. The AlAs layers on either side of the LT-GaAs act as diffusion barriers for point defects during the subsequent thermal annealing (Figure 27). The control sample differs only in that the top thin layers of AlAs and GaAs were grown at 600°C. After growth, the LT-GaAs capped sample was cleaved into pieces, covered with a GaAs wafer, and then separately annealed in a rapid thermal annealer (RTA) at 600°C for 30 s in a forming gas environment (90% N_2 and 10% H_2). Annealed samples together with control samples, were exposed to the Ar^+ ion beam at room temperature in a RBIBE system for 3 minutes at beam energies of 500 eV or 600 eV with an ion beam current of 50 $\mu A/cm^2$ at normal incidence. As shown in Figure 28, samples covered

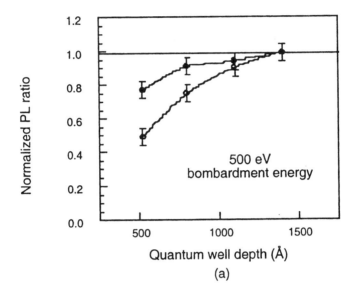

(a)

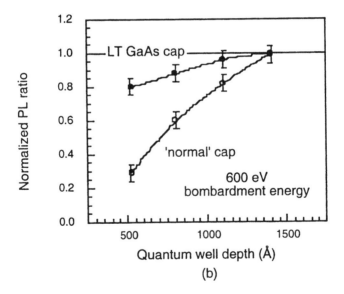

(b)

Fig. 28. Damage profiles for samples with a thin capping layer of LT-GaAs (dots) or regular GaAs (circles), following 3 minute argon ion beam exposure at ion energies of (a) 500 eV and (b) 600 eV, 50 μA/cm^2 at normal incidence.

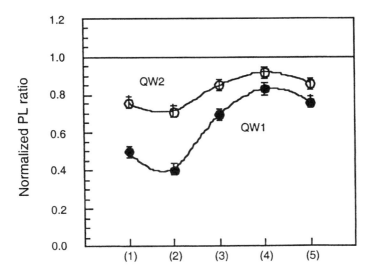

Fig. 29. The variation of QW PL efficiency, after 500 eV Ar ion bombardment for 3 minutes at 50 μA/cm^2. MQW samples were either (1) as-grown without a LT cap, (2) as-grown with a LT cap, and annealed at (3) 500°C (4) 600°C and (5) 700°C.

with a thin layer of LT-GaAs showed a great improvement in the quantum well luminescence by about 70% for a 600 V bombarded sample, indicating that the annealed LT GaAs layer can effectively reduce the deep penetration of ion damage.

To better understand the role played by the LT GaAs, we attempted to systematically vary the microstructure of the LT GaAs through post-growth anneals at various temperatures: 500°C, 600°C, and 700°C for 30 s in a Rapid Thermal Annealer (RTA). LT-GaAs layers annealed at different temperatures produced varying degrees of improvement in the PL of the underlying QWs. As shown in Figure 29 the best improvement in PL efficiency of QW1 and QW2 was observed for samples annealed at 600°C. The data also show a reduction of photoluminescence, compared to the control for the LT-GaAs capped sample that did not undergo a post-growth thermal annealing. This may be due to the out-diffusion of point defects from LT-GaAs capping layer into the underlying QW material, a process that may in turn be accelerated by the ion bombardment process. Figure 29 also shows a correlation of damage profiles with the microstructure of the capping LT-GaAs layer. Increased annealing temperatures will increase the precipitate size, at the same time decreasing the precipitate density in LT-GaAs layers, that is, increasing the separation between precipitates. Table II summarizes the average sizes and spacings between the arsenic precipitates at the various

TABLE II
Characteristics of As precipitates in the thin layer (~210 Å) of LT-GaAs annealed at
various temperatures from 500–700°C for 30 s, assuming the individual
precipitates to be spherical with a diameter of d.

Annealing Temperature (°C for 30 s)	Density (10^{16} cm^{-3})	Diameter (Å)	Spacing (Å)	Precipitates/ GaAs interface (10^5 cm^2)	Volume fraction
500	152	17	86	1.38	0.004
600	22	55	166	2.1	0.019
700	5.7	90	267	1.45	0.02

annealing temperature, as determined by analysis of plan-view transmission electron microscopy (TEM) images.

We have seen that a rather thin layer of LT GaAs obviously alters the damage profile of the underlying substrate, even at incident ion energies of 600 eV. What is the basic underlying mechanism of damage mitigation? Stoffel has noted that his SCHLEICH simulations suggest that 'an amorphous GaAs surface layer at least 8 nm thick is required to effectively prevent the subsequent channeling of 1 keV Ar ions' [64]. Whatever the nature of the microstructure, the 21 nm thick LT GaAs is clearly not a fully disordered amorphous layer, and the angled incidence data of Figure 30 indicate that the LT GaAs does *not* substantially block direct channeling. If the main effect of the LT GaAs is not to affect channeling, might it influence the diffusion of ion-induced defects? These effects are best observed as the channeling range corresponding to a particular incident ion energy straddles the region occupied by the LT GaAs layer. We see from Figure 31 that the relative improvement offered by the LT GaAs layer diminishes as the ion energy increases.

For 550 or 600 eV incident ions, the majority of defects would be expected to be formed deeper in the material than the LT GaAs layer, rendering the material ineffective as a damage barrier. Yet some small enhancement persists, even at these higher ion energies. The data of Figure 29 perhaps provide an answer, showing the maximum enhancement for normally incident ions occurring for cap layers annealed at 600°C. Reference to Table II shows that this is the annealing temperature that results in maximum common precipitate-GaAs interfacial area. This in turn could be related to a scattering cross section — with the role of the LT GaAS layer to alter the statistics of ion scattering into channeling directions.

These studies have shown how the microstructure of a material can influence the propagation of ion-induced defects into the surrounding material. The interesting aspect of working with LT-GaAs is the range of

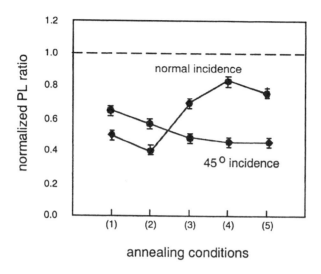

Fig. 30. Normalized PL of the first QW for ion bombardment at normal incidence or 45° incidence. Samples were either (1) as-grown without a LT cap, (2) as-grown with a LT cap, and annealed at (3) 500°C (4) 600°C and (5) 700°C. Ar ion bombardment was carried out at 500 eV for 3 minutes at 50 μA/cm^2.

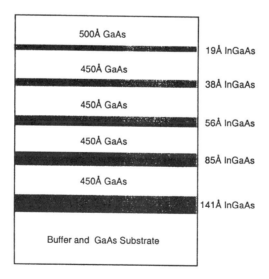

Fig. 31. Normalized PL of first quantum well as a function of argon ion bombardment energy. Samples either did (dots) or did not (circles) have a 200 Å LT GaAs capping layer, annealed at 600°C. The first QW is located $\sim$600 Å below the surface.

control that can be exercised in modifying the microstructure. Pragmatically, LT-GaAs may be incorporated into structures to achieve various advantages in device performance; in such cases, there may be additional attendent processing-related benefits of the LT-GaAs.

5.3.   Strained Layer Materials

All the material structures discussed thus far have involved lattice-matched heterostructures. Nonlattice, matched, or strained heterostructures might be expected to display dramatically different damage profiles: the change in lattice constant could alter channeling probabilities, and the strained interfaces could serve even more effectively than the lattice matched SL material as gettering sites for defects. A set of preliminary experiments were carried out using the strained-layer, $In_{0.2}Ga_{0.8}As/GaAs$ structure shown in Figure 32. The structure has a 500 Å GaAs surface barrier region and five $In_{0.2}Ga_{0.8}As$ quantum wells of widths 19, 38, 56, 85, and 141 Å separated by 450 Å GaAs barriers[1]. Samples of this strained-MQW structure and the lattice-matched GaAs/AlGaAs MQW structure were exposed for 3 minutes to an Ar ion beam at 500 V, 50 $\mu A/cm^2$, 30°C substrate temperature, and at either normal or 45° ion beam angle of incidence. The damage profiles for the two structures are compared in Figure 33(a) for normal incidence and in Figure 33(b) for 45° incidence. Clearly, for the normal incidence exposure, the damage measured by the strained-MQW was observed to be much greater than that of the unstrained material. The damage to the two structures is apparently similar for the 45° incidence exposure.

On the one hand, modulation of the lattice constant in the strained MQW samples might promote dechanneling of the incident ions. However, the regions of strained material are sufficiently thin that they may not prove to be very effective in dechanneling. This point seems to be borne out by Figure 33(b), where the ranges of ions (hence defects) in both strained and lattice-matched materials appear to be comparable for ions directed along the ⟨011⟩ channeling direction. (The greater-than-unity normalized CL intensity shown in Figure 33(b) is probably a result of incorrect normalization of the CL intensities: without an explicitly-grown, deeply lying reference QW, we used the 141 Å quantum well as a reference, situated only ∼2568 Å from the surface. This QW may not have been situated deeply enough not to have been itself affected by the ion bombardment; therefore all data points for the strained MQW in Figure 33(b) might more accurately be translated to lower values). Therefore, the lower luminescence, at each corresponding QW for the two samples, might reflect some enhanced dechanneling in the strained material. Detailed simulations of our structures will need to

---

[1]This material was kindly provided to us by Professor C. Stanley of Glasgow University.

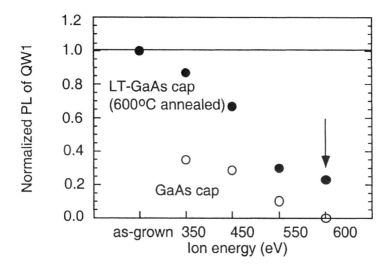

Fig. 32. $In_{0.2}Ga_{0.8}As$ strained multiple quantum well sample.

be carried out to verify if substantial dechanneling could indeed occur. In addition, the strained interfaces may serve as more effective gettering sites for ion-generated defects; i.e., the strained MQW material may be more efficient collectors of surrounding defects. However, looking back to the SCHLEICH calculations of ion channeling range as a function of incident ion energy, shown in Figure 13, the channeling range alone for 500 eV argon ions is much smaller than the almost 2000 Å depth of damage shown in Figure 33(a). It is highly likely that strain-enhanced diffusion plays an important role in ion processing of these materials. Other instances of such strain-enhanced diffusion have been reported [7, 12].

5.4.   Quantum Dot Structures

One of the most tantalizing and critical applications of ion-enhanced dry etching techniques is for the fabrication of sub-100 nm structures in which quantum confinement and other unusual properties are believed to apply. However, it is those very structures, by virtue of high surface area and small volume, which may be the most susceptible to ion-induced damage. A number of research groups have attempted to form quantum wire and quantum dot structures through a combination of high spatial resolution lithography, such as electron beam, together with dry-etch transfer of the pattern into compound semiconductor substrates [2, 30, 61]. Most attempts found that

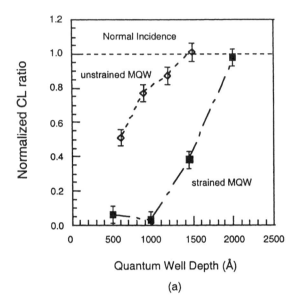

(a)

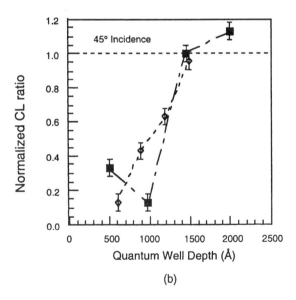

(b)

Fig. 33. Comparison of profiles of strained and lattice-matched MQW samples subject to 500 eV Ar ion beam bombardment at either (a) normal incidence or at (b) 45° to substrate. Samples were subject to 3 minutes bombardment at 50 $\mu A/cm^2$.

the limitations of sidewall and other damage severely limited the optical and electrical efficiency of those materials. Without post-etch processing such as a wet-etch 'clean up', or regrowth to passivate the etched sidewalls [2], very low optical performance resulted. In addition to the issues of processing damage, questions of the uniformity and absolute dimensions achievable called into question the practicality of the combination of lithography and ion-assisted etching to fabricate quantum structures. However, an exciting and effective alternative has been provided by the growth of 'self-assembled' InAs quantum dots (QDs) [38]. Formed *in situ* through strain-mediated epitaxial growth processes, these self-assembled dots yield a uniformity and control over size not achievable using previous, more conventional approaches. The natural overgrowth/capping possible with this technique results in highly efficient optical structures. In fact, such QDs have demonstrated improved resistance to damage by dislocations and surface recombination centers over quantum wells [21, 22]. These characteristics were attributed to efficient carrier capture and exciton localization in the QDs. The following experiments describe some initial assessments of the relative robustness of such QD structures to argon ion bombardment [56].

The quantum dot samples were grown by MBE on (100) semi-insulating GaAs wafers. These structures are composed of a short period superlattice (SPS) of AlAs/GaAs (20 Å /20 Å , 40 times), followed by a 500 Å GaAs buffer layer. A 10ML reference $In_{0.2}Ga_{0.8}As$ QW was grown to allow for normalization of the PL spectra between samples. This was followed by a 4000 Å GaAs barrier layer to isolate the reference well from point defects introduced by radiation exposure. Depending on the sample, either an 18 $In_{0.2}Ga_{0.8}As$ QW or 1.7ML InAs QDs layer (QD density of 10^{10} cm^{-2}) was then grown, followed by a 500 Å GaAs top layer. The sample structures (designated as either QW or QD) are shown in Figure 34. The procedure for the InAs QDs layer growth has previously been described elsewhere [19]. Samples were exposed to an Ar plasma (300 eV, 35 μA/cm^2) for times between 10–60 seconds, yielding ion doses of 2×10^{15}/cm^2 to 1.3×10^{16}/cm^2. PL data were taken using both an Ar ion laser and a Titanium-sapphire (TSL) laser ($\lambda = 840$ nm), allowing for optical pumping both above and below GaAs energy gap, respectively. We will refer to PL taken with the TSL as selective PL (SPL) since it selectively probes only the reference QW and QW or QDs layers, and not the surrounding GaAs.

The dependence of the PL intensity of the QW and QDs layers on ion dose is shown in Figure 35. The PL spectra were obtained by pumping with an Ar ion laser having 10 mW power, and a spot size of about 100 μm. The quantity R is the ratio of the area under the QW or QD PL peak normalized to that for the as-grown samples. Before the quantity R is computed, all the spectra for each ion exposure are normalized to the reference QW in each of the samples. As Figure 35 quite clearly shows, although there is a

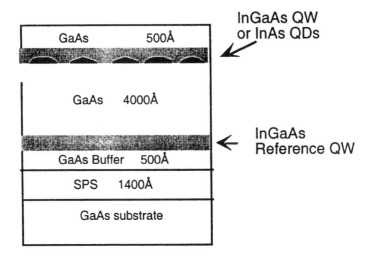

Fig. 34. Schematic diagram of the quantum dot (quantum well) material structures.

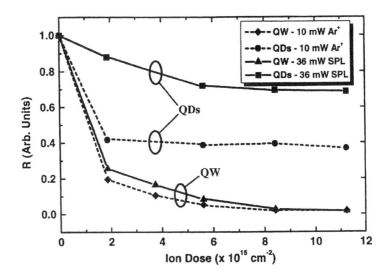

Fig. 35. Dependence of PL intensity of QD and QW samples on Ar ion beam dose.

degradation in luminescence efficiency for both types of materials, as the ion dose is increased, the QD samples appear to suffer substantially less degradation. Moreover, this difference is accentuated for the SPL data, with the QD samples exhibiting even higher relative optical efficiency.

How are we to understand these preliminary, but tantalizing results? On the one hand, we expect that the wave functions associated with electron-hole pairs (and excitons — the coupled electron-hole pairs) generated in the QD material will be more tightly confined, or localized within those structures, with less extension out into the barrier layers. Therefore the optical output from the QDs may be sampling a more limited extent of material than that given from the QW material. The stronger confinement, or localization, may result in greater optical output, since the photo-generated carriers in the QDs are not sampling as many defect-generated, non-radiative sites. On the other hand, quantum dots fabricated through a combination of lithography and etching, discussed earlier, showed high sensitivity to ion damage. In many cases, such quantum dots were far less optically efficient than the corresponding quantum well material. These structures would be expected to similarly confine and localize excitons. An argument that invokes exciton localization alone does not account for the profound differences in optical efficiency of the two kinds of quantum dots. Is there some key feature relating to the nature of the quantum dot materials themselves? The quantum dots in this most recent study have been formed through strain-mediated growth. The sizes and distributions of these dots are influenced by the strain in the material, and we expect that there is residual strain remaining. Our earlier considerations of the effects of strain on damage propagation suggested that defect accumulation and propagation would be aggravated by strain. However, in this case, the comparison of QD PL and SPL data suggests that there is a high density of defects clustered at or immediately outside of the QD-barrier interface, as if defects were clustered *outside* of the dot region.

Yoon *et al.* have carried out a study of the reliability of lasers utilizing strained InGaAs quantum wells, with differing amounts of In, and hence differing amount of strain [72]. Their studies indicated a reduced rate of laser degradation for more highly strained material, and they suggest that the strain must play a role in hindering defect propagation into the active (strained) layers of the device. They have also carried out some analysis that indicates, under certain conditions, it may be energetically unfavorable for defects to surmount a 'strain energy barrier' and propagate further into a given material. Much more will have to be determined about the nature of the strain fields in these quantum dot structures and about the nature of the ion-generated defects themselves. It is interesting, however, to contemplate yet another means of controlling defect placement and propagation through the engineering of the material structure.

6. SUMMARY

This chapter has reviewed some recent progress in the understanding of the propagation of defects generated through low-energy ion etching processes. Focussing primarily on the optical characterization of specially engineered multiple quantum well substrates, we have traced a steady decrease in the discrepancies between experimental assessments of ion damage and simulations of the damage profiles. Although we begin in a largely phenomenological manner, the steady application of simulation to different experimental conditions, and the addition of damage data for a wider ranging category of materials will allow us to reach a more fundamental understanding of the damage mechanisms. This in turn will allow us to design etch processes or utilize material structures that will minimize damage. There is still a tremendous amount of information to be learned about the precise nature and behavior of the ion-created defects. Ultimately, such studies will be critical not only in understanding the as-processed damage, but also in extrapolations to lifetimes and reliability of ion-etched devices.

Acknowledgements

The authors gratefully acknowledge the contributions of the host of researchers at UCSB who have contributed to this work over the years. These include H.F. Wong, D.G. Lishan, J.A. Skidmore, D.G. Yu, and P.M. Petroff. We are also grateful to N.G. Stoffel for making SCHLEICH available to us. This work has been supported by DARPA, under the Optoelectronic Technology Center, and through QUEST, an NSF Science and Technology Center, through Grant No. DMR 91-20007.

References

1. K. Akita, M. Taneya, Y. Sugimoto and H. Hidaka, *Journ. Vac. Sci. & Technol.*, **A8**(4), 3274 (1990).
2. H.E.G. Arnot, M. Watt, C.M. Sotomayor-Torres, R. Glew, R. Cusco, J. Bate and S.P. Beaumont, *Superlattices and Microstructures*, **5**, 459 (1989).
3. J.L. Benton, B.E. Weir, D.J. Eaglesham and R.A. Gottscho, *Journ. Vac. Sci. Technol.*, **B10**, 540 (1992).
4. R. Bäuerlein, 'Displacement Thresholds in Semiconductors', Radiation damage in solids, *Proceedings of the International School of Physics*, XVIII Course Sept. 5–24, 1960, Varese, IT. (ed. Billington), p.358 (1962).
5. C. Bourgoin and J.W. Corbett, *Phys. Lett.*, **38A**, p.135 (1972).
6. P.D. Brewer, G.M. Reksten and R.M. Osgood, Jr., *Solid State Technology*, **28**(4), 273 (1985).
7. Y.-L. Chang, M. Krishnamurthy, I.-H. Tan, E.L. Hu, J.L. Merz, P.M. Petroff, A. Frova and V. Emiliani, *Journ. Vac. Sci. Technol.*, **B11**(4), 1702 (1993).
8. C.-H. Chen, D.L. Green and E.L. Hu, *Journ. Vac. Sci. Technol.*, **B13**, 2355 (1995).

9. C.-H. Chen, D.L. Green, E.L. Hu, J.P. Ibbetson and P.M. Petroff, *Appl. Phys. Lett.*, **69**, 58 (1996).
10. C.-H. Chen, D.G. Yu, E.L. Hu and P.M. Petroff, *J. Vac. Sci. Technol.*, **B14**, 3684 (1996).
11. C.-H. Chen, Y.-J. Chiu and E.L. Hu, *J. Vac. Sci. Technol.*, **B15**, 2648 (1997).
12. C.-H. Chen and E.L. Hu, to be published in *Journ. Vac. Sci. Technol.*, **B**, Nov./Dec. 1998.
13. Y.-J. Chiu, S.B. Fleischer and J.E. Bowers, MWP'97, International Topical Meetings on Microwave Photonics, PDP-1 (1997).
14. E.M. Clauson, H.G. Craighead, J.P. Harbison, A. Scherer, L.M. Schiavone, B. Van der Gaag and L. Florez, *Journ. Vac. Sci. Technol.*, **B11**, 2011 (1989).
15. T.A. Strand, B.J. Thibeault and L.A. Coldren, *Journ. Appl. Phys.*, **81**(8), 3377 (1997).
16. R.J. Davis and P. Jha, *J. Vac. Sci. Technol.*, **B13**, 242 (1995).
17. L. Deng, M. Rahman, S.K. Murad, A. Boyd and C.D.W. Wilkinsen, *Journ. Vac. Sci. Technol.*, **B16**, 334 (1998).
18. G.F. Doughty, C.L. Dargan and C.D.W. Wilkinson, IEE Colluquium on 'Towards Semiconductor Integrated Optoelectronics' London, UK: IEE, p.8 (1985).
19. M. Fricke, A. Lorke, J.P. Kotthaus, G. Medeiros-Ribeiro and P.M. Petroff, *Appl. Phys. Lett.*, **36**, 197 (1996).
20. K. Gamo, H. Miyake, Y. Yuba, S. Namba, H. Kashara, H. Swaragi and R. Aihara, *Journ. Vac. Sci. Technol.*, **B6**, 2124 (1988).
21. M. Garcia, G. Medeiros-Ribeiro, K. Schmidt, T. Ngo, J.L. Feng, A. Lorke, J. Kotthaus and P.M. Petroff, *Appl. Phys. Lett.*, **71**(14), 2014 (1997).
22. M. Gerard, O. Cabrol and B. Sermage, *Appl. Phys. Lett.*, **68**, 3123 (1996).
23. R. Germann, A. Forchel, M. Bresch and H.P. Meier, *Journ. Vac. Sci. Technol.*, **B7**, 1475 (1989).
24. S.K. Ghandi, *VLSI Fabrication Principles* (Wiley-Interscience, New York, 1983) p.139.
25. H.P. Gillis, D.A. Choutov, P.A. Steiner, IV, J.D. Piper, J.H. Crouch, P.M. Dove and K.P. Martin, *Appl. Phys. Lett.*, **66**(19), 2475 (1995).
26. D.L. Green, E.L. Hu, P.M. Petroff, V. Liberman, M. Nooney, R. Martin, *Journ. Vac. Sci. Technol.*, **B11**, 2249 (1993).
27. D.L. Green, E.L. Hu and N.G. Stoffel, *Journ. Vac. Sci. Technol.*, **B12**, 3311 (1994).
28. V.J. van Gurp and J.M. Jacobs, *Phillips Journal of Research*, **44**, 211 (1989).
29. J.P. Ibbetson, J.S. Speck, A.C. Gossard and U.K. Mishra, *Appl. Phys. Lett.*, **62**, 169 (1993).
30. A. Izrael, B. Sermage, J.Y. Marzin, A. Ougazzaden, R. Azoulay, J. Etrillard, V. Thierry-Mieg and L. Henry, *Appl. Phys. Lett.*, **56**(9), 830 (1990).
31. J.E. Johnson and C.L. Tang, *Electronics Letters*, **28**(21), 2025 (1992).
32. T. Kanayama, T. Wada and Y. Sugiyama, *Mater. Sci. Forum*, **196–201** (18th International Conference on Defects in Semiconductors, Sendai, Japan, 23–28 July 1995), 1939 (1995).
33. M. Kawabe, N., Kanzaki, K. Masuda and S. Namba, *Appl. Optics*, **17**(16), 2556 (1978).
34. ?. Kirillov, C.B. Cooper III and R.A. Powell, *Journ. Vac. Sci. Technol.*, **B4**, 1316 (1986).
35. H. Kudo, H. Takiguchi, C. Sakane, S. Sugahara, S. Yano and T. Hijikata, *Sharp Technical Journal*, **41**, 15 (1989).
36. Lang, D.V. and L.C. Kimmerling, *Phys. Rev. Lett.*, **33**, 489 (1974).
37. M. Lax, *J. Appl. Phys.*, **48**, 3919 (1977).
38. D. Leonard, M. Krishnamurthy, C.M. Reaves, S.P. Denbaars and P.M. Petroff, *Appl. Phys. Lett.*, **63**, 3203 (1993).
39. Z. Liliental-Weber, W. Swider, K.M. Yu, I. Kortright, F.W. Smith and A.R. Cawala, *Appl. Phys. Lett.*, **58**, 2153 (1991).
40. Z. Liliental-Weber, *Mater. Res. Sec. Symp. Proceedings*, **241**, 101 (1992).
41. Lindhard and M. Scharff, *Phys. Rev.*, **124**, 128 (1961).
42. Lindhard, M. Scharff, and H.E. Schiott, Kongelige Danske Videnskabernes Selskab, *Matematis.-Fysiske Medd.*, **33**, (1963).
43. C. Look, *J. of Appl. Phys.*, **70**, 3148 (1991).
44. C. Look, *Thin Solid Film*, **231**, 61 (1993).
45. B.E. Maile, A. Forchel and R. Germann, *Appl. Phys. Lett.*, **54**(16), 1552 (1989).
46. B.E. Maile, A. Forchel, R. Germann, D. Grützmacher, H.P. Meier and J.-P. Reithmaier, *Journ. Vac. Sci. Technol.*, **B7**(6), 2030 (1989).

47. A. Matsutani, F. Koyama and K. Iga, Fabrication and Characterization of Semiconductor Optoelectronic Devices and Integrated Circuits, *AIP Conference Proceedings*, **227**, 112 (1991).
48. J.W. Mayer, L. Eriksson and J.A. Davies, Ion Implantation in Semiconductors, Silicon and Germanium,' Academic Press, NY (1970).
49. U.K. Mishra, *Proceedings of the European Materials Research Society 1993 Spring Meeting, Symposium B: Low Temperature Molecular Beam Epitaxy of III–V Materials: Physics and Applications*, Strasbourg, France, 72 (1993).
50. T. Murotani, T. Shimanoe and S. Mitsui, *J. of Crystal Growth*, **45**, 302 (1978).
51. S.W. Pang, M.W. Geis, N.N. Efremow and G.A. Lincoln, *Journ. Vac. Sci. Technol.*, **B3**, 398 (1985).
52. S.W. Pang, *Journ. Electrochem. Soc.*, **133**, 784 (1986).
53. P.M. Petroff, R.C. Miller, A.C. Gossard and W. Wiegmann, *Appl. Phys. Lett.*, **44**, 217 (1984).
54. M. Rahman, N.P. Johnson, M.A. Foad, A.R. Long, M.C. Holland and C.D.W. Wilkinson, *Appl. Phys. Lett.*, **61**, 2335 (1992).
55. M. Rahman, *J. Appl. Phys.*, **82**, 2215 (1997).
56. W.V. Schoenfeld, C.-H. Chen, P.M. Petroff and E.L. Hu, *Appl. Phys. Lett.*, **73**(20), 2935 (1998).
57. J.A. Skidmore, D.L. Green, D.B. Young, J.A. Olsen, E.L. Hu, L.A. Coldren and P.M. Petroff, *J. Vac. Sci. Technol.*, **B9**(6), 3516 (1991).
58. J.A. Skidmore, G.D. Spiers, J.H. English, Z. Xu, C.B. Prater, L.A. Coldren, E.L. Hu and P.M. Petroff, *Proceed. SPIE*, **1671**, 268 (1992).
59. W. Smith, A.R. Cawala, C.-L. Chen, M.J. Manfra and L.J. Mahoney, *IEEE Electron Device Lett.*, **9**, 77 (1988).
60. W. Smith, H.Q. Le, V. Diadiuk, M.A. Hollis, A.R. Cawala, S. Gupta, M. Frankel, R. Dykaar, G.A. Mouton and T.Y. Hsiang, *Appl. Phys. Lett.*, **54**, 890 (1989).
61. C.M. Sotomayor-Torres, W.E. Leitch, D. Lootens, P.D. Wang, G.M. Williams, S. Thomas, H. Wallace, P. Van Daele, A.G. Cullis, C.R. Stanley, P. Demeester and S.P. Beaumont, *Proceed. of the International Symposium on Nanostructures and Mesoscopic Systems*, Santa Fe, NM, May 20–24, 1991, p.455, ed. Reed and Kirk (1991).
62. O.M. Steffensen, D. Birkedal, J. Hanberg, O. Albrektsen and S.W. Pang, *Journ. Appl. Phys.*, **78**(3), 1528 (1995).
63. D. Stievenard and J.C. Bourgoin, *Phys. Rev.*, **B33**, 8410 (1986).
64. N.G. Stoffel, *Journ. Vac. Sci. Technol.*, **B10**, 651 (1992).
65. M. Taneya, Y. Sugimoto and K. Akita, *Journ. Appl. Physics*, **66**(3), 1375 (1989).
66. V.L. Vinetskii and G.A. Kholodar, in *Physics of Radiation Effects in Crystals*, R.A. Johnson and A.N. Orlov, Editors. 1986: New York.
67. C. Warren, J.M. Woodall, I.L. Reeouf, D. Grischkowsky, D.T. McInturff, M.R. Melloch and N. Otsuka, *Appl. Phys. Lett.*, **57**, 1331 (1990).
68. W. Wesch, E. Wendler, T. Bachmann and O. Herre, *Nucl. Instrum. and Methods in Physics Research*, **B96**, 290 (1995).
69. R.S. Williams, *Sol. State Communicat.*, **41**, 153 (1982).
70. H.F. Wong, D.L. Green, T.Y. Liu, D.G. Lishan, M. Bellis, E.L. Hu, P.M. Petroff, P.O. Holtz and J.L. Merz, *Journ. Vac. Sci. Technol.*, **B6**, 1906 (1988).
71. L.-Y. Yin, Y. Hwang, J.L. Lee, R.M. Kolbas, R.J. Trew and U.K. Mishra, *IEEE Electron Device Lett.*, **11**, 561 (1990).
72. H. Yoon, Y.C. Chen, L. Davis, H.C. Sun, K. Zhang, J. Singh and P.K. Bhattacharya, *Proceedings IEEE/Cornell Conference on Advanced Concepts in High Speed Semiconductor Devices and Circuits*, Ithaca, NY, 348 (1993).
73. Y. Yuba, T. Ishida, K. Gamo and S. Namba, *Journ. Vac. Sci. Technol.*, **B6**, 253 (1988).

CHAPTER 4

Dry Etch Damage in Widegap Semiconductor Materials

S.J. PEARTON[1] AND R.J. SHUL[2]

[1] *Department of Materials Science and Engineering,*
University of Florida, Gainesville, FL 32611, USA
[2] *Sandia National Laboratories, Albuquerque, NM 87185, USA*

1. INTRODUCTION

Wide bandgap semiconductors are being developed for two basic classes of applications, namely [1–4]:

(i) Blue/green/UV emitters — the main candidates here are the GaN/InN/AlN system and the ZnSe/ZnS system [1]. While SiC blue light-emitting diodes have been fabricated and indeed are commercially available, the inefficient nature of the emission process due to the indirect gap of SiC means that these devices cannot compete with III-nitride or II–VI based devices. Laser diode lifetimes are currently >10,000 hours for InGaN/AlGaN devices and >100 hours for ZnSe/ZnSSe structures, and both blue and green InGaN LEDs are commercially available from a number of companies in Japan and the United States.

(ii) High power, high temperature electronics — the main candidates here are SiC, GaN and diamond. SiC is by far the most developed materials and device technology, while GaN looks extremely promising based on its excellent transport properties and the availability of hetero-structures such as InGaN/GaN and AlGaN/GaN. Diamond suffers from the lack of reproducible n-type doping and interest in diamond-based electronics has waned somewhat in recent years.

In this chapter we will give some details of the dry etching processes for III-nitrides, SiC and II–VI compounds, and the types of damage induced by various etching methods. Since all of these materials are compounds, there is the additional issue of avoiding creation of non-stoichiometric, near-surface regions through preferential loss of one of the elements during etching. Therefore "damage" to the semiconductor may take numerous forms, all of which lead to changes in the electrical and optical properties of the materials. These changes may be classesd into the following categories [5]:

1. Ion-induced creation of lattice defects which generally behave as deep level states and thus produce compensation, trapping or recombination in the material. Due to channeling of the low energy ions that strike the sample, and rapid diffusion of the defects created, the effects can be measured as deep as 1000 Å from the surface, even though the projected range of the ions is only ≤ 10 Å.

2. Unintentional passivation of dopants by atomic hydrogen. The hydrogen may be a specific component of the plasma chemistry, or may be unintentionally present from residual water vapor in the chamber or from sources such as photoresist mask erosion. The effects of the hydrogen deactivation of the dopants is a strong function of substrate temperature, but may occur to depths of several thousand angstroms.

3. Deposition of polymeric films from plasma chemistries involving CH_x radicals, or from reaction of photoresists masks with Cl_2-based plasmas.

4. Creation of non-stoichiometric surfaces through preferential loss of one of the lattice elements. This can occur because of strong differences in the volatility of the respective etch products, leading to enrichment of the less volatile species, or by preferential sputtering of the lighter lattice

element if there is a strong physical compared to the etch mechanism. Typical depths of this non-stoichiometry are ≤ 100 Å.

It is not always easy to delineate the damage mechanism. Clearly surface analysis by Auger Electron Spectroscopy (AES) or X-Ray Photoelectron Spectroscopy (XPS) can measure the stoichiometry of the etched surface and the presence of deposited films, but electrical methods are generally unable to differentiate between the compensation caused by hydrogen and by true etch damage. In both cases, annealing around 400–500°C may largely restore the initial properies of the near-surface region.

2. DAMAGE IN THE InGaAlN SYSTEM

2.1. Background

The binary nitrides are relatively strongly bonded compared to normal III–V's such as GaAs. GaN has a bond energy of 8.92 eV/atom, AlN 11.52 eV/atom and InN 7.72 eV/atom compared to the GaAs value of 6.52 eV/atom [6]. Therefore, fairly high ion energies (or high ion fluxes) are necessary to provide efficient bond breaking that will allow the etch products to form. Typical Ar^+ ion milling rates for binary nitrides are shown in Figure 1. The values for the $In_{0.5}Ga_{0.5}N$ ternary alloy falls between those for the component binaries. These ion milling rates are approximately a factor of 2–3 lower than for GaAs and InP under the same conditions [7]. This essentially precludes use of ion milling in GaN-based device structures because extremely thick masks would be required even for relatively shallow etch depths.

A number of different dry etch methods involving a chemical component to the etching in addition to physical sputtering. Figure 2 shows schematics of three different reactors [8–10]. In simple capacitively-coupled Reactive Ion Etching (RIE) the plasma is generated by application of rf power to the sample electrode via a coupling capacitor. The sample position sustains a negative potential with respect to the body of the plasma, and thus produces ion bombardment of the sample in addition to the impingement of neutral gas atoms and molecules. Increasing the applied rf power increases both the ion density and energy. Two different high density plasma systems are shown in the lower part of Figure 2. In both configurations, ion density is controlled by the rf chuck power. Decoupling these two parameters allows one to have a large ion flux, but a low ion energy and thus etch rate can be enhanced without creation of excessive lattice damage. In the system at center an Electron Cyclotron Resonance (ECR) source operating at microwave frequency couples power into the chamber and magnetic confinement of the plasma induces a high ion density through efficient electron-gas molecule collisions. Similar high ion densities ($\sim 5 \times 10^{11}$ cm^{-3}) can be achieved using

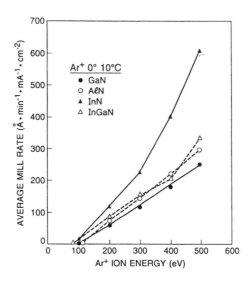

Fig. 1.   Ion milling rates for nitride materials at 10°C, as a function of Ar$^+$ ion energy. The beam had vertical incidence.

an Inductively Coupled Plasma (ICP) source (lower part of Figure 2) that has the advantage of more mature automatic tuning technology and superior uniformity over large areas. Some of the typical characteristics of these three sources (and another variation, the Helicon wave source) are shown in Table I.

A typical nitride device structure, the laser diode, is shown in Figure 3 [11]. The n-GaN side of the junction is exposed by dry etching, since wet etching is not practical in this materials system. The mirror facets are also formed by dry etching, although cleaving, polishing and selective/crystallographic crystal growth have all been demonstrated to produce working devices [12]. A typical GaN sidewall produced by dry etching is shown in Figure 4. Note the presence of vertical striations and micro-roughness, which can originate from roughness on the edge of the initial photoresist mask that is used to pattern the dielectric that then serves as a mask during the GaN etching [13]. This roughness can lead to loss of laser output intensity through scattering as the light traverses the waveguide [14].

A large number of plasma chemistries have been reported for etching GaN, as shown in Table II [15]. The chemistries are similar to those for other III–V compounds, i.e. they are based on Cl_2, Br_2, I_2, or CH_4/H_2. Notice the etch rates are much higher for the ECR and ICP reactors, at lower dc chuck self-biases. This is emphasized in Figure 5, where GaN etch rates are plotted as a function of rf chuck power in $Cl_2/H_2/CH_4$ plasmas at 1 mTorr

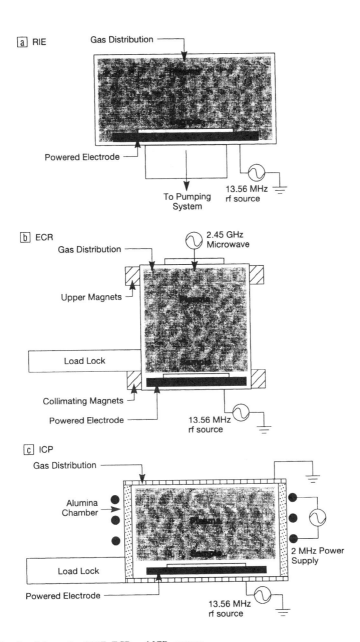

Fig. 2. Schematic of RIE, ECR and ICP reactors.

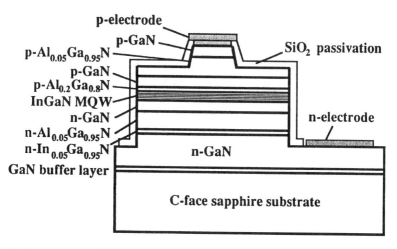

Fig. 3. Schematic of GaN-based laser diode.

for three different reactors [10]. The high density tools produce higher rates due to their improved bond-breaking efficiency. The etch yields are almost the same in both RIE and high density tools, but the higher fluxes in the latter systems lead to the higher etch rates [9, 10].

A problem in nitride etching is the difference in volatility of the group III etch products. This can be illustrated by the data of Table III, which shows the boiling points of potential etch products for Ga, Al and In, and also for N. Note that most of the products are more volatile than their P or As counterparts and much more volatile than the group III products. Therefore, preferential loss of nitrogen is difficult to avoid when dry etching of all of the nitrides.

TABLE I

Characteristics of different high efficient plasma sources, compared with conventional rf discharges used for RIE.

Type	Pressure (mTorr)	Fractional ionization	Frequency (MHz)	Magnetic field (G)
RIE	50	10^{-3}	13.56	0
ECR	1	10^{-1}	2450	875
ICP	1	10^{-1}	13.56	0
Helicon	1	10^{-1}	13.56	200

RIE: reactive ion etching; ECR: electron cyclotron resonance; ICP: inductively coupled plasma.

TABLE II

Plasma chemistries for dry etching of GaN (compiled from ref. 2, 7–10).

Plasma	Technique	GaN etch rate (Å/min)	dc bias
Cl_2/H_2	ECR	1,100	−150
Cl_2/SF_6	ECR	900	−150
$CH_4/H_2/Ar$	ECR	400	−250
$Cl_2/CH_4/H_2 Ar$	ECR	3,100	−125
$Cl_2/H_2/Ar$	ECR	2,200	−100
CCl_2F_2	ECR	400	−250
BCl_3	RIE	500	−230
BCl_3	MIE	3,500	−75
$SiCl_4$ (Ar, SiF_4)	RIE	500	−400
HBr (H_2/Ar)	RIE	600	−400
HI/H_2	ECR	1,100	−150
HBr/H_2	ECR	900	−150
Ar	ion milling	250	−400
Cl_2/Ar	CAIBE	2,000	−600
H_2	LEEEE	75	0
IBr	ECR	3,000	−150
ICl	ECR	13,000	−150
$Cl_2/H_2/Ar$	ICP	6,875	−280
Cl_2	RIBE	240	−1000
$SiCl_4/Ar$	ECR	950	−280
CHF_3, C_2ClF_5	RIE	450	−500
HCl	CAIBE	2,000	−500
H_2/Cl_2	LEEEE	2,500	0

TABLE III

Boiling points of III–V etch products.

Species	Boiling Point (°C)	Species	Boiling Point (°C)
$GaCl_3$	201	NCl_3	<70
$GaBr_3$	279	NI_3	explodes
GaI_3	sub 345	NF_3	−129
$(CH_3)_3Ga$	55.7	NH_3	−33
		N_2	−196
$InCl_3$	600	$(CH_3)_3N$	2.9
$InBr_3$	>600		
InI_3	210	PCl_3	76
$(CH_3)_3In$	134	PBr_5	106
		PH_3	−88
$AlCl_3$	183		
$AlBr_3$	263	$AsCl_3$	130
AlI_3	191	$AsBr_3$	221
$(CH_3)_3Al$	126	AsH_3	−55
		AsF_3	−63

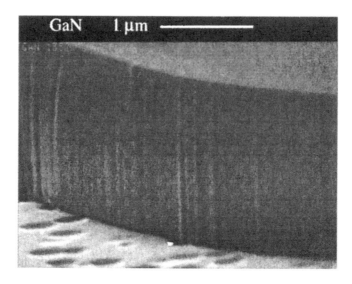

Fig. 4. Sidewall of feature formed in GaN by ICP dry etching in $Cl_2/H_2/Ar$.

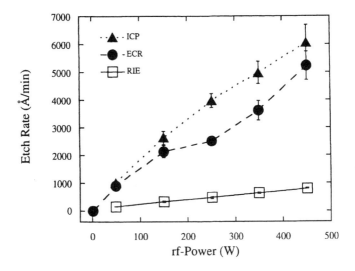

Fig. 5. GaN etch rates as function of radio-frequency (rf) power in an ECR-, RIE- and ICP-generated $Cl_2/H_2/CH_4$ Ar plasma. The plasma conditions were 10 sccm of Cl_2, 15 sccm of H_2, 10 sccm of Ar, 3 sccm of CH_4, 25°C electrode temperature, and 1 mTorr total pressure. For the ECR and ICP plasmas, the rf power ranged from 1 to 450 W with corresponding direct-current (dc) biases of -10 to -400 ± 25 V. Reactive-ion-etching rf powers ranged from 50 to 450 W with corresponding dc biases of -270 to -950 ± 50 V.

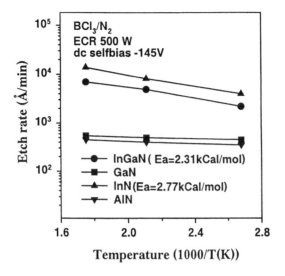

Fig. 6. Arrhenius plot of nitride etch rates in BCl₃/N₂ ECR discharges.

Since the etching requires a strong ion component, the etch products need not be fully coordinated. For example one could represent the etching of GaN in Cl_2-based plasmas by the reaction

$$2GaN + 6Cl \rightarrow 2GaCl_3 + N_2$$

However, with a high ion bombardment flux incident on the surface, one could expect ion-assisted desorption of sub-chlorides. This behavior is reflected in the non-Arrhenius dependence of etch rate on sample temperature, as shown in Figure 6 for four different nitride materials. For example, $GaCl_3$ has an enthalpy of reaction of ~ 18 kCal·mol^{-1}, but clearly desorption of this species is not limiting the etch rate.

2.2. Effect of Plasma Parameters

As discussed above, it is expected that preferential loss of nitrogen would readily occur in dry etching of nitrides. Figure 7 shows AES surface scans of GaN before and after ECR $Cl_2/Ar/CH_4$ plasmas (2 mTorr, 750 W source power, 65 W rf chuck power) at different rf chuck powers. Note that the Ga:N ratio increases as incident ion energy increases. At 65 W chuck power, dc self-bias is approximately -80 V, while at 275 W it increases to approximately -250 V. Since ion energy is directly proportional to this chuck self-bias, the data of Figure 7 shows that Ga-enrichment occurs even at relatively low energies [16, 17].

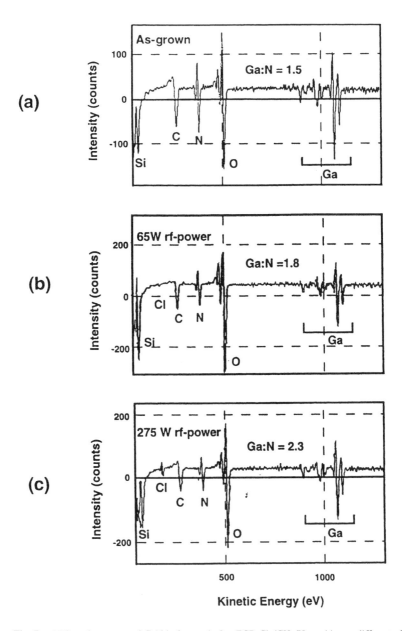

Fig. 7. AES surface scans of GaN before and after ECR $Cl_2/CH_4/H_2$ etching at different rf chuck powers.

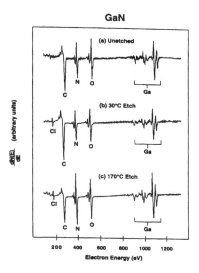

Fig. 8. AES surface scans of GaN before and after ECR $Cl_2/CH_4/H_2$ etching at different substrate temperatures.

The near-surface stoichiometry of dry etched GaN was shown to much less sensitive to substrate temperature at fixed plasma conditions [16]. Figure 8 shows AES surface scans of GaN before and after dry etching in ECR $Cl_2/Ar/CH_4$ plasmas (2 mTorr, 750 W source power, 65 W rf chuck power).

The effects of ion bombardment on the electrical properties of the nitrides can be measured using Van der Pauw Hall data on sample resistance before and after exposure to a pure Ar discharge [18]. This is a worst-case scenario because damage is allowed to accumulate and is not simultaneously removed by etching as it would be in a real etching process. Figure 9 shows the ratio of final-to-initial sheet resistivity ratio of different nitride materials exposed for 1 min to ECR Ar discharges (1000 W source power, 1.5 mTorr) as a function of rf chuck power. There are several features of this data that are important. First, InGaN is more resistant to damage introduction than pure InN, as expected from its higher bond strength. Second, damage is more readily apparent in lightly-doped materials (the InAlN was n-type, 10^{17} cm^{-3}, while the other two materials were n-type, 10^{20} cm^{-3}). Third, at constant ion flux, increasing the ion energy produces more damage. Since the material resistance increases, the ion bombardment clearly creates deep acceptor states and reduces carrier mobility, leading to higher resistivity [18].

Similar data is shown in Figure 10 for RIE exposures to Ar plasmas, as a function of rf chuck power. Comparison of this data to that in Figure 9 also

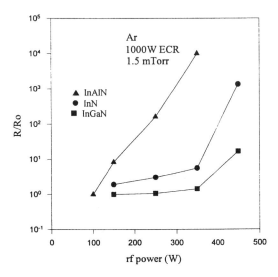

Fig. 9. Sheet resistance ratio of nitrides before and after exposure to ECR Ar plasmas for 1 min, as a function of rf power.

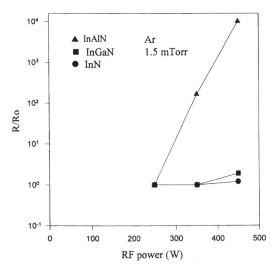

Fig. 10. Sheet resistance ratio of nitrides before and after exposure to RIE Ar plasmas for 1 min, as a function of rf power.

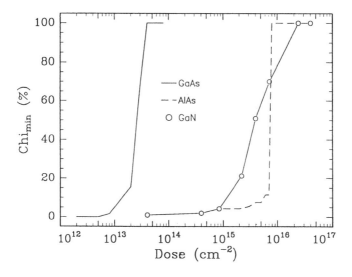

Fig. 11. Backscattering yield of Si^+ implanted III–V's, as a function of ion dose.

shows that ion flux plays an important role in damage accumulation once ion energy is above the atom displacement threshold.

Some idea of how resistant GaN is to ion damage introduction can be obtained from measurements of backscattering yield in ion-implanted material [19]. Figure 11 shows that the onset of amorphization for room temperature Si^+ implantation in GaN occurs at a similar dose to that in AlAs, another material with high bond strength and well-established resistance to damage. Note that both materials have amorphization dose thresholds more than two orders of magnitude higher than that of GaAs.

Returning to the Ar plasma damage experiments, Figure 12 shows that damage accumulates rapidly under high ion flux conditions, indicating that over-etch times should be minimized in real etching processes. This applies to all of the nitride materials we investigated.

At fixed ECR source power and rf chuck power, the damage was found to worsen with process pressure (Figure 13). This is due to the increase in Ar^+ flux, even though increased collision rates in the sheath at higher pressure will reduce the average ion energy.

A common method for reducing the effects of plasma-induced damage is post-etch annealing. Figure 14 shows that Ar damage could be essentially completely removed by annealing in N_2 at $\sim 650°C$, while the same treatment still left some residual electrical effects in InN. The activation energy for damage repair was calculated to be > 2.7 eV, and probably corresponds to

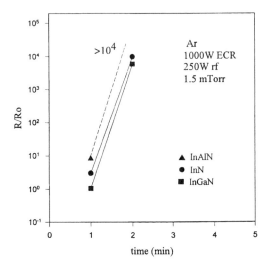

Fig. 12. Sheet resistance ratio of nitrides before and after exposure to ECR Ar plasmas, as a function of exposure time.

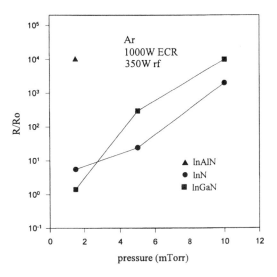

Fig. 13. Sheet resistance ratio of nitrides before and after exposure to ECR Ar plasmas, as a function of process pressure.

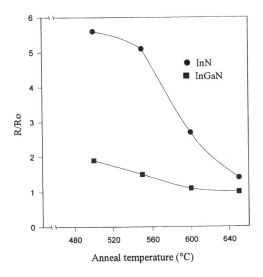

Fig. 14. Sheet resistance ratio of plasma damaged nitrides subjected to post-etch annealing.

point defect diffusion in the nitrides leading to dissolution of stable defect complexes that have levels in the gap [18].

2.3. Damage in Electronic Devices

There has been much recent attention on development of AlGaN/GaN heterostructure field effect transistors (HFETs) for high frequency and high power application [20–25]. Both enhancement and depletion mode devices have been demonstrated, with gate lengths down to 0.2 μm. Excellent dc performance has been reported up to 360°C [25], and the best devices have a maximum frequency of oscillation f_{max} of 77 GHz at room temperature [22]. Even better speed performance could be expected from InAlN channel structures, both because of the superior transport properties and the ability to use highly doped $In_xAl_{1-x}N$ ($x = 0 \rightarrow 1$) graded contact layers which should produce low specific contact layers which should produce low specific contact resistivities. We have previously demonstrated that nonalloyed Ti/Pt/Au metal on degenerately doped InN ($n = 5 \times 10^{20}$ cm^{-3}) has $P_C \sim 1.8 \times 10^7$ Ωcm^2 [26]. While metalorganic chemical vapor deposition (MOCVD) has generally been employed for growth of nitride-based photonic devices and for most of the prototype electronic devices [27, 28], the ability of the molecular beam techniques to control layer thickness and incorporate higher In concentrations in the ternary alloys is well suited to growth of HFET structures [29, 30].

The exceptional chemical stability of the nitrides has meant that dry etching must be employed for patterning [10, 31–33]. To date most of the work in this area has concentrated on achievement of higher etch rates with minimal mask erosion, in particular because a key application is formation of dry etched layer facets. In that case etch rate, etch anisotropy, and sidewall smoothness are the most important parameters, and little attention has been paid to the effect of dry etching on the stoichiometry and electrical properties of the nitride surface.

In these experiments, we used an InAlN and GaN FET structure as a test vehicle for measuring the effect of electron cyclotron resonance (ECR) BCl_3-based dry etching on the surface properties of InAlN and GaN. Preferential loss of N leads to roughened morphologies and creation of a thin n^+ surface layer which degrades the rectifying properties of subsequently deposited metal contacts.

The InAlN samples were grown by MOMBE on 2" diameter GaAs substrates using a Wavemat ECR N_2 plasma and metalorganic group III precursors (trimethylamine alane, triethylindium). A low temperature ($\sim$ 400°C) AlN nucleation layer was followed by a 500 Å thick AlN buffer layer grown at 700°C. The $In_{0.3}Al_{0.7}N$ channel layer ($\sim 5 \times 10^{17}$ cm^{-3}) was $\sim$500 Å thick, and then an ohmic contract layer was produced by grading to pure InN over a distance of $\sim$500 Å .

The GaN layer structure was grown on double-side polished c-Al_2O_3 substrates prepared initially by $HCl/HNO_3/H_2O$ cleaning and an *in situ* H_2 bake at 1070°C [7]. A GaN buffer <300 Å thick was grown at 500°C, and crystallized by ramping the temperature to 1040°C, where trimethylgallium and ammonia were again used to grow $\sim$ 1.5 μm of undoped GaN ($n < 3 \times 10^{16}$ cm^{-3}), a 2000 Å channel ($n = 2 \times 10^{17}$ cm^{-3}) and a 1000 Å contact layer ($n = 1 \times 10^{18}$ cm^{-3}).

FET surfaces were fabricated by depositing TiPtAu source/drain ohmic contacts, which were protected by photoresists. The gate mesa was formed by dry etching down to the InAlN or n-GaN channel using an ECR BCl_3 or BCl_3/N_2 plasma chemistry. During this process, we noticed that the total conductivity between the ohmic contacts did not decrease under some conditions. CH_4/H_2 etch chemistry was also studied. To simulate the effects of this process, we exposed the FET substrates to D_2 plasma, we saw strong reductions in sample conductivity. The incorporation of D_2 into the InAlN was measured by secondary ion mass spectrometry. Changes to the surface stoichiometry were measured by Auger electron spectroscopy (AES). All plasma processes were carried out in a Plasma Therm SLR 770 System with an Astex 4400 low profile ECR source operating at 500 W. The samples were clamped to an rf-powered, He backside cooled chuck, which was left at floating potential (about -30 V) relative to the body of the plasma.

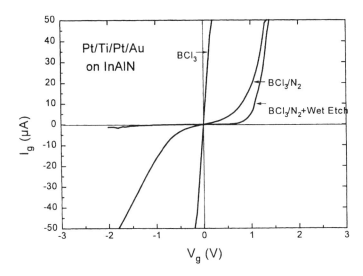

Fig. 15. I–V characteristics of Pt/TiPtAu contacts on InAlN exposed to to different ECR plasmas.

Upon dry etch removal of the InAlN capping layer, a Pt/Ti/Pt/Au gate contact was deposited on the exposed InAlN to complete the FET processing. If pure BCl_3 was employed as the plasma chemistry, we observed ohmic and not rectifying behavior for the gate contact. If BCl_3/N_2 was used, there was some improvement in the gate characteristics. A subsequent attempt at a wet-etch clean-up using either H_2O_2/HCl or H_2O_2/HCl produced a reverse breakdown in excess of 2 V (Figure 15). These results suggest that the InAlN surface becomes nonstoichiometric during the dry etch step, and that addition of N_2 retards some of this effect.

Figure 16 shows the I_{DS} values obtained as a function of dry etch time in ECR discharges of either BCl_3 or BCl_3/N_2. In the former case the current does not decrease as material is etched away, suggesting that a conducting surface layer is continually being created. By contrast BCl_3/N_2 plasma chemistry does reduce the drain-source current as expected, even though the breakdown characteristics of gate metal deposited on this surface are much poorer than would be expected.

From the atomic force microscopy (AFM) studies of the InAlN gate contact layer surface after removal of the $In_xAl_{1-x}N$ contact layer surface in BCl_3, BCl_3/N_2 or BCl_3/N_2 plus wet etch, we can see the mean surface roughness is worse for the former two chemistries, indicating that preferential loss of N is probably occurring during the dry etch step (Figure 17). To try to

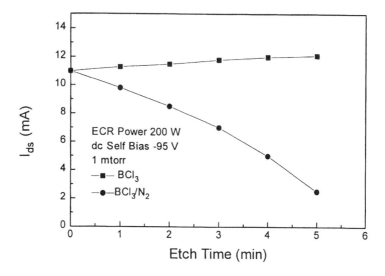

Fig. 16. I_{DS} values at 5 V bias for InAlN FETs etched for various times in BCl_3 or BCl_3/N_2 ECR plasmas.

remove the group III enriched region, the samples were rinsed in a 1:1 HCl/H_2O_2 solution at 50°C for 1 min. This step did increase the reverse breakdown voltage (2.5 V) compared to dry etched only samples, (<1 V), but did not produce a completely damage-free surface. This is because the HCl/H_2O_2 does not remove the InAlN immediately below the surface which is closer to stoichiometry but is still defective. Auger electron spectroscopy surface scans from the as-grown and BCl_3/N_2 plus wet etch samples are shown in Figure 18. In the former sample, only the peaks from the InN at the top of the structure, plus adventitious C and O from the native oxide are present. In the etched structure, the Al is now visible, along with some chlorine etch residue (probably in the form of $InCl_3$). However, the etched surface is clearly more In-rich, as shown in Table IV where we have listed the ratio of raw counts for In/N. While BCl_3/N_2 produces less enrichment than pure BCl_3, there is clearly the presence of a defective layer that prevents achievement of acceptable of rectifying contacts. From the I–V measurements, we believe this defective layer is probably strong n-type, in analogy with the situation with InP described earlier. At this stage, there is no available wet etch solution for InAlN that could be employed to completely remove the nonstoichiometric layer in the type of clean-up step commonly used in other III–V materials. Other possible solutions to this problem include use of a higher Al concentration in the stop layer, which should be more

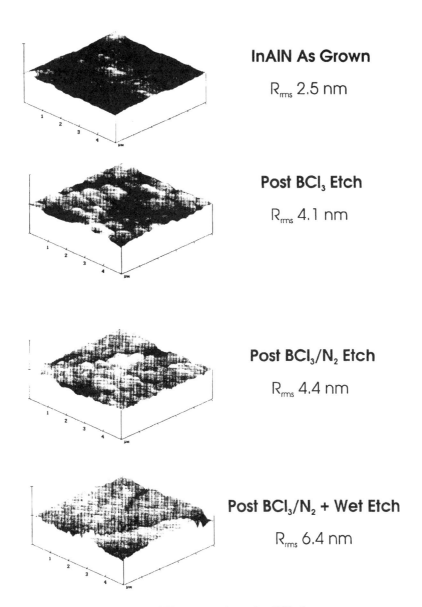

Fig. 17. AFM scans of InAlN FET structures after various ECR plasma exposures.

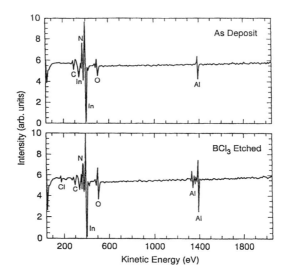

Fig. 18. AES surface scans of InAlN FET structures before and after exposure to a BCl₃ ECR plasma.

resistant to nitrogen loss, or employment of a layer structure that avoids the need for gate recess [34].

A study of simulating the CH_4/H_2 etch chemistry was conducted on the full FET structure of Figure 19, which requires etching of the InAlN contact layer to expose the gate contact layer for deposition of the gate metallization. Figure 20 shows the I_{DS} as a function of applied bias before and after a D_2 plasma treatment at 200°C for 30 min. The loss of conductivity could be due to two different mechanisms. The first is hydrogen forming neutral complexes $(D-H)°$, with the donors extra electrons taken up in forming a bond with the hydrogen [35]. The second mechanism is creation of deep acceptor states

TABLE IV

In/N ratio measured in raw counts from AES analysis of dry etching In AlN samples. No correction for relative sensitivity factor was applied.

Sample	In/N
As-grown	2.86
BCl₃ etch	3.64
BCl₃/N₂ etch	3.49
BCl₃/N₂ + wet etch	3.25

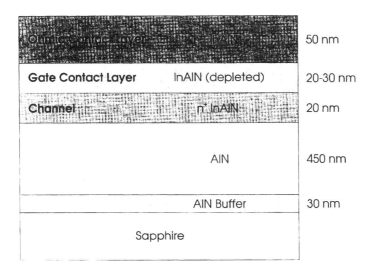

Fig. 19. Schematic of InAlN FET structure.

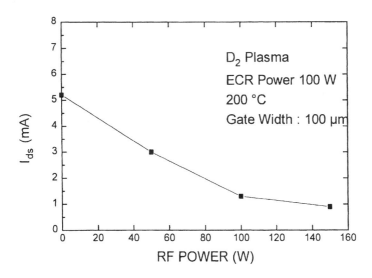

Fig. 20. I_{DS} values at 5 V bias, as a function of the rf power used during D_2 ECR plasma exposures at 200°C for 30 min.

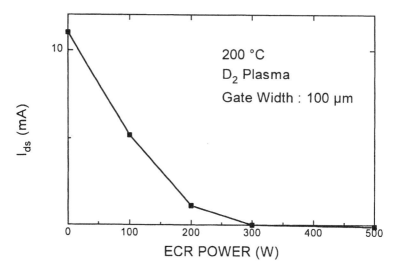

Fig. 21. I_{DS} values at 5 V bias as a function of the ECR source power used during D_2 plasma exposure at 200°C for 30 min.

that trap the electrons and remove them from the conduction process. These states might be formed by the energetic D^+ or D_2^+ ion bombardment from the plasma.

The decrease in I_{DS} in the InAlN FET was a strong function of both the ion energy in the plasma, and of the active neutral ($D°$) and ion density (D^+, D_2^+). Figure 21 shows the dependence of I_{DS} on ECR microwave power. As this power is increased, both dissociation of D_2 molecules into atoms, and ionization of atomic and molecular species will increase and thus it is difficult to separate out bombardment and passivation effects. At fixed ECR power, the I_{DS} values also decrease with rf power, which controls the ion energy. At 150 W, the ion energy increases to ~200 eV, with a corresponding decrease in I_{DS}. This is good evidence that creation of deep traps is playing at least some role, and the fact that hydrogen is found to diffuse all the way through the sample also implicates passivation as contributing to the loss of conductivity.

Figure 22 shows SIMS depth profiles of ^2H in the FET after a 500 W D_2 ECR plasma treatment at 250°C for 30 min. A high concentration of ^2H ($\sim 10^{20}$ cm^{-3}) has permeated the entire layer structure under these conditions. Subsequent annealing at 700°C reduces the deuterium concentration by an order of magnitude, by loss of the deuterium to the surface. The fact that the ^2H profile has a plateau shape indicates it is trapped at defects or impurities with this concentration. This is a common phenomenon in highly doped

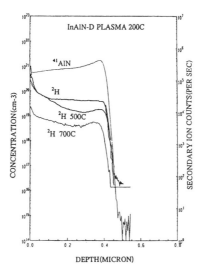

Fig. 22. SIMS profiles of ^{2}H in InAlN treated in D_2 ECR plasmas and then annealed at $500°$C or $700°$C.

semiconductors containing hydrogen [35]. Thus, the results of Figure 21 and Figure 22 lead us to believe that both mechanisms are contributing to the decrease in sample conductivity.

Figure 23 shows the gate current-voltage characteristic when the gate metal is deposited on the as-etched GaN surface. The Schottky contact is extremely leaky, with poor breakdown voltage. We believe this is caused by the presence of a highly conducting N deficient surface, similar to the situation encountered on dry etching InP where preferential loss of P produces a metal-rich surface which precludes achievement of rectifying contacts. Auger electron spectroscopy analysis of the etched GaN surface showed an increasing the Ga-to-N ratio (from 1.7 to 2.0 in terms of raw counts) upon etching. However, a 5 min. anneal at $400°$C under N_2 was sufficient to produce excellent rectifying contacts, with a gate breakdown of $\sim$25 V (Figure 24). We believe the presence of the conducting surface layer after etching is a strong contributing factor to the excellent ρ_c values reported by Lin *et al.* [30] for contacts on a reactively ion etched n-GaN.

The drain I–V characteristics of the 1×50 μm^2 MESFET are shown in Figure 25. The drain-source breakdown was -20 V, with a threshold voltage of -0.3 V. The device displays good pinch-off and no slope to the I–V curves due to gate leakage, indicating that the anneal treatment is sufficient to restore the surface breakdown characteristics. We believe these devices

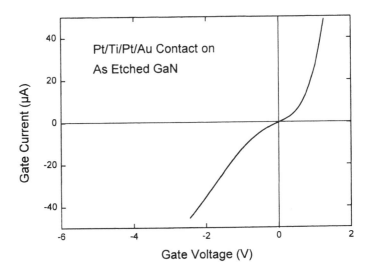

Fig. 23.   I–V characteristic on ECR BCl$_3$-etched GaN.

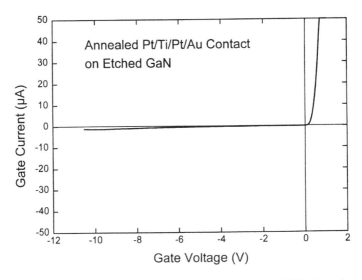

Fig. 24.   I–V characteristic on ECR BCl$_3$-etched GaN annealed at 400°C prior to deposition of the gate metal.

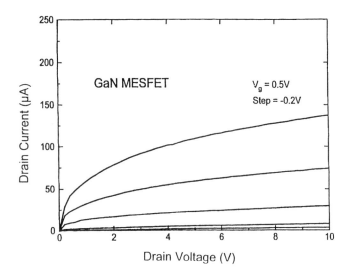

Fig. 25. Drain I–V characteristics of a 1 × 50 μm^2 MESFET.

are well-suited for high power applications since GaN is a robust material and the contact metallizations employed are also very stable [36].

III-nitrides FET structures are sensitive to several effects during dry etching of the gate mesa. Firstly, if hydrogen is present in the plasma there can be passivation of the doping in the channel layer. Secondly, the ion bombardment from the plasma can create deep acceptor states that compensate the material. Thirdly, even when these problems are avoided through use of H-free plasma chemistries and low ion energies and fluxes, preferential loss of N can produce poor rectifying gate characteristics for metal deposited on the etched surface. Ping *et al.* [37] observed that pure Ar etching produced more damage in Schottky diodes than SiCl$_4$ RIE. The diodes characteristics were strongly dependent on plasma self-bias, while annealing at 680°C removed much of the damage.

2.4.   **Effect on Optical Properties**

Increasing the ion energy in a plasma etch system typically results in highly anisotropic, high rate etching due to the physical sputter desorption of the etch products. However, bombardment of semiconductor surfaces with energetic ions generated during plasma etching can damage the near surface region and produce lattice damage if the ions energy are greater than the displacement energy of the host atoms. As these energetic ions strike the sample, damage as deep as 100 nm can occur [38], causing degradation of device performance. This damage can include simple Frenkel pairs consisting of a vacancy and the displaced atom, implanted etch ions, broken bonds, formation of dangling bonds, or deposition. Attempts to minimize the damage by reducing the ion energy below the damage threshold for compound semiconductors (<40 eV) [39] or by increasing the chemical component of the etch results in more isotropic profiles, significantly limits minimum dimensions, and reduces the etch rate. It is therefore necessary to develop plasma etch processes which couple high etch rates, anisotropy, and sidewall profile control with low-damage for optimum device performance.

Since GaN is more chemically inert that GaAs and has higher binding energies, higher ion energies may be used during the etch with potentially less damage to the material. However, reports of plasma-etch-induced-damage of the group-III nitrides has been limited. Pearton and co-workers have reported plasma-induced-damage results for InN, InGaN, and InAlN in an ECR-generated plasma where the damage increased as a function of ion flux and energy [40]. ICP etching offers an attractive alternative etch technique which may be easier to scale-up than ECR sources, and may be more economical in terms of cost and power requirements. ICP plasmas are formed in a dielectric vessel encircled by an inductive coil into which rf-power is applied. A strong magnetic field is induced in the center of the chamber which generates a high-density plasma due to the circular region of the electric field that exists concentric to the coil. At low pressures (≤ 10 mTorr), the plasma diffuses from the generation region and drifts to the substrate at relatively low ion energy. Thus, ICP etching is expected to produce low damage while achieving high etch rates. In this section, we discuss ICP and ECR plasma-induced-damage of GaN as a function of rf-power and source power. Pure Ar plasmas were used to simulate the ion bombardment conditions created during plasma etching of the group-III nitrides.

The GaN samples used in this study were grown by metal organic chemical vapor deposition (MOCVD) on a c-plane sapphire substrate in a multiwafer rotating disk reactor at 1040°C with a 20 nm GaN buffer layer grown at 530°C [41]. The GaN film was approximately 1.8 μm thick. The ECR plasma reactor used in this study was a load-locked Plasma-Term SLR 770 etch system with a low profile Astex 4400 ECR source in which the upper magnet was operated at 165 A. Energetic ion bombardment was provided by superimposing an rf-bias (13.56 MHz) on the sample. Etch gases were introduced through an annular ring into the chamber just below the quartz window. To minimize field divergence to optimize plasma uniformity and ion density across the chamber, an external secondary collimating magnet was located on the same plane as the sample and was run at 25 A. The ICP reactor was a load-locked Plasma-Therm SLR 770 etch system with a Plasma-Therm ICP source. The reactor was a cylindrical coil configuration with a ceramic vessel encircled by a 3-turn inductive coil into which the rf-power (2 MHz) was applied. Identical to the ECR, energetic ion bombardment was provided by superimposing an rf-bias (13.56 MHz) on the sample. Etch gases were introduced through an annular region at the top of the chamber. Unless otherwise mentioned, ECR and ICP etch parameters used in this study were: 40 sccm of Ar, 30°C electrode temperature, 1 mTorr total pressure, 500 W of applied source power, and 1 to 250 W rf-power with corresponding dc-biases of -10 to -300 ± 25 V.

All samples were mounted using vacuum grease on an anodized Al carrier that was clamped to the cathode and cooled with He gas. Samples used to measure PL intensity were 5 mm × 5 mm and unpatterned. Samples used to calculate etch rates were patterned using AZ 4330 photoresist. Etch rates were calculated from the depth of etched features measured with a Dektak stylus profilometer after the photoresist was removed with an acetone spray. Each sample was approximately 1 cm^2 and depth measurements were taken at a minimum of three positions. Standard deviation of the etch depth across the sample was nominally less than $\pm 10\%$ with run-to-run variation less than $\pm 10\%$. Root-mean square (rms) surface roughness was quantified using a Digital Instruments Dimension 3000 atomic force microscope (AFM) system operating in tapping mode with Si tips.

PL measurements were made at liquid helium temperature (10 K) in a continuous flow cryostat. A HeCd laser (325 nm) was used as the excitation source and the typical excitation power was 5 mW. The detection system consisted of a 0.275 meter spectrometer in conjunction with a thermoelectrically cooled UV-enhanced CCD detector. Measurements were taken using a 100 line/mm grating for high resolution data and a 150 line/mm grating for low resolution, broad spectral range data. To evaluate the effect of high temperature on the PL, samples were annealed in an Addax AET rapid thermal annealer in flowing Ar, preceded by a three cycle pump/purge sequence to reduce the background oxygen level. In the annealer, samples were contained in a SiC coated graphite crucible with thermocouples monitoring the temperature at two points on the crucible. Anneal times were 30 s at the prescribed set point $\pm 10°C$.

Prior to the etch experiments, the two inch GaN wafer was mapped out to examine the uniformity of the PL emission. In Figure 26 we show the PL spectrum at the center of the wafer taken with the low resolution grating. The spectrum consisted of two distinct features. The dominant near band-edge resonance was seen at 3.472 eV (as verified by spectra taken with 1 meV resolution). Emission resonances in this spectral region have been identified with recombination of a neutral-donor-bound excition [42, 43], was not clearly resolved. The broad spectral feature centered at approximately 2.21 eV was associated with emission from deep level impurities. The oscillations in the deep level emission were due to optical interference effects

S.J. PEARTON and R.J. SHUL

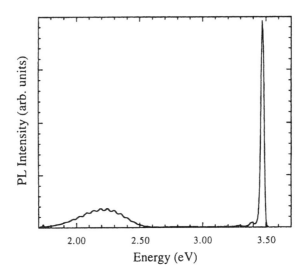

Fig. 26. PL spectrum from GaN films at $T = 10$ K.

in the film. We focused on the near-band-edge emission for the majority of our etch studies.

In Figure 27, we plot the peak intensity of the near-band-edge emission as a function of radial position on the two inch GaN wafer. The intensity dropped by approximately 15% at a radial position of 0.5 inches and approximately 30% at a radial position of 0.8 inches. Toward the edge of the wafer, the intensity drop was significantly more rapid. The samples that were used in the etch studies were diced from the center part of the wafer, within the 0.6 inch radial position. The PL spectra were taken before and after etching for each sample. The PL spectrum of a reference sample from the wafer was compared during each experiment to ensure consistency of the excitation conditions. The PL band-edge emission is expected to be accurate to ±10%.

The first study evaluated the effect of rf-power on the peak near band-edge PL intensity. GaN samples were exposed to ICP- and ECR-generated Ar plasmas for 1 minute under identical plasma conditions while the rf-power was increased. The dc-bias was approximately 10 to 65% higher in the ECR under comparable conditions. In Figure 28, the percent change in the peak PL intensity versus rf-power is plotted for both ECR and ICP etching. For the ICP case, at relatively low rf-powers (1 and 50 W) the PL intensity slightly degraded, and as the rf-power was increased up to 250 W increasing degradation in PL intensity was seen. Depth profiling of similar films at a rf-power of 1 W (~10 V dc-bias) revealed no detectable material removed

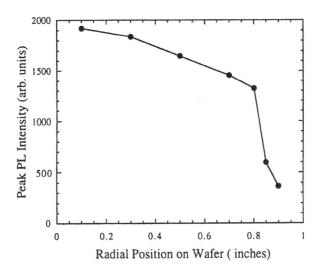

Fig. 27. Variation of the peak PL intensity as a function of radial position on the two inch GaN/Al$_2$O$_3$ wafer.

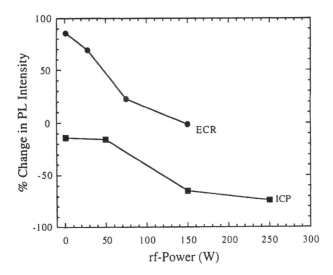

Fig. 28. Percent change in the peak PL intensity as a function of rf power for ECR and ICP exposures.

whereas the 250 W etch (-300 V dc-bias) resulted in GaN sputter loss of approximately 770 Å during a 1 minute exposure.

Distinctly different results were obtained for etching in the ECR plasma system. As seen in the figure, etching with very low rf-power (1 W) resulted in an over 80% increase in the PL intensity and virtually no sputter loss of GaN. Etching at higher rf-powers also improved the PL intensity, but to a lesser degree as the rf-power was increased. The highest power (150 W) etch resulted in a very slight decrease in PL intensity and the sputter rate under these conditions was determined to be approximately 820 Å /min.

We also studied the effect of plasma density on the peak near band-edge PL intensity. GaN samples were exposed to ICP- and ECR-generated Ar plasmas for 1 minute. The dc-bias was held approximately constant at -65 ± 15 V by varying the rf-power. The data was more scattered than the rf-power data for both ICP and ECT conditions. The ICP showed virtually no change in PL intensity at 250 W source power and then decreased by 30% as the ICP power was increased to 750 W. The PL intensity at 250 W source power was then decreased by 30% as the ICP power was increased to 750 W. The PL intensity decreased by only 10% at 1000 W ICP power which was an improvement of almost 20% over 750 W. In the ECR, we observed an increase of $\sim 115\%$ in PL intensity at 250 W ECR power. Similar to the trend observed as a function of rf-power, the PL intensity also improved at higher ECR powers but at a lower rate. Sputter rates for GaN were approximately 30 ± 10 Å /min at 250 W source power and 225 ± 25 Å /min at 1000 W. Further studies are underway to identify the effect of plasma density.

The PL emission efficiency of other III–V bulk semiconductors (e.g. GaAs) has been shown to be strongly dependent on surface conditions, specifically the presence of a native oxide, the surface recombination velocity, as well as band bending at the surface [44, 45]. Previous work on hydrogen plasma passivation in GaAs films with *in situ* PL monitoring has demonstrated that both PL intensity enhancement as well as degradation can occur in different etching regimes. In particular, the reduction of surface As concentration in the initial stages of the etch can result in an increase in the PL efficiency whereas extended exposure to ion bombardment can create damage that reduces the PL efficiency.

While the exact nature of surface oxides and surface states in GaN are not well understood, our data show that the PL efficiency can be strongly affected by exposure to an Ar plasma and is highly sensitive to the exact plasma conditions. A common result for both plasma environments was the decrease in PL intensity with increasing rf-power which suggested that plasma-induced damage can occur in GaN films under moderate ion energies. Our initial data suggests that GaN surfaces are considerably more sensitive to process-inducted changes than is widely recognized. Brief low bias exposures to ECR discharges almost doubled the band-edge PL, suggesting that the

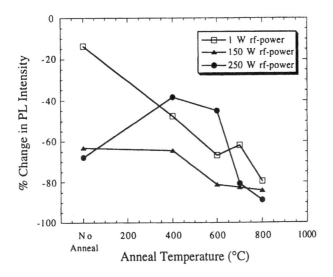

Fig. 29. Percent change in the peak PL intensity as a function of post-etch anneal temperature for ICP etched samples, for different rf power conditions.

native oxide has a strong effect on the surface recombination velocity. Work is in progress to more closely examine the changes in GaN surface conditions under various plasma conditions.

Post-etch anneals in Ar were performed on selected samples to investigate the effects of high temperature on the PL intensity. In Figure 29, we show the change in peak PL intensity as a function of anneal temperature for samples etched under 1 W, 150 W and 250 W rf-power in the ICP reactor. The post-etch condition, specifically the initial change in PL intensity due to the etch, is indicated at the "no anneal" condition. The data show that the effect of the anneal is strongly dependent on the initial (post-etch) sample conditions. In particular, the sample etched under very low power (1 W) degraded with increasing anneal temperature, whereas the sample etched under moderate power (250 W) showed an initial enhancement in the post-etch PL intensity, followed by degradation as the anneal temperature was increased beyond 400°C. These results suggested that relatively low temperature annealing may reduce the (non-radiative) damage induced under the 250 W etching conditions. In all cases, however, the effect of post-etch annealing in Ar for $T > 700$°C resulted in a degradation of the PL intensity. At higher rf-powers we would expect a greater initial PL degradation because of the greater damage depth. The low temperature (± 400°C) annealing stage for the 250 W sample may result from the individual damage sites being closer together,

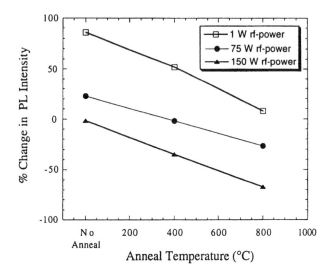

Fig. 30. Percent change in the peak PL intensity as a function of post-etch anneal temperature for ECR etched samples, for different rf power conditions.

as is seen in ion-implanted material, where annealing is actually easier in moderately damaged sample. However the key result is that overall, the etch damage is stable to >800°C, much higher than in other III–V materials.

Similar post-etch annealing experiments were performed on selected samples etched in the ECR reactor, and the effect on PL intensity is shown in Figure 30. For all rf-power conditions, a degradation in the PL was seen after anneals at 400°C and 800°C. The loss of PL intensity with anneal temperature was similar for all of the samples, in strong contrast to the ICP results. This result was not unexpected, due to the large differences in the PL intensities (post-etch) at the start of the anneal experiments. It is interesting to note that for the highest rf-power conditions, annealing at temperatures as high as 800°C resulted in a similar reduction of pre-etch PL intensity (75–90%) for both ICP and ECR etches, despite the large discrepancy in post-etch PL intensities.

In summary, peak PL intensity was strongly affected by exposure to an Ar ECR and ICP plasma, with both enhancement and degradation seen under various etch conditions, as is the case with GaAs. For both plasma environments the PL intensity decreased with increasing rf-power. Exposure to an ICP Ar plasma resulted in decreased PL intensity, whereas the PL intensity increased following exposure to the ECR. The effect of post-etch annealing in Ar varies, depending on initial film conditions. For all etch conditions examined in this work, annealing at temperatures above 400°C resulted in a reduction in the PL intensity. Possible degradation mechanisms may be due to defect migration to form stable non-radiative centers, loss of passivating hydrogen initially in the GaN, or the formation of a non-stoichiometric surface oxide. We are currently investigating all of these mechanisms using time-resolved PL and surface recombination velocity measurements. Although the nature of surface states and oxides in GaN is not entirely understood, our results suggest that surface conditions can significantly affect radiative recombination efficiency in GaN films. Further work is in progress to determine the nature and effect of plasma-induced surface passivation in the group-III nitride materials.

Usikov et al. [46] studied the effect of Ar and Ar/N_2 collimated ion beam etching on GaN p-n junctions. On undoped GaN epilayers, Ar^+ ion beams etching increased the near-surface electron concentration from 10^{16} cm^{-3} to 10^{18} cm^{-3}. In the PL spectrum an intense, broad band appeared at 3.05 eV, and there was an increase in intensity of the yellow band at 2.20 eV. The latter is thought to involve defects such as Ga_v in some models. Use of Ar^+/N_2^+ ion beams produced less degradation of both optical and electrical properties.

Saotome et al. [47] studied the effects of RIBE/ECR etching with pure Cl_2 on GaN properties. Etch rates up to 1,000 Å · min^{-1} at 500 V beam voltage were obtained. He-Cd (325 nm) laser irradiation was used to measure PL from RIBE GaN samples before and after photo-assisted wet etching in a 85% KOH:H_2O (1:3) solution. The RIBE treatment decreased near band-edge PL intensity by a factor of approximately five, whereas subsequent photo-assisted wet etching restored this to about half of the original value.

2.5. Effects of Hydrogen

Hydrogen plays an important role in nitrides because of its effect on passivating acceptor dopants during cool-down after epitaxial growth and during processing steps [48]. Table V shows some examples of the amount and depth of hydrogen incorporation in GaN measured by Secondary Ion Mass Spectrometry (SIMS) after different processing steps. When the hydrogen encounters a dopant atom, it may form neutral complexes of the type shown in Figure 31, where it is in an antibonding or bond-entered location and passivating the electrical activity of the dopants. Acceptor passivation is extremely efficient in GaN, but one might also expect somewhat less efficient donor passivation based on past results in other compound semiconductors.

Hydrogen readily enters GaN and the other nitrides during dry etching in CH_4-containing plasma chemistries. Although most attention on III-V nitrides has been focused on GaN [49], the In-based materials InN and InGaN are expected to play a significant role in future blue/UV lasers and wide bandgap electronics. For example, InN is generally auto-doped n-type

TABLE V
Processing steps in which hydrogen is found to be incorporated into GaN.

Process	Temperature (°C)	Max. [H] (cm^{-3})	Incorporation Depth (μm)
H$_2$O Boil	100	10^{20}	1.0
PECVD SiN$_x$	125	3×10^{19}	0.6
Dry Etch	170	10^{19}–10^{20}	>0.2
Implant Isolation	25	dose dependent	2.0
Wet Etch	85	2×10^{17}	0.6

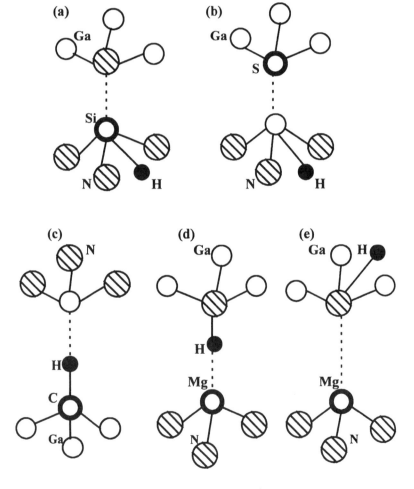

Fig. 31.   Schematic of dopant-hydrogen complexes in GaN.

by the presence of native defects with shallow ($\sim$30 meV) energy levels. By grading through $In_xGa_{1-x}N$ to an n^+ InN layer, we can therefore produce low resistance ohmic contacts to GaN. Similarly, InGaN is also a potential channel material for heterostructure field effect transistor structures capable of high temperature operation. The problem of unintentional hydrogen passivation of dopants in GaN and related alloys has previously been observed during cool-down after MOCVD growth [49], and during dry etching processes [50]. An understanding of the role of hydrogen in both InN and InGaN is necessary for control of the conductivity of these materials during steps such as annealing in H_2, dielectric deposition with SiH_4, wet etching in acids, or dry etching with CH_4/H_2 discharges, all of which are known to introduce hydrogen into other III–V semiconductors [51].

To study these effects, layers of $\sim$ 0.3 μm thick InN or $In_{0.5}Ga_{0.5}N$ were grown on semi-insulating GaAs layers at 500°C in an Intevac Gas-Source Gen II system [51]. The nitrogen source was an electron cyclotron resonance discharge (2.45 GHz, 200 W), and triethylgallium and trimethylindium transported by He provided the group-III flux. The films contain both cubic and hexagonal phases and were n-type with carrier concentrations of $\sim 2 \times 10^{20}$ cm^{-3} for InN and 5×10^{19} cm^{-3} for InGaN. Ohmic contacts were placed at the four corners of 5×5 mm^2 sections using alloyed (420°C, 5 min) HgIn eutectic. Hydrogen plasma exposures were performed in a Plasma-Therm SLR 770 reactor using an ECR source and separate RF (13.56 MHz) biasing of the sample position. The RF power was held fixed at 100 W, corresponding to a DC bias of $\sim$125 V, while the microwave power was varied for 0–1000 W. The process pressure was varied form 1.5 to 10 mTorr. Van der Pauw geometry Hall measurements were used to obtain the sheet resistance of the layers after various hydrogenation treatments.

Figure 32 shows the ratios between the post-hydrogenation resistance of InN and InGaN films and their initial values, against the microwave power used to create the hydrogen discharge. In the case of InGaN, the passivation efficiency of the native donor states increases with microwave power, which is consistent with the increased concentration of atomic hydrogen in the plasma. This was confirmed by monitoring the 656 nm emission line of this species using optical emission spectroscopy. For InN however, there is a fall-off in passivation efficiency at higher microwave powers. This coincides with the observation of severe degradation of the InN surface. Preferential loss of N leads to a very rough morphology and its subsequent effects are two-fold: first, it prevents efficient permeation of further atomic hydrogen into the InN and secondly, the additional nitrogen vacancy concentration may increase the apparent n-type conductivity of the material.

The dependence of the layer resistance on the process pressure during hydrogenation is shown in Figure 33. For both InN and InGaN, the effect of hydrogen is decreased as the presence is increased. At higher pressures,

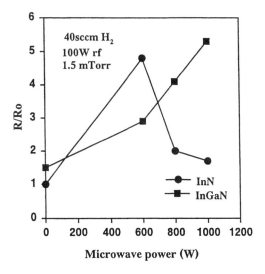

Fig. 32. Sheet resistance ratio of nitrides before and after ECR H_2 plasma exposure for 1 min, as a function of ECR source power.

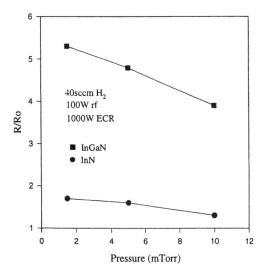

Fig. 33. Sheet resistance ratio of nitrides before and after ECR H_2 plasma exposure for 1 min, as a function of process pressure.

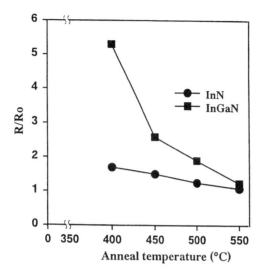

Fig. 34. Sheet resistance ratio of ECR H_2 plasma exposed nitrides as a function of subsequent annealing temperature.

the efficiency of plasma dissociation under ECR conditions is subsequently reduced and recombination of hydrogen atoms into molecules decreases the passivation efficiency. We emphasize that the changes in the layer sheet resistance are dominated by hydrogen passivation effects since exposure of the samples to Ar plasmas under the same conditions produced no significant changes. This indicates that chemical (passivation) effects are much more important than physical (damage) induced changes in resistance.

Hydrogen passivation of the shallow donors in both materials can be reversed by subsequent annealing. Figure 34 shows the annealing temperature dependence of the reactivation (1 min anneals). The initial conductivity in both materials is recovered with similar characteristics, with slightly higher stability for the wider bandgap InGaN. This would be expected if the native donors have a common origin, either as defects or impurities, whose electrical properties are altered slightly by the different matrixes in which they are incorporated. Similar thermal stabilities have been reported for passivated shallow native donors in InAlN and InGaAlN, with reactivation energies of ~2.4 eV [52].

In conclusion, shallow native donors in InN and InGaN are passivated by atomic hydrogen, but with much lower efficiencies, than acceptors in the III–V nitrides where changes in resistance of > 6 orders of magnitude can be obtained.

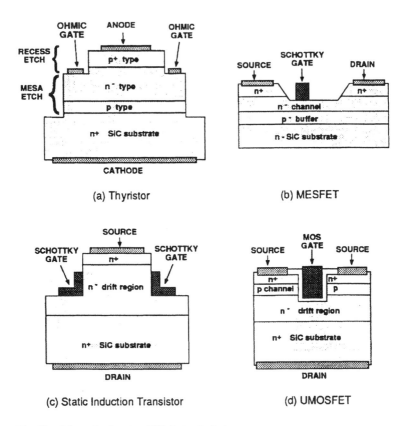

Fig. 35.   Schematic of various SiC electronic devices.

3. DAMAGE IN SIC

3.1. Background

There has been a revival of interest in SiC-based high power, high temperature (>250°C) devices and circuits for applications ranging from advanced avionics, automobiles and space exploration to bore-hole logging [53–72]. SiC is the most mature of the candidate semiconductors, which include diamond and GaN, and has the advantage of high thermal conductivity and availability in both bulk, single-crystal and thin-film form. The two most common polytypes are 6H and 4H, although cubic material (3C) is also available. There are a wide variety of device structures that have been fabricated in 6H, including thyristors, static induction transistors, Schottky diodes, metal-semiconductor field effect transistors (MESFETs) and carious vertical Metal-Oxide-Semiconductor (MOS) devices. Schematics of some of these structures are shown in Figure 35.

In all of these structures there is a need for pattern transfer capability. While some success has been obtained with photo-chemical etching in electrolytes that oxidize the SiC surface and subsequently dissolve this oxide [73, 74], it is generally agreed that conventional wet chemical etching is not possible at practical temperatures.

The absence of practical wet chemical etches for SiC has focused attention on the development of dry etching for device fabrication [75]. Some of the particular process steps performed using dry etching include deep mesa isolation ($\sim 10\ \mu$m) and shallow recess etching to expose a buried n-ohmic contact layer in high-voltage thyristors, formation of narrow trenches in vertical MOSFET devices and exposure of gate contact regions in MESFETs or static induction transistors. Both Si and C form volatile etch products in either F_2 [54–71] or Cl_2-based [63, 72] plasma chemistries and some of the mixtures employed to date include CHF_3, $CBrF_3$, SF_6, CF_4 or NF_3 in the former category and Cl_2, BCl_3 and $SiCl_4$ in the latter. Addition of O_2 is often found to improve the removal efficiency of C (as CO or CO_2) [66] and to reduce surface residues when using CHF_3, which is known to form polymers on some etched surface. There have been two general classes of plasma reactor reported for pattern transfer in SiC. The first is conventional reactive ion etching (RIE) at pressures from 10–225 mTorr. Under these conditions there is often a relatively high dc self-bias, which can cause problems with mask erosion, damage to the semiconductor lattice and micromasking through back-sputtering of the electrode materials. The second general class is the so-called high density plasma sources, including Electron Cyclotron Resonance (ECR) [83–87, 90, 91], magnetron-enhanced RIE [87, 91] and Inductively Coupled Plasma (ICP) [95, 96]. These tools operate at lower pressure (1–2 mTorr) and with ion fluxes several orders of magnitude higher than for RIE. Etch rates of SiC with RIE tools are quite low (typically <1,000 Å /min), even though the etch products are volatile. We expect that the main factor limiting etch rate is actually the initial bond-breaking which must precede etch product formation and the fact the Si etch rates are typically 5–10 times faster than those of SiC under the same conditions is most likely a manifestation of the differences in bond strength (78 kCal·mol^{-1} for Si, 111.5 kCal·mol^{-1} for 6H-SiC). Etch rates for SiC in the range 1500 Å /min have been reported for both ECR and ICP tools and fluorine plasma chemistries produced faster rates than either Cl_2- or Br_2-based mixtures. It is also generally found the NF_3 produces somewhat faster rates than SF_6 and much faster rates than carbon-containing fluorines.

One attribute of the RIE technique is the high ion energy (typically (200 eV), which is useful in breaking the bonds in the SiC. However a downside to high ion energies is mask erosion and residual lattice damage in the semiconductor. The etch products with fluorinated plasma chemistries are SiF_x and CF_x species, and under high bias conditions (i.e. physically-dominated process) these probably do not need to be fully fluorinated (i.e. $x = 4$) to be desorbed from the surface by ion assistance. Alternatively plasma chemistries include Cl_2-, Br_2- or I_2-based gases, but these produce slower etch rates than the F_2-based mixtures. Rather than rely simply on high ion energy to stimulate etching of the SiC, another approach is to employ a high ion flux with lower ion energy. This is the basis of the newer high density plasma tools in vogue for pattern transfer in Si. Etching of SiC in Electron Cyclotron Resonance (ECR) and Inductively Coupled Plasma (ICP) reactors has been reported by several groups, with fairly good etch rates and good anisotropy. The operating pressure (1–2 mTorr) of these tools is much lower than in RIE systems (10–300 mTorr), with much higher ion fluxes ($\geq 10^{11}$ cm^{-3} compared to ($\geq 10^9$ cm^{-3}). A major advantage with the newer reactors is the ability to separately control ion flux and ion energy, leading to increased flexibility in designing etch products [97].

3.2. Effect of Plasma Parameters

In the work of Flemish *et al.* [83], the Schottky barrier height of Pd/SiC diodes degraded from 1.05 eV on control samples to 0.64 eV on samples etched under RIE conditions. By contrast, ECR-etched diodes displayed barrier heights of 0.87 eV and excellent ideality factors (1.09 vs. 1.34 for control samples). The current-voltage measurements used for the barrier height evaluations are strong functions of the perfection and stoichiometry of the immediate surface region, providing a sensitive method for monitoring disruption to these quantities by the plasma processing.

The intrinsic resistance of SiC to the introduction of ion-induced damage can be measured on samples exposed to pure Ar plasmas, which induce little change in near-surface stoichiometry. Previous measurements of this type have shown, for example, that InGaN is much more resistant to damage introduction than a more common semiconductor with similar chemical makeup, InGaP [98]. Moreover, the thermal stability of plasma-induced damage in SiC has not been established, and this is an important consideration in designing the process sequence for device fabrication. In this section, we show that n-type SiC is strongly resistant to introduction of deep level compensating centers by ion bombardment, and a significant annealing of these centers occurs at ⟨700°C with an activation energy of 3.4 eV.

The β-SiC layers were grown by low pressure chemical vapor deposition on ⟨111⟩Si substrates at ∼ 1000°C. The thickness of the films as 1.2 μm, and the tensile stress was 250 MPa. Hall measurement showed a net electron mobility at 300 K of ∼50 cm^2/V s. This residual n-type conductivity may be

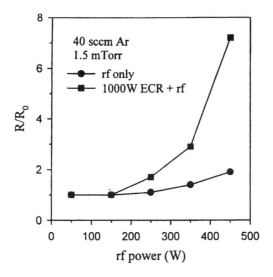

Fig. 36. Sheet resistance ratio of SiC before and after ECR Ar plasma exposure for 1 min, as a function of rf power.

due to unintentional nitrogen doping during growth. Van der Pauw geometry Hall samples were prepared with e-beam evaporated Ti annealed at 450°C to form Ohmic contacts [99], and exposed to ECR Ar discharges created in a Plasma-Therm SLR 770 system using an Astex 440 microwave (2.45 GHz) source operating at either 1000 or 0 W with additional rf (13.56 MHz) biasing of the sample position (50–450 W). The process pressure was varied from 1.5 to 10 mTorr and the plasma exposure time set at 1–10 min. The contacts retained sufficiently good Ohmic nature for the Hall measurements.

Figure 36 shows the ratio of the sample sheet resistance (R) after 1 min Ar plasma exposures, relative to the initial control value (R_0), for different rf powers applied to the sample chuck. When no additional microwave power is employed (which corresponds to a reactive ion etch configuration) there is little change in the average resistance of the epitaxial layers until a rf power of ≥ 250 W. The corresponding negative dc bias induced by this rf power is approximately -275 V, which acts to accelerate ions from the edge of the edge of the plasma sheath to the sample surface. Much larger changes in sheet resistance ratio are observed under ECR-rf conditions due to the much higher ion density ($\sim 5 \times 10^{11}$ cm^{-3} vs. $\sim 5 \times 10^9$ cm^{-3} without microwave power). The threshold rf power for damage introduction is ≤ 150 W, corresponding to dc biases of ≤ -170 V. It would be expected that the increased ion current for the combined microwave rf plasmas would induce more damage when a

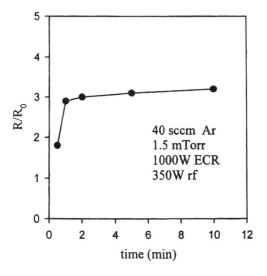

Fig. 37. Sheet resistance ratio of SiC before and after ECR Ar plasma exposure, as a function of exposure time.

nonetching gas such as Ar is used, since the damage is able to accumulate in the SiC. If an etch gas such as CF_4 or SF_6 were employed, then the near-surface damage would be removed as it was introduced, and one would not observe much change in the average conductivity of the film. Therefore, the current experiment is a worst-case scenario for plasma-induced damage because the SiC removal rate is negligible (sputter rate in Ar of $\leq$ 150 Å /min for 750 W microwave power and 450 W rf power). A plausible explanation for the resistance increases is the introduction of deep level states which trap some of the conduction electrons in the SiC. This is the mechanism for the creation of high resistance regions in ion-implanted semiconductors [100].

The time dependence of the resistance increases for exposure to a 1.5 mTorr, 1000 W ECR, 350 W rd Ar discharge is shown in Figure 37. The effect saturates after $\sim$ 2 min, which we assume means that there is an approximate balance between defect creation and recombination. The impinging Ar^+ ions create simple vacancy-interstitial pairs upon impact with the SiC surface. While the microscopic nature of the defects causing the deep level states is unknown, they are likely to consists of complexes between point defects and impurities or lattice atoms. If the vacancies and interstitials initially created annihilate by recombination then the deep states will not form. It appears that after several minutes of deep state formation is saturated due to a limited supply of at least one of its constituents.

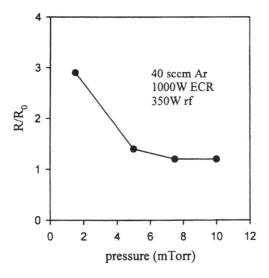

Fig. 38. Sheet resistance ratio of SiC before and after ECR Ar plasma exposure, as a function of process pressure.

As the process pressure is increased, the resistance change decreases (Figure 38). This is consistent with a reduction in the average ion energy through the additional collisions occurring at higher pressure. The ECR coupling efficiency also decreases with increasing pressure, so that the ion density is also reduced above ~5 mTorr in our system.

Once introduced, the damage can be removed by subsequent annealing. Figure 39 shows the annealing temperature dependence of the resistance changes for a sample into which damage was created by exposure to a 1.5 mTorr, 1000 W microwave, 350 W rf Ar discharge for 1 min. If we assume the annealing mechanism is a break-up of the defect complex by the separation and diffusion of its constituents, then we can estimate the activation energy for damage removal E_a, from the relationship:

$$E_a = -kT \ln[(-1/tv) \ln(1 - R/R_0)], \tag{1}$$

where the annealing time t is 60 s at temperature T, k is Boltzmann's constant, and the attempt frequency v is assumed to be 10^{14} s^{-1}, a typical value for these type of processes. The data in Figure 39 yield a value of $E_a \sim 3.4$ eV, somewhat higher than reported for dry etch damage removal in the wide-band-gap nitrides (~2.7 eV).

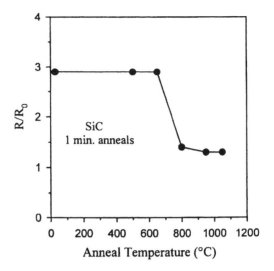

Fig. 39.   Sheet resistance ratio of plasma-damaged SiC, as a function of subsequent annealing temperature.

In conclusion, n-type SiC is quite resistant to changes in its conductivity upon exposure to Ar^+ ion fluxes. Under conventional RIE conditions rf powers ≤ 250 W induce no significant changes in resistance, while under ECR conditions it is necessary to limit the additional rf power to ≤ 150 W to avoid measurable damage introduction. A major annealing stage occurs in ion-damaged SiC at $\geq 700°C$, although there is still some remnant disorder remaining even at $1050°C$. The best strategy is obviously to operate under etching conditions below the threshold where damage is introduced.

The surface cleanliness and stoichiometry of SiC after dry etching in high density plasmas with F_2-based chemistries is generally very good, as shown in the AES surface scans of SiC after ICP etching in NF_3/O_2 discharges of various compositions (Figure 40).

3.3.   Damage in Electronic Devices

Flemish *et al.* studied the effect of RIE or ECR plasma etching on the I–V characteristics of Pd Schottky diodes formed in SiC surfaces. Schottky diode I–V characteristics are very sensitive to the metal-semiconductor interface quality. Assuming a thermionic emission model, the barrier height (Φ) was determined from the measured diode saturation current. The effective Richardson constant for SiC was taken to be 72 A-K^2/cm^2 based on an electron effective mass of 0.60. For each sample approximately

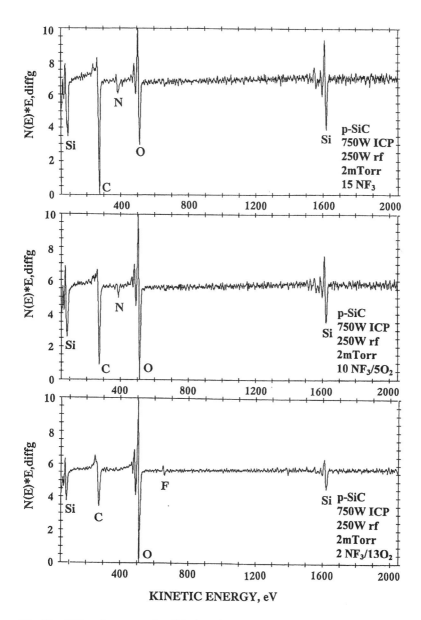

Fig. 40. AES surface scans of *p*-SiC after ICP etching in NF_3/O_2 discharges of different compositions.

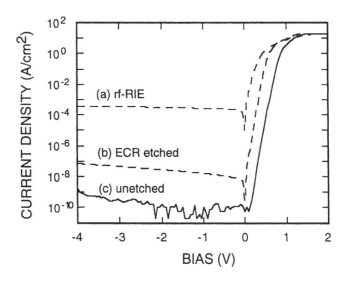

Fig. 41. I–V characteristics of Pd Schottky diodes on dry etched SiC.

15 to 20 diodes were measured. Out of these a small number on each sample exhibited unusually high leakage current which were presumed to be related to defects in the as-grown SiC layers. Disregarding the outlying characteristics, Figure 41 shows I–V characteristics 0.59 W/cm^2) and ECR (-100 V, 800 W) etched surfaces. For these samples approximately 500 nm of material was etched away, and for rf-RIE H_2 was added to obtain a smooth surface. Most notable is the large increase in the reverse bias leakage current for the etched samples. This increase is approximately two orders of magnitude for ECR etched samples and approximately six orders of magnitude for the rf-RIE samples. The Schottky diodes on the as-grown sample show average values of $\Phi = 1.05 \pm 0.02$ eV and $n = 1.34 \pm 0.08$, with the errors representing the standard deviations. The high ideality factor may arise from the presence of a thin interfacial layer between the Schottky contact and the SiC. The existence of such a layer is consistent with AES results showing O and N on the RCA-cleaned SiC surface. Schottky diodes on the as-grown sample show average values of $\Phi = 0.637 \pm 0.004$ and $n = 1.27 \pm 0.03$. The very high leakage current and low barrier height are consistent with the occurrence of significant lattice damage to the SiC caused by rf-RIE. In contrast, Schottky diodes on the ECR etched surface show values of $\Phi = 0.874 \pm 0.002$ and $n = 1.09 \pm 0.05$. The barrier height is lower than that of the as-grown sample, but much higher than that of the rf-RIE sample. This result together with the fact that less degradation is seen in the leakage

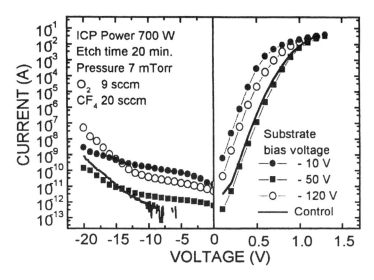

Fig. 42. I–V characteristics of Au Schottky diodes on ICP dry etched SiC.

currents suggests that the ECR etching process is far less damaging to the underlying SiC than the rf-RIE process.

Cao *et al.* [96] performed similar experiments on ICP CF_4/O_2 etched SiC surfaces. Figure 42 shows I–V characteristics (Ni rear contacts, annealed at 1050°C, and Au front Schottky contacts) from diodes etched at different substrate bias voltages. For biases of −50 V, the characteristics are similar to those for control samples that were not etched. Interestingly, diodes etched at −10 V had higher leakage currents, perhaps due to residues that are not removed under these conditions. ICP etching was employed to fabricate 4H-SiC gate turn-off thyristors with blocking voltage of ~460 V without gate current.

3.4. Effects of Hydrogen

Hydrogen is a component of virtually every gas or chemical involved in the growth and processing of SiC. For example, silane (SiH_4) and propane (C_3H_8) diluted with hydrogen can be employed as the sources for SiC epitaxial growth [101], and alternative organo-silanes are also gaining attention [102]. Boron doping can be accomplished with diborane (B_2H_6) diluted with hydrogen [101]. The extremely good chemical stability of SiC requires that dry etching be employed for patterning, and hydrogen is generally added to the plasma chemistry to avoid rough surfaces [103–105]. High temperature

(1500–1700°C) annealing in H_2, followed by quenching to room temperature produces hydrogen passivation (up to $\sim$75% decrease in carrier density) in both n-type (N-doped) and p-type (Al, B) SiC [106]. This effect was reversed by sybsequent annealing in He at the same temperatures. Annealing at 1700°C of chemical vapor deposited 6H-SiC epilayers increased the net hole concentration by a factor of approximately three and was accompanied by outdiffusion of hydrogen, and was consistent with residual passivation of B acceptors in the as-grown films. Theoretical studies of the properties of hydrogen in SiC have suggested that a tetrahedral interstitial site is the lowest energy site in undoped 3C-SiC, with a barrier to diffusion to the bond centered site of 1.5 eV [107–109]. In 6H-SiC, it was suggested that another interstitial position (the R-Site) was the lowest energy site, with the diffusivity being slightly slower in Si [107–109].

Thus, unintentional incorporation of hydrogen may occur at virtually any stage during the growth and processing of SiC, as it does in Si [110–112]. This will produce uncontrolled changes in the near-surface resistivity of the material through dopant passivation. The resistivity and breakdown voltage of this surface region are inversely dependent on doping density and therefore device characteristics may be strongly affected by the hydrogenation effects. In Si, acceptor passivation is a more efficient and thermally stable process than donor passivation because hydrogen occupies a more strongly bound bond-centered position between the acceptor and the nearest-neighbor Si atom, whereas in n-type material the hydrogen is at a less strongly bonded, anti-bonding interstitial position. It is, therefore, also of interest to examine differences in susceptibility of hydrogen passivation of n- and p-type material.

Nominally undoped SiC ($n \sim 8 \times 10^{17}$ cm^{-3}) substrates cut 3.5° off (0001) towards the middle of [1010] and [1120] were implanted into the polished Si face with 100 keV $^2H^+$ ions to a dose of 2×10^{15} cm^{-2}. Sections from these substrates were annealed under N_2 for 20 mins at temperatures up to 950°C in a tube furnace, or for 1 min at 1000°C in a Heatpulse 410T system. Other substrates were exposed to an Electron Cyclotron Resonance ^2H plasma (500 W microwave power, 10 mTorr Pressure) for 30 min at 300°C. The deuterium profiles were measured by Secondary Ion Mass Spectrometry (SIMS) using a Cs^+ ion beam [113]. The concentrations were quantified using implant standards, while the depth scales were established from stylus profilometry measurements of the analysis craters.

In Figure 43 are shown the ^2H atomic profiles in the plasma exposed sample before and after annealing at temperatures up to 1000°C. For as-treated samples, the incorporation depth is ≤ 0.1 μm and the peak ^2H density at the surface is $\sim 10^{20}$/cm^3. The deuterium concentration in the plasma-treated material is reduced by $\sim$90% with annealing at 800°C, and <1% remains after a 1000°C, 1 min anneal. These profiles lead to an effective diffusivity

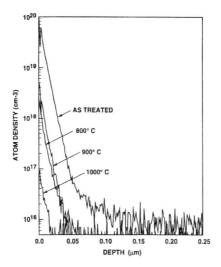

Fig. 43. SIMS profiles of ^{2}H in SiC exposed to a plasma at 300°C for 30 min, and subsequently annealed at 800, 900°C for 20 min or at 1000°C for 1 min.

of $\leq 6 \times 10^{-16}$/cm^2/s, which is much lower than that found in Si or in wide bandgap materials such as GaN and AlN [114]. The fact that the deuterium is present at a high concentration in a shallow region near the surface is indicative that it is present in platelet form or other extended clusters [115]. We have not performed transmission electron microscopy on the samples to confirm this hypothesis. In B-doped SiC containing hydrogen, the profile follows the B profile because of the presence of neutral (B-H)^0 complexes. The intrinsic solubility of hydrogen in undoped SiC material is so low that isolated hydrogen is essentially undetectable, and only other complexes (e.g., H$_2$ and larger clusters, Si-H) are present at significant concentrations.

The atomic distribution of ^{2}H in implanted samples has a much higher thermal stability. In Figure 44 are shown SIMS profiles of implanted samples before and after 20 min anneals at temperatures up to 950°C. Under these conditions there is no measurable redistribution of the deuterium. These profiles cannot be described by a single Gaussian, but require a Pearson IV-type distribution. This is fairly typical for implanted hydrogen in semiconductors [116]. The peak ^{2}H density is $\sim 10^{20}$/cm^3 and occurs at $\sim 0.75 \ \mu$m.

We expect that the deuterium in the implanted material is located in isolated interstitial positions, and that part of the ^{2}H population is trapped at the implant damage sites that overlap the ^{2}H profile. Therefore, with annealing

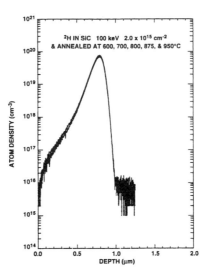

Fig. 44. SIMS profiles of ^2H in SiC implanted with $^2H^+$ ions (100 keV, 2×10^{15} cm^{-2}) and subsequently annealed at various temperatures for 20 min at each of the indicated temperatures.

there is remnant deuterium that remains in the initial implanted distribution shape. The ^2H redistribution is controlled by the annealing of the implant damage, which releases trapped deuterium. Apparently, up to 950°C very little of the implant damage has been annealed.

The data in Figure 44 indicate a much higher thermal stability for implanted deuterium (or hydrogen) than in any other semiconductor we have examined, including AlN [117, 118]. We believe that the difference in apparent thermal stability between the plasma treated and implanted materials is due to the presence of lattice damage in the latter which traps the deuterium. It is well established that hydrogen is attracted to any regions of strain or disorder in semiconductors [114], and thus the outdiffusion is retarded. By contrast, in the plasma treated material there is no major impediment to the deuterium motion once it is released from its initial state.

In conclusion, high concentrations ($> 5 \times 10^{19}$ cm^{-3}) of deuterium can be incorporated into undoped SiC by exposure to plasmas at 300°C, but the incorporation depth is shallow (≤ 0.10 μm). Most of this ^2H is removed by annealing at temperatures above 800°C. In contrast, implanted ^2H is thermally stable at higher annealing temperatures ($\geq 950°C$) without any loss from the crystal. Therefore, different methods of ^2H incorporation lead to different apparent thermal stabilities, as has been reported for III-N crystals [117]. Control of the hydrogen concentration in SiC structures and devices thus requires careful understanding of the growth and processing steps have the potential of introducing it.

Acceptor passivation is much more efficient in 6H-SiC than is donor passivation. This is a fairly general phenomenon in semiconductors and is related to the more stable bond centered position for hydrogen passivating acceptors compared to the antibonding interstitial position for hydrogen passivating donors. Reactivation of passivated acceptors occurs at $\sim 700°C$ in p-SiC, corresponding to a reactivation energy of ~ 3.3 eV. The diffusivity of ^2H is much slower in SiC at 200°C than it is in Si of the same doping level. The incorporation distance can be influenced by the average ion energy in the plasmas used to introduce hydrogen or deuterium, which is probably related to a decrease in self-trapping efficiency at the immediate surface of the SiC. One needs to be aware of acceptor passivation effects than can occur when hydrogen is part of the growth or processing ambient for SiC, in particular since the passivation can occur at many different stages. Hydrogen is a common component of most of the gases and liquids used during device fabrication and may induce significant changes in the electrical properties of the near-surface of SiC.

4. DAMAGE IN II–VI COMPOUNDS

4.1. Background

Zn-based II–VI semiconductors, especially ZnSe, has been a focus of attention for blue-green lasers and LEDs. Though plagued by reliability problems due to defect migration during device operation there is still great potential if substrate issues can be solved, ZnS and ZnSSe have also found application as low loss waveguide layers in AlGaAs/GaAs laser diodes. Cd-based materials have been employed in applications ranging from solar cells (CdS and CdTe), photo-electrochemical cells (CdSe), nuclear radiation detectors and electro-optical modulators. Hg-based materials have been used in infra-red detectors operating in the 3–5 and 8–12 μm ranges and for optical communication devices in the near 1R (1–3 μm). All of these materials are relatively susceptible to plasma-induced damage because of the different volatilities of the etch products.

4.2. Effect of Plasma Parameters

ZnSe, ZnCdSe, ZnMnSe, ZnS, ZnTe and ZnSSe have been dry etched in either chlorine-based (BCl$_3$, Cl$_2$) [119, 120] or methane-hydrogen (CH$_4$/H$_2$, H$_2$) [121, 122] plasma chemistries. In the former it is expected that Zn is ejected by largely physical means, whereas the group VI element can be removed as species such as Se_2Cl_2 which are quite volatile. In CH_4/H_2 etching, typical reductions in PL intensity of ~20% relative to unetched materials has been reported, and etch-induced non-radiative recombination is observed. Hydrogen passivation of both donors and acceptors is typical, and this is not reversed even for N_2 annealing at 400°C for 2 mins [123].

The Cd-based materials have been almost exclusively etched in CH_4/H_2. CdTe/CdMnTe quantum dots fabricated with this process showed compressive strain due to etch damage, with annealing removing most of this strain [124]. The Hg-based materials have only been etched in H_2 or CH_4/H_2, and there are usually problems maintaining surface stoichiometry [125].

Some of the issues encountered in high density plasma etching of the II–VI compounds include incorporation of Cl into the near-surface region (this is an n-dopant in these materials) [126] and surface roughening due to preferential loss of the group VI element. Typical depths for S, Se or Te loss are ≤ 100 Å , in general. Etch damage in a material such as HgCdTe is expected to include lattice defects, type conversion and stoichiometry disturbances resulting from sample heating, ion bombardment and possible impurity introduction. Ion sputtering with 260 eV Ar^+ ions has been reported to produce severe degradation of PL efficiency in ZnSe and CdZnSe, and dislocation loops are produced in the near-surface through agglomeration of interstitials [127]. CdS surfaces become slightly S-rich during ion milling due to differences in sputter yield for Cd and S [128]. In CdTe, dislocation loops were also reported in ion-milled material, with a damage layer extending well beyond the projected range of bombarding ions. Ion milling of HgCdTe produced preferential depletion of HgTe to a depth of ~50 Å (1–3 keV Ar^+ ion beam). Some ion-free techniques, such as laser-assisted etching, appear to produce less lattice disruption [129].

5. SUMMARY

An extensive amount of work has been performed on dry etching and related damage in wide bandgap semiconductors because of the absence of practical wet etch processes for GaN and SiC, in particular. Currently all commercial nitride LEDs and laser diodes employ dry etching for facet or mesa formation, and essentially damage-free pattern transfer has been demonstrated in SiC electronic device fabrication. High density plasma systems display much faster etch rates and generally produce cleaner, smoother semiconductor etched surfaces than conventional RIE methods. The remnant damage in dry etched structures depends strongly on many factors, including ion energy, ion flux and etch rate, because slow rates allow damage to accummulate.

Acknowledgements

The work at UF was performed in conjunction with C.R. Abernathy, F. Ren, J.W. Lee, J.J. Wang, C.B. Vartuli, J.M. Grow, M. Ostling, C.-M. Zetterling and J.R. Flemish. The work was partially supported by a DARPA/EPRI grant (D. Radack/J. Melcher) and USF grant, DMR 94-21109. Sandia is operated by Sandia Corporation, a subsidiary of Lockheed Martin, for DOE under grant DEAC04-94-AL85000.

References

1. see for example MRS Bulletin Special Issues; *GaN and Related Materials for Device Applications*, **22**(2) (1997); *SiC Electronic Materials and Devices*, **22**(3) (1997); *Blue-Green Laser Diodes*, **20**(7) (1995).
2. *GaN and Related Materials*, ed. S.J. Pearton (Gordon and Breach, New York 1997).
3. S.N. Mohammed, A.A. Salvador and H. Morkoc, *Proc. IEEE*, **83**, 1306 (1995).
4. S. Nakamura, M. Senoh, S. Nagahama, N. Iwasa, T. Yamada, T. Matsushitu, H. Kiyoku and Y. Sugimoto, *Jap. J. Appl. Phys.*, **35**, L74 (1996).
5. S.J. Pearton, *Appl. Surf. Sci.*, **117/118**, 597 (1997).
6. W.A. Harrison, *Electronic Structure and Properties of Solids* (Freeman, San Francisco 1980).
7. S.J. Pearton, C.R. Abernathy, F. Ren and J.R. Lothian, *J. Appl. Phys.*, **76**, 1210 (1994).
8. R.J. Shul, A.J. Howard, S.J. Pearton, C.R. Abernathy and C.B. Vartuli, *ECS Symp. Proc.*, **95–21**, 217 (1995).
9. J.C. Zolper and R.J. Shul, *MRS Bulletin*, **22**, 36 (1997).
10. R.J. Shul, in *GaN and Related Materials*, ed. S.J. Pearton (Gordon and Breach, New York 1997).
11. S. Nakamura, *MRS Bulletin*, **22**, 29 (1997).
12. S. Nakamura, in *GaN*, ed. J.I. Pankove and T.D. Moustakas (Academic Press, San Diego, 1998).
13. S.J. Pearton, R.J. Shul, C. Constantine and G.F. McLane, *Solid State Electron*, **41**, 159 (1997).
14. R.J. Shul, C.T. Sullivan, M.B. Snipes, G.B. McClellan, M. Hatich, C.T. Fuller, C. Constantine, J.W. Lee and S.J. Pearton, *Solid State Electron*, **38**, 2047 (1995).
15. S.J. Pearton and R.H. Shul, in *GaN*, ed. J.I. Pankove and T.D. Moustakas (Academic Press, San Diego 1998).
16. R.J. Shul, S.D. Kilcoyne, M.H. Crawford, J.E. Burmeter, C.B. Vartuli, C.R. Abernathy and S.J. Pearton, *Appl. Phys. Lett.*, **66**, 1761 (1995).
17. R.J. Shul, A.J. Howard, S.J. Pearton, C.R. Abernathy, C.B. Vartuli, P.A. Barnes and M.J. Bozack, *J. Vac. Sci. Technol.*, **B13**, 2016 (1995).
18. S.J. Pearton, J.W. Lee, J.D. MacKenzie, C.R. Abernathy and R.J. Shul, *Appl. Phys. Lett.*, **67**, 2329 (1995).
19. H.H. Tan, J.S. Williams, J. Zou, D.J.H. Cockayne, S.J. Pearton and R.J. Stall, *Appl. Phys. Lett.*, **69**, 2364 (1996).
20. M.A. Khan, J.N. Kuznia, D.T. Olson, W. Schaff, J. Burm and M.S. Shur, *Appl. Phys. Lett.*, **65**, 1121 (1994).
21. S.C. Binari, L.B. Rowland, W. Kruppa, G. Kelner, K. Doverspike and D.K. Gaskill, *Electron. Lett.*, **30**, 1248 (1994).
22. M.A. Kahn, M.S. Shur, J.N. Kuznia, J. Burm and W. Schuff, *Appl. Phys. Lett.*, **66**, 1083 (1995).
23. M.A. Kahn, Q. Chen, C.J. Sun, J.W. Wang, M. Blasingame, M.S. Shur and H. Park, *Appl. Phys. Lett.*, **68**, 514 (1996).

140 S.J. PEARTON and R.J. SHUL

24. W. Kruppa, S.C. Binari and K. Doverspike, *Electron. Lett.*, **31**, 1951 (1995).
25. S.C. Binari, L.B. Rowland, G. Kelner, W. Kruppa, H.B. Dietrich, K. Doverspike and D.K. Gaskill, *Inst. Phys. Conf. Ser.*, **141**, 459 (Bristol, UK: Inst. of Physics, 1995).
26. F. Ren, C.R. Abernathy, S.N.G. Chu, J.R. Lothian and S.J. Pearton, *Appl. Phys. Lett.*, **66**, 1503 (1995).
27. S. Strike, M.E. Lin and H. Morkoc, *Thin Solid Films*, **231**, 197 (1993).
28. H. Morkoc, S. Strite, G.B. Bao, M.E. Lin, B. Sverdlov and M. Burns, *J. Appl. Phys.*, **76**, 1363 (1994).
29. C.R. Abernathy, J.D. MacKenzie, S.R. Bharatan, K.S. Jones and S.J. Pearton, *Appl. Phys. Lett.*, **66**, 1632 (1995).
30. M.E. Lin, Z.E. Fan, Z. Ma, L.H. Allen and H. Morkoc, *Appl. Phys. Lett.*, **64**, 887 (1994).
31. I. Adesida, A.T. Ping, C. Youtsey, T. Dow, M.A. Khan, D.T. Olson and J.A. Kuznia, *Appl. Phys. Lett.*, **65**, 889 (1994).
32. G.F. McLane, L. Casas, S.J. Pearton and C.R. Abernathy, *Appl. Phys. Lett.*, **66**, 3328 (1995).
33. S.J. Pearton, C.R. Abernathy and F. Ren, *Appl. Phys. Lett.*, **64**, 3643 (1994).
34. F. Ren, J.R. Lothian, S.J. Pearton, C.R. Abernathy, C.B. Vartuli, J.D. MacKenzie, R.G. Wilson and R.F. Karlicek, *J. Electron. Mater.*, **26**, 1287 (1997).
35. S.J. Pearton, J.W. Corbett and M. Stavola, *Hydrogen in Crystalline Semiconductors* (Springer-Verlag, Heidelberg, 1992).
36. F. Ren in *GaN and Related Materials*, ed. S.J. Pearton (Gordon and Breach, New York, 1997).
37. A.T. Ping, A.C. Schmitz, I. Adesida, M.A. Khan, O. Chen and Y.W. Yang, *J. Electron. Mater.*, **26**, 266 (1997).
38. S.W. Pang, *J. Electrochem. Soc.*, **133**, 784 (1986).
39. C. Constantine, D. Johnson, S.J. Pearton, U.K. Chakrabarti, A.B. Emerson, W.S. Hobson and A.P. Kinsella, *J. Vac. Sci. Technol.*, **B8**, 596 (1990).
40. S.J. Pearton, J.W. Lee, J.D. MacKenzie, C.R. Abernathy and R.J. Shul, *Appl. Phys. Lett.*, **67**, 2329 (1995).
41. C. Yuan, T. Salagaj, A. Gurary, P. Zawadzki, C.S. Chern, W. Kroll, R.A. Stall, Y. Li, M. Schurman, C.-Y. Hwang, W.E. Mayo, Y. Lu, S.J. Pearton, S. Krishnankutty and R.M. Kolbas, *J. Electrochem. Soc.*, **142**, L163 (1995).
42. W. Shan, T.J. Schmidt, X.H. Yang, J. Hwang, J.J. Song and B. Goldenberg, *Appl. Phys. Lett.*, **66**, 985 (1995).
43. G.D. Chen, M. Smith, J.Y. Lin, H.X. Jiang, M. Asif Kahn and C.J. Sun, *Appl. Phys. Lett.*, **67**, 1653 (1995).
44. R.A. Gottscho, B.L. Preppernau, S.J. Pearton, A.B. Emerson and K.P. Giapis, *J. Appl. Phys.*, **68**, 440 (1990).
45. E.S. Aydil and R.A. Gottscho, *Mat. Sci. For.*, **148/149**, 159 (1994).
46. A.S. Usikov, W. L. Lundin, U.I. Ushakov, B.V. Pushnyi, N.M. Schmidt, Y.M. Zadiranov and T.V. Shubtra, *Proc. ECS*, **98-14**, 57 (1998).
47. K. Satore, A. Matsutani, T. Shirasawa, M. Mori, T. Honda, T. Sakaguchi, F. Koyama and K. Iga, *Mat. Res. Soc. Symp. Proc.*, **449**, 1029 (1997).
48. S.J. Pearton in *GaN and Related Materials*, ed. S.J. Pearton (Gordon and Breach, New York, 1997).
49. S. Nakamura, T. Mukai, M. Senoh and N. Iwasa, *Jap. J. Appl. Phys.*, **31**, L-139–L142 (1992).
50. S.J. Pearton, C.R. Abernathy, C.B. Vartuli, J.D. MacKenzie, R.J. Shul, R.G. Wilson and J.M. Zavada, *Electron. Lett.*, **31**, 836–837 (1995).
51. C.R. Abernathy, *J. Vac. Sci. Technol. A.*, **11**, 869–875 (1993).
52. S.J. Pearton, C.R. Abernathy, J.D. MacKenzie, R.G. Wilson, F. Ren and J.M. Zavada, *Electron. Lett.*, **31**, 327–328 (1995).
53. C.E. Weitzel, J.W. Palmour, C.H. Carter, Jr., K. Moore, K.J. Nordquist, A. Allen, C. Thero and M. Bhatnagar, *ICCC Trans. Electron. Dev.*, **43**, 1732 (1996).
54. K.Z. Xie, J.H. Zhao, J.R. Flemish, T. Burke, W.R. Buchwald, G. Lorenzo and H. Singh, *IEEE Electron Dev. Lett.*, **17**, 142 (1996).
55. B.J. Baliga, *IEEE Trans. Electron. Dev.*, **43**, 1717 (1996).

56. A.K. Agarwal, G. Augustine, V. Balakrishna, C.D. Brandt, A.A. Burke, L.S. Chen, R.C. Clarke, P.M. Esker, H.M. Hobgood, R.H. Hopkins, A.W. Morse, L.B. Rowland, S. Seshadri, R.R. Siergiej, T.J. Smith, Jr. and S. Siriam, *Tech. Dig. Inst. Electron. Dev. Meeting*, 9.1.1–9.1.6, Dec. 1996.

57. D.M. Brown, E. Downey, M. Ghezzo, J. Kretchmer, V. Krishnamurthy, W. Hennessy and G. Michon, *Solid State Electron.*, **39**, 1531 (1996).

58. V.E. Chelnokov, *Mat. Sci. Eng.*, **B11**, 103 (1992).

59. T.P. Chow and M. Ghezzo, *Mat. Res. Soc. Symp. Proc.*, **423**, 9 (1996).

60. S. Siriam, R.C. Clarke, A.A. Burk, Jr., H.M. Hobgood, P.G. McMullin, P.A. Orphanos, R.R. Siergiej, T.J. Smith, C.D. Brandt, M.C. Driver and R.H. Hopkins, *IEEE Electron. Dev. Lett.*, **15**, 458 (1994).

61. K.E. Moore, C.E. Weitzel, K.J. Nordquist, L.L. Pond, III, J.W. Palmour, S. Allen and C.H. Carter, Jr., *IEEE Electron. Dev. Lett.*, **18**, 69 (1997).

62. J.W. Palmour, J.A. Edmond, H.S. Kong and C.H. Carter, Jr., *Physica*, **B185**, 461 (1993).

63. J.B. Casady, D.C. Sheridan, W.C. Dillard and R.W. Johnson, *Mat. Res. Soc. Symp. Proc.*, **423**, 105 (1996).

64. S.T. Sheppard, M.R. Melloch and J.A. Cooper, *IEEE Trans. Electron. Dev.*, **41**, 1257 (1994).

65. J.B. Casady, J.D. Cressler, W.C. Dillard, R.W. Johnson A.K. Agarwal and R.R. Siergiej, *Solid State Electron.*, **39**, 777 (1996).

66. J.N. Sheroy, J.A. Cooper, Jr. and M.R. Melloch, *IEEE Electron. Dev. Lett.*, **18**, 93 (1997).

67. C.D. Brandt and R.H. Hopkins, *Mat. Res. Soc. Symp. Proc.*, **423**, 87 (1996).

68. P.G. Neudeck, D.J. Larkin, C.S. Salupo, J.A. Powell and L.G. Matus, *Inst. Phys. Conf. Ser.*, **137**, 475 (1994).

69. O. Kordina, J.P. Bergman, A. Henry, E. Fanzen, S. Savage, J. Andre, L.P. Ramberg, U. Lindfelt, W. Hermanasson and K. Bergman, *Appl. Phys. Lett.*, **67**, 1561 (1995).

70. M. Bhatnagar, P.K. McLarty and B.J. Baliga, *IEEE Electron. Dev. Lett.*, **13**, 501 (1992).

71. W. Xie, J.A. Cooper, Jr. and M.R. Melloch, *IEEE Electron Dev. Lett.*, **15**, 455 (1994).

72. J.B. Casady and R.W. Johnson, *Solid State Electron.*, **39**, 1409 (1996).

73. J.S. Shor, A.D. Kurtz, I. Grimberg, B.Z. Weiss and R.M. Osgood, *J. Appl. Phys.*, **81**, 1546 (1997).

74. D.H. Collins, G.L. Harris, K. Wongchotigul, D. Zhang, N. Chen and C. Taylor, *Inst. Phys.*

75. see for example, *Properties of SiC*, ed. G.L. Harris (Inspec, London, UK, 1995) pp. 131–149; J.R. Flemish, in *Processing of Wide Bandgap Semiconductors*, ed. S.J. Pearton (Noyes Publications, Park Ridge, NJ, 1997); P.H. Yih and A.J. Steckl, *J. Electron. Soc.*, **142**, 2853 (1995); C.E. Weitzel, J.W. Palmour, C.H. Carter, Jr., K. Moore, K.J. Nordquist, S. Allen, C. Thero and M. Bhatnagar, *IEEE Trans Electron. Dev.*, **43**, 1732 (1996).

76. J.B. Casady, E.D. Luckowski, M. Bozack, D. Sheridan, R.W. Johnson and J.R. Williams, *J. Electron. Soc.*, **143**, 1750 (1996).

77. P.H. Yih and A.J. Steckl, *J. Electron. Soc.*, **142**, 312 (1995).

78. J.W. Palmour, R.F. Davis, T.M. Wallett and K.B. Bhasin, *J. Vac. Sci. Technol.*, **A4**, 590 (1986).

79. A.J. Steckl and P.H. Yih, *Appl. Phys. Lett.*, **60**, 1966 (1992).

80. B.P. Luther, J. Rozyllo and D.L. Miller, *Appl. Phys. Lett.*, **63**, 171 (1993).

81. W.S. Pan and A.J. Steckl, *J. Electrochem. Soc.*, **137**, 212 (1990).

82. R. Padiyath, R.L. Wright, M.I. Chaudry and S.V. Babu, *Appl. Phys. Lett.*, **58**, 1053 (1991).

83. J.R. Flemish, K. Xie and J. Zhao, *Appl. Phys. Lett.*, **64**, 2315 (1994).

84. J.R. Flemish, K. Xie, W. Buchwald, L. Casas, J.H. Zhao, G.F. McLane and M. Dubey, *Mat. Res. Soc. Symp. Proc.*, **339**, 145 (1994).

85. G.F. McDaniel, J.W. Lee, E.S. Lambers, S.J. Pearton, P.H. Holloway, F. Ren, J.M. Grow, M. Bhaskaran and R.G. Wilson, *J. Vac. Sci. Technol.*, **A14**, 885 (1997).

86. J.R. Flemish and K. Xie, *J. Electrochem. Soc.*, **143**, 2620 (1996).

87. K. Xie, J.R. Flemish, J.H. Zhao, W.R. Buchwald and L. Casas, *Appl. Phys. Lett.*, **67**, 368 (1995).

88. see for example, J.B. Casady in *Processing of Wide Bandgap Semiconductors*, ed. S.J. Pearton (Noyes Publications, Park Ridge, NJ, 1997).

89. J.B. Casady, E.D. Luckowski, M. Bozack, D. Sheridan, R.W. Johnson and J.R. Williams, *Inst. Phys. Conf. Ser.*, **142**, 625 (1996).

90. F. Ren, J.M. Grow, M. Bhaskaran, J.W. Lee, C.B. Vartuli, J.R. Lothian and J.R. Flemish, *Mat. Res. Soc. Symp. Proc.*, **421**, 251 (1996).
91. J.R. Flemish, K. Xie and G.F. McLane, *Mat. Res. Soc. Symp. Proc.*, **421**, 153 (1996).
92. F. Lanois, P. Lassagne and M.L. Locatelli, *Appl. Phys. Lett.*, **69**, 236 (1996).
93. J. Wu, J.D. Parsons and D.R. Evans, *J. Electrochem. Soc.*, **142**, 669 (1995).
94. E. Niemann, A. Boos and D. Leidich, *Inst. Phys. Conf. Ser.*, **137**, 695 (1994).
95. L. Cao, B. Li and J.H. Zhao, presented at SiC and Related Conf., Stockholm Sweden, Sept. 1997.
96. L. Cao and H.J. Zhao, *IEEE Electron. Dev. Lett.* (in press).
97. see for example, M.A. Liebermann and A.J. Lichtenburg, *Principles of Plasma Discharges and Materials Processing* (Wiley & Sons, New York, 1994).
98. J.W. Lee, S.J. Pearton, C.R. Abernathy, W.S. Hobson and F. Ren, *Appl. Phys. Lett.*, **67**, 3129 (1995).
99. J.S. Shor, R.A. Weber, L.G. Provost, D. Goldstein and A.D. Kurtz, *J. Electrochem. Soc.*, **141**, 579 (1994).
100. S.J. Pearton, *Mat. Sci. Rep.*, **4**, 313 (1990).
101. D.J. Larkin, S.G. Sridhara, R.P. Devanty and W. P. Choyke, *J. Electron. Mater.*, **24**, 289 (1995).
102. J.M. Grow, *Proc. SOTAPOCS XXIV*, ed. F. Ren, S.J. Pearton, S.N.G. Chu, R.J. Shul, W. Pletschen and T. Kamijoh (Pennington, NJ: *Electrochem. Soc.*, 1996), **96-2**, 60 (1996).
103. A.J. Steckl and P.H. Yih, *Appl. Phys. Lett.*, **60**, 1966 (1992).
104. P.H. Yih and A.J. Steckl, *J. Electrochem. Soc.*, **140**, 1813 (1993); **142**, 312 (1995).
105. J.R. Flemish, *Wide Bandgap Semiconductors and Devices*, ed. F. Ren (Pennington, NJ: *Electrochem. Soc.*, 1995), **95–21**, 231 (1995).
106. F. Gendron, L.M. Porter, C. Poole and E. Bringuier, *Appl. Phys. Lett.*, **67**, 1253 (1995).
107. S.K. Estreicher, *Mater. Sci. Eng.*, **R14**, 319 (1995).
108. S.K. Estreicher, *Wide Bandgap Semiconductors and Devices*, ed. F. Ren (Pennington, NJ: *Electrochem. Soc.*, 1995), **95–21**, 78 (1995).
109. M.A. Robertson and S.K. Estreicher, *Phys. Rev.*, **B44**, 10578 (1991).
110. S.J. Pearton, C.R. Abernathy, R.G. Wilson, F. Ren and J.M. Zavada, *Electron. Lett.*, **31**, 496 (1995).
111. S.J. Pearton, J.W. Lee, J.M. Grow, M. Bhaskaran and F. Ren, *Appl. Phys. Lett.*, **68**, 2987 (1996).
112. J.I. Pankove and N.M. Johnson, *Hydrogen in Semiconductors* (San Diego: Academic Press, 1991).
113. R.G. Wilson, F.A. Stevie and C.M. Magee, *Secondary Ion Mass Spectrometry: A Practical Guide for Depth Profiling and Bulk Impurity Analysis* (Wiley, 1989).
114. J.M. Zavada, R.G. Wilson, C.R. Abernathy and S.J. Pearton, *Appl. Phys. Lett.*, **64**, 2724 (1994).
115. F.A. Ponce, N.M. Johnson, J.C. Tramontana and J. Walker, in *Microscopy of Semiconducting Materials*, ed. A.G. Cullis (IOP, Bristol, 1987), p.49.
116. R.G. Wilson, *J. Appl. Phys.*, **61**, 2826 (1987).
117. J.M. Zavada and R.G. Wilson, in *Hydrogen in Compound Semiconductors*, ed. S.J. Pearton (Trans Tech, Zurich, 1994).
118. R.G. Wilson, S.J. Pearton, C.R. Abernathy and J.M. Zavada, *J. Vac. Sci Technol.*, **A13**, 719 (1995).
119. K. Ohkawa, T. Karasawa and T. Mitsuyu, *J. Vac. Sci. Technol.*, **B9**, 1934 (1991).
120. E.M. Blausen, H.G. Craighead, M.C. Tamargo, J.L. Miguel and L.M. Schiavone, *Appl. Phys. Lett.*, **53**, 690 (1988).
121. G.J. Orloff, J.L. Elkind and D. Koch, *J. Vac. Sci. Technol.*, **A10**, 1371 (1992).
122. M.A. Foad, C.D.W. Wilkinson, C. Dunscomb and R.H. Williams, *Appl. Phys. Lett.*, **60**, 2531 (1992).
123. K. Ohtsuka, M. Imaizumi, H. Sugimoto, T. Isu and Y. Endoh, *Appl. Phys. Lett.*, **60**, 3025 (1992).
124. Y.S. Tang, P.D. Wang, C.M. Sotomayor-Torres, B. Lunn and D.E. Ashenford, *J. Appl. Phys.*, **77**, 6481 (1995).
125. J.L. Elkind and G.J. Orloff, *J. Vac. Sci. Technol.*, **A10**, 1106 (1992).

126. M. Tomouchi and T. Miyasato, *J. Appl. Phys.*, **70**, 3367 (1991).
127. J.B. Malherbe, *Conf. Rev. Solid State Mater. Sci.*, **19**, 55 (1994).
128. E.K. Chieh and Z.A. Munir, *J. Mater. Sci.*, **26**, 4268 (1991).
129. M. Rothschild, C. Arrone and D.J. Ehrlich, *J. Mater. Res.*, **2**, 244 (1987).

CHAPTER 5

Generation, Removal, and Passivation of Plasma Process Induced Defects

STELLA W. PANG

Department of Electrical Engineering and Computer Science,
The University of Michigan, Ann Arbor,
Michigan 48109-2122, USA

1. INTRODUCTION

In plasma etching, energetic particles in the discharge can cause damage in devices [1–10]. These high energy particles, which include ions, electrons, and photons, can introduce radiation damage in materials. Often, defects induced by dry etching penetrate deeply into the devices, way beyond the typical ion penetration range [11, 12]. In addition, contamination from materials coming off the plasma system or etch mask, deposition from the reactive species in the discharge, or stoichiometry changes due to preferential etching or layer intermixing in the compound semiconductors can also result in device degradation [13–16]. Therefore, it is important to understand the mechanisms for plasma process induced damage and to develop plasma etching conditions with minimal or no surface damage. However, there are multiple requirements that need to be satisfied for dry etching electronic or optoelectronic devices. These include controllable etch rate, selectivity, profile, surface morphology, uniformity, reproducibility, etch stop, and low damage. In order to meet most of these needs, some surface defects could be generated by dry etching. Therefore, sensitive techniques to analyze these surface defects are important to identify their origins and their influence to device performance. In addition, surface passivation and damage removal techniques are also critical to restore the device performance after plasma processing.

Among different evaluation techniques, electrical measurements on simple device structures such as diode, capacitor, and transmission line are particularly sensitive to surface damage. Since only a few processing steps are needed to form these devices, plasma induced defects are easier to identify without the complications of extensive processing. The high sensitivity of the electrical measurements also allows defect density as low as 10^{10} cm^{-2} to be detected [17–19]. Optical measurements are also sensitive to surface defects. Using multiple quantum well (MQW) layers, the depth of defect penetration can be derived from the photoluminescence (PL) signals off wells at different depths below the surface. Device degradation obtained from electrical evaluations could be different from those obtained using optical measurements [20, 21]. In addition, surface analyses are needed to identify the changes in structure, composition, and impurity after dry etching [6, 15].

In this chapter, surface damage induced by dry etching in III–V devices will be reviewed. Surface defects induced by dry etching are balanced by their generation and removal rates. The degree of damage depends strongly on the etch conditions, such as the ion energy, ion density, and etch species used for etching. In general, it is important to use low ion energy to prevent extensive damage. Therefore, high-density plasma systems are desirable since they can generate low energy ions with high density to reduce damage while still maintaining fast etch rates. Typical high-density plasma systems consist of an electron cyclotron resonance (ECR) source or an inductively coupled plasma (ICP) source. These plasma systems will be compared to conventional parallel plate reactive ion etching (RIE) systems. The effects of dry etching induced defects on the electrical and optical properties of a number of devices will be shown. These include changes in device characteristics of Schottky diodes, transmission lines, conducting wires, heterostructure bipolar transistors (HBTs), in-plane gated (IPG) quantum wire transistors, quantum dots, waveguides and mirrors. Techniques to minimize or remove etch-induced damage will be discussed. Improvements can be observed after thermal annealing, two-step etching, damage removal by low energy reactive species, and by surface passivation or coating.

2. DRY ETCHING SYSTEMS

Three commonly used dry etching systems for electronic and optoelectronic devices will be briefly described. The RIE systems are simpler and consist of a single rf power supply. To increase the etch rates, higher rf power and pressure are used but they often lead to higher device damage, lower selectivity, and less vertical profile. Recently, high-density plasma systems such as ECR and ICP systems become more popular since they provide better control and more flexibility. These high-density plasma systems usually have 2 separate power supplies to control the density and energy of the reactive species. In addition, plasma can be maintained at lower pressure. The low pressure helps to minimize the scattering of reactive species which results in vertical etch profile. Since the high-density plasma sources are more efficient in generating reactive species, high etch rates can be obtained even at low pressure.

2.1. Reactive Ion Etching Systems

Most of the devices are dry etched in parallel plate RIE systems as shown in Figure 1. In a RIE system, devices are placed on the bottom electrode that is typically powered by a 13.56 MHz rf power supply [22, 23]. The top electrode and the rest of the chamber are at ground potential. A self-induced dc bias ($|V_{dc}|$) is developed at the bottom electrode across the dark space sheath. The $|V_{dc}|$ is related to the area ratio between the top and bottom electrodes. It increases with rf power and it is directly related to the energy of ions bombarding the samples. To increase the etch rates, which improves the throughput, higher rf power is often applied. At higher rf power, the density of the charged and neutral species increases, but it also increases the ion energy. While it is desirable to have higher concentration of reactive species to provide faster etching, the high-energy ions decrease etch selectivity and they can also cause more damage to the devices.

To provide more flexibility and better control for dry etching, the density and energy of the reactive species should be adjusted independently. Ideally, the flux of the reactive species should be high to provide fast etch rates, while the energy of the ions should be low to avoid low selectivity and high damage

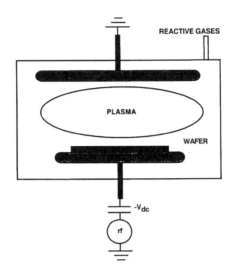

Fig. 1. Schematic of a parallel plate reactive ion etching system with rf power applied to the bottom electrode.

caused by bombardment of energetic ions. Unfortunately, in a typical RIE system, the ion density and ion energy cannot be controlled independently since there is only one rf power supply that controls both at the bottom electrode. Therefore, alternate dry etching systems have been used to provide this independent control. They are typically high-density plasma systems with two separate power supplies to allow independent control of concentration and energy of the reactive species. Examples of these high-density plasma systems will be described next.

2.2. Electron Cyclotron Resonance Plasma Systems

A schematic of the system which consists of an ECR source and an rf-coupled stage is shown in Figure 2. Microwave power at a frequency of 2.45 GHz is coupled into the ECR cavity through a rectangular waveguide and a copper probe [24, 25]. The cavity is a hollow cylindrical structure whose top height is adjustable and whose bottom consists of a quartz dome. The quartz dome is transparent to microwave power which allows the plasma to be ignited in the vacuum chamber. The cavity mode is tuned to minimize the reflected power and provide a stable or resonant electromagnetic mode by adjusting the top height of the cavity and the length of the copper probe in the cavity. The electromagnetic mode can be determined by measuring the microwave power radially and axially along the cavity. Electron cyclotron resonance

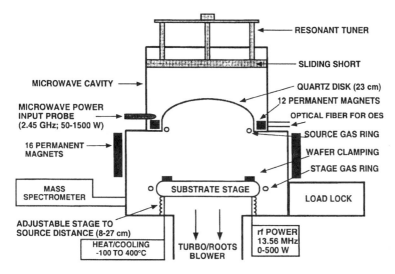

Fig. 2. Plasma etching system with an electron cyclotron resonance source in a tunable cavity. Microwave power is applied to the source for gas ionization and dissociation, while rf power is applied to the stage to control mainly the ion energy.

occurs due to the matching of the microwave frequency to the magnetic field that is generated by the permanent magnets surrounding the quartz disk. At resonance, the electrons are rotating in phase with the electric field. This allows the electrons to gain sufficient energy to ionize the gases in the chamber and create the plasma efficiently.

The advantage of using an ECR source to generate the plasma is that the use of two separate power sources allows separation of the plasma generation and wafer biasing. Additionally, due to the high efficiency provided when the resonance condition is met, a plasma can be generated at low pressure. Chamber pressure as low as 0.1 mTorr has been used to fabricate structures with extremely high anisotropy. A separate rf power supply is capacitively coupled to the wafer stage at 13.56 MHz. This rf power mainly controls the $|V_{dc}|$ which determines the energy of the incident ions. On the other hand, the microwave power coupled to the ECR source mostly controls the density of the reactive species. With high microwave power to the ECR source and low rf power to the stage, high concentration of charged and neutral reactive species can be generated with low ion energy for the charged particles. This is desirable for etching of devices since the high density but low energy ions can provide fast etch rate, high etch selectivity, and low surface damage.

A drift tube surrounded by permanent magnets is often used below the ECR source. These magnets can provide further plasma confinement, reduce

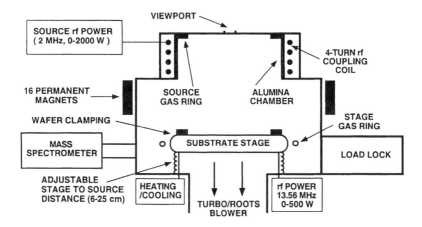

Fig. 3. An inductively coupled plasma source with an rf-powered stage. The source is driven by 2 MHz rf power and the stage is biased with 13.56 MHz rf power.

the recombination of reactive species at the chamber wall, and improve the discharge uniformity. Wafers are clamped to the rf-powered stage and the stage temperature can be controlled by a combination of resistive heating and liquid nitrogen cooling. Helium is flown at the backside of the wafer to improve the thermal conductivity between the wafer and the stage. Effective temperature control for the wafers is very important since the wafers can be easily heated up during etching by the power applied to generate the plasma. The temperature raise in the wafers can affect the etch characteristics as well as the degree of surface damage on the wafers.

2.3. Inductively Coupled Plasma Systems

As shown in Figure 3, the ICP source uses an inductive element adjacent to a discharge region to couple energy from an rf power source to gases in the chamber. The inductive circuit element is typically a planar or solenoid rf coil which acts like primary winding of a transformer and induces current in the plasma as a single-turn secondary coil [26, 27]. A matching network is used between the rf power source and the coupling coil to maximize the forward power and minimize the reflected power. The resonant circuit causes large rf currents to flow in the inductive coils. The rf magnetic flux generated by these currents penetrates into the adjacent discharge region. This time-varying rf magnetic flux induces a solenoidal rf electric field. It is this 'inductive' electric field which accelerates free electrons in the discharge and sustains the plasma.

The main advantage of an ICP source is its simplicity. Unlike the ECR plasma sources, no static magnetic field is required and only rf power rather than microwave power is needed. Coupling of rf power through a cable is much easier than coupling microwave power to the system using waveguides. Due to their flexibility in antenna design and lack of dc magnetic field, ICP sources are capable of generating uniform high-density plasmas over large areas and are well suited for scale-up to larger size substrates. Typically, inductive coupling of rf power can produce ion densities in excess of 10^{12} cm^{-3} even at sub-millitorr pressure, much higher density compared to capacitively coupled plasma sources with parallel plate electrodes. Since the coupling rf coil is not in direct contact with the plasma, there is no contamination due to sputtering of the coil. The ICP source shown in Figure 3 is coupled to a 2 MHz rf power supply. The ion density is primarily controlled by the 2 MHz rf power supply to the ICP source while the ion energy is controlled by the 13.56 MHz rf power supply that is capacitively coupled to the stage. The nearly independent control of the ion energy and density makes it easier to optimize processes.

3. PLASMA PROCESS INDUCED DAMAGE

During dry etching, a number of parameters can be controlled to provide the desired characteristics. For the high-density plasma systems, the controllable parameters include power to the source, power to the stage, gas chemistry, pressure, temperature, and distance between source and wafer. As these parameters are changed, they affect the concentration, energy, and directionality of the reactive species, the chemical vs. physical etch species, and the reaction rates. A number of requirements such as fast etch rate, vertical etch profile, smooth surface morphology, high etch selectivity, and low surface damage need to be satisfied. While it is highly desirable to meet all the requirements simultaneously, it often becomes necessary to compromise some of the criteria in order to achieve high device performance and low production cost. For example, high etch rate and vertical profile require the devices to be etched with high-energy ions. However, high ion energy also induces more surface damage in the devices. Therefore, techniques such as 2-step etching and damage removal or passivation become very useful. Initially, high ion energy may be used to maintain high etch rate and vertical profile. In 2-step etching, the etch condition can be adjusted near the end to reduce the ion energy so that low surface damage can be obtained. In addition, by removing or passivating the surface damage after etching is completed, high device performance can be restored.

In the following sections, the dependence of surface damage on etch conditions will be described. The changes in electrical and optical characteristics of high-speed devices after dry etching will be presented. Examples will be provided in HBTs, IPG quantum wire transistors, and quantum well waveguides and mirrors.

3.1. Effects of Ion Energy, Ion Density, Etch Species, and Etch Temperature

To evaluate the effects of etch conditions on the electrical characteristics of the dry etched materials, simple device structures such as Schottky diodes, transmission lines, and conducting wires were used [17, 18]. These devices require only a few fabrication steps without high temperature treatment. This allows dry etching induced surface damage to be identified without the complication of complex fabrication processes. Most of the devices were dry etched to a constant etch depth of 100 to 200 nm and their electrical responses were compared to those fabricated without dry etching.

GaAs Schottky diodes were formed on n-type epitaxial layer with doping concentration of 2×10^{16} cm^{-3} over n$^+$-substrate. After dry etching, 50/300 nm Ti/Au Schottky contact was deposited on the etched surface using a liftoff process. The samples were etched at 8 cm below the ECR source using a Cl$_2$/Ar mixture with 10% Cl$_2$ at 0.5 mTorr and 35 W microwave power. The diodes became more leaky with higher rf power. As the rf power was increased from 20 to 200 W, the barrier height (ϕ_B) was reduced from 0.74 to 0.66 eV and the ideality factor (n) was increased from 1.03 to 1.16. Since $|V_{dc}|$ increased from 46 to 383 V as rf power was increased from 20 to 200 W, higher ion bombardment energy could cause more defects to be generated at higher rf power. The reverse breakdown voltage (V_{BR}) also decreased as rf power was increased.

The capacitance-voltage (C-V) characteristics under the same etch condition are shown in Figure 4. The extracted intercept voltage (V_i) increased with rf power from 0.9 to 1.0 V. This could be explained by the presence of a non-conducting damaged layer for samples etched at high rf power. The series capacitance of this non-conducting damaged layer would cause the $1/C^2$ vs. V curve to shift up at higher rf power. A similar upward shift of the $1/C^2$ plot has been observed by others on GaAs after being etched by RIE in CH$_4$/H$_2$ [28] or CCl$_2$F$_2$/He [29]. At high rf power, the etch-induced damage is quite serious and could result in depletion of carriers near the surface. This surface depleted zone could be viewed as a non-conducting damaged layer, which increases V_i because of the series capacitance effects. The barrier height, on the other hand, continues to decrease because of the high concentration of induced defects. The ideality factor under these conditions was high and a decrease in the effective doping near the surface was observed.

The effects of ion energy on defect generation were further investigated by transmission electron microscopy (TEM). With 20% Cl$_2$ in a Cl$_2$/Ar plasma,

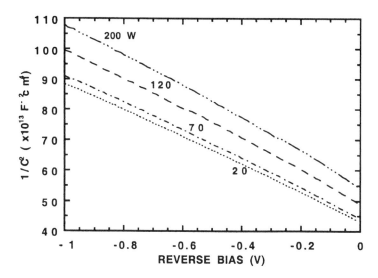

Fig. 4. Variation of $1/C^2$ vs. V for GaAs etched using 20, 70, 120 and 200 W rf power. The pressure was 0.5 mTorr with 35 W microwave power and 8 cm source distance. A Cl_2/Ar mixture with 10% Cl_2 was used.

GaAs was etched at a rf power of 20, 100, and 200 W. Smooth surface morphology was observed for all the samples. The defects are found to be mostly dislocation loops that are 2.4 nm in diameter. The defect density increased from 9.6×10^9 to 5.0×10^{10} cm^{-2} as the rf power was increased from 20 to 200 W. This agrees with the degradation of the Schottky diodes at higher rf power. The depth of the damaged layer decreased from 133 nm at 20 W to 53 nm at 200 W rf power. The decrease in damage depth at higher rf power could be due to the higher defect density at the surface and/or the faster defect removal due to the higher etch rate. Therefore, at higher rf power, even though the defect depth is shallower, the higher defect density still results in more leaky diodes.

The effects of ion flux were investigated by varying the microwave power at fixed $|V_{dc}|$. The changes in the unalloyed contact resistance extracted from GaAs transmission lines are shown in Figure 5. The samples were etched in a Cl_2/Ar plasma at 0.5 mTorr while $|V_{dc}|$ was fixed at 150 and 320 V. To maintain $|V_{dc}|$ at 150 V, rf power was varied from 53 to 63 W as microwave power was increased from 0 to 500 W. The contact resistance increased from 0.4 to 2.0 kΩ as the microwave power was increased from 0 to 500 W, indicating that more defects are induced on the surface etched with a higher ion flux. A similar increase of the contact resistance from 1.1 to 36.3 kΩ corresponding to 0 to 250 W microwave power was measured with $|V_{dc}|$

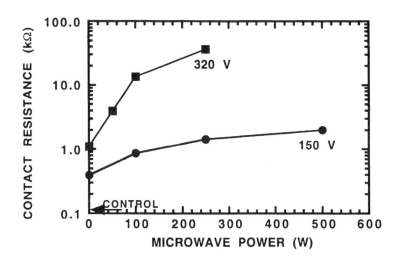

Fig. 5. Changes in the unalloyed contact resistance for samples etched at various microwave power using 150 V and 20% Cl$_2$ (●) and 320 V and 10% Cl$_2$ (■) in a Cl$_2$/Ar plasma generated at 0.5 mTorr.

fixed at 320 V. The high ion flux can cause more surface damage because of the faster defect generation rate.

Auger electron spectroscopy (AES) was used to study the changes in the stoichiometry of GaAs etched at different ion flux. It is found that the atomic concentration profiles for samples etched between 0 and 500 W microwave power are all similar to the control sample, and there is no deviation in the stoichiometry after etching. This shows that even though the electrical characteristics of the samples degrade at higher ion flux, the surface compositions of the etched GaAs remain the same. In addition, no Cl or Ni mask deposits can be detected on the etched surface, showing that etching with a Cl$_2$/Ar plasma is free of residue. However, Cl can be detected on the Ni mask with an atomic concentration ranging from 3.5 to 18.7%.

Different etch species could induce various degree of surface damage. Comparisons have been made between etching using Ar and N$_2$ addition in Cl$_2$. The current-voltage (I-V) characteristics after etching with different gas compositions are shown in Figure 6. The samples were etched at 15 cm below the ECR source with 50 W rf power and 35 W microwave power at 0.5 mTorr. |V_{dc}| was 170 V and was basically independent of the gases used in the plasma. A pure Ar or N$_2$ plasma caused the diode leakage current to increase substantially. The ideality factor was 1.57 and 1.28 after being etched with a pure Ar and N$_2$ plasma, respectively. The high n suggests the

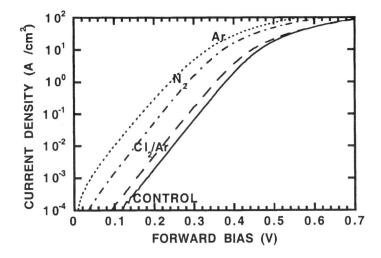

Fig. 6. Forward I–V characteristics using different gas compositions including Ar, N_2, 10% Cl_2 in Cl_2/Ar and control sample. Both the rf and microwave power were 50 W and the pressure was 0.5 mTorr. The samples were located at 15 cm below the ECR source.

formation of a heavily damaged layer on the surface. Evidently, Ar causes more physical damage to the surface than N_2. Samples etched by Ar sputtering also tend to have a lower V_{BR} than those etched in a N_2 plasma. Since Ar is an inert gas and has higher mass than N_2, this may cause more etch-induced surface damage. A 10% Cl_2 addition in Ar or N_2 resulted in a significant improvement of the I-V characteristics similar to that of the control sample. The $|V_{dc}|$ remained the same when Cl_2 was added to the Ar or N_2 plasma. This shows the importance of having reactive etch species to reduce surface damage.

The C-V characteristics for the samples etched under the same conditions show that a higher V_i was obtained with a pure Ar plasma compared to pure N_2. Similar to before, a 10% Cl_2 addition to either the Ar or N_2 reduces V_i to about the same value as the control sample. The higher V_i obtained with pure Ar or N_2 could be explained by the series capacitance caused by a damaged non-conducting surface layer. This is further verified by the carrier concentration profiles obtained from the C-V measurements. The carrier density is depleted near the surface after etching with either a pure Ar or N_2 plasma. Addition of Cl_2 as little as 10% recovered the effective doping concentration to the same level as the control sample. The significant reduction in surface damage when Cl_2 is added could be caused by the enhanced etch rate. The etch rates were 4.2, 0.4, and 91.9 nm/min for

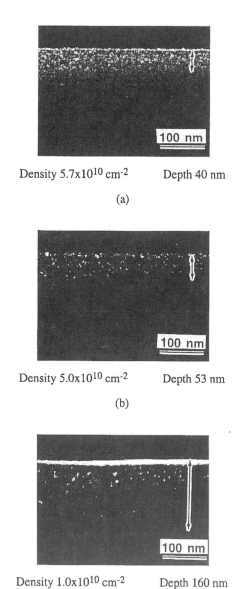

Density 5.7×10^{10} cm^{-2} Depth 40 nm

(a)

Density 5.0×10^{10} cm^{-2} Depth 53 nm

(b)

Density 1.0×10^{10} cm^{-2} Depth 160 nm

(c)

Fig. 7.   Cross-sectional transmission electron micrographs of GaAs etched at (a) $-130°$C, (b) $25°$C and (c) $350°$C. The samples were etched with 50 W microwave power and 200 W rf power at 0.5 mTorr in a Cl$_2$/Ar plasma with 10% Cl$_2$.

Ar, N_2, and 10% Cl_2 in Ar plasma, respectively. The increased etch rate could partially remove the generated surface damage. In addition, the adsorption of Cl_2 and its reactive species on the surface could act as a protective layer to reduce or passivate ion-induced damage.

The diffusion of the penetrated ions and the generated defects at higher etch temperature were investigated using TEM. Cross-sections of GaAs etched at -130, 25, and $350°C$ using 20% Cl_2 in a Cl_2/Ar plasma are shown in Figure 7. As the etch temperature is increased from -130 to $350°C$, the defect density decreases from 5.7 to 1.0×10^{10} cm^{-2} and the depth of defects increases from 40 to 160 nm. This shows that even though defects are distributed further below the etched surface at higher etch temperature, the defect density is still lower because of the annealing effect. From the measurements of the unalloyed contact resistance of GaAs transmission lines and conducting wires, however, higher etch temperature improves both the contact resistance of the transmission lines as well as reduces sidewall damage depth from the conducting wires. This suggests that the electrical characteristics of GaAs are more sensitive to the defect density than to the depth of defects under these etch conditions.

The surface stoichiometry at various etch temperature was studied using AES. From the depth profiles, the stoichiometry for the etched GaAs surface is found to be similar to the control sample for etch temperature as high as $350°C$. This shows that there is no preferential etching of either Ga or As under these etch conditions even at high etch temperature. No Cl was detected on the etched surface for etch temperature ranging from 25 to $350°C$. The concentration of Cl on the Ni etch mask decreased with increasing etch temperature, suggesting that higher temperature helps to desorb the deposited Cl.

3.2.   Time and Material Dependent Defect Generation

The reduction in feature size typically reduces the etch times for pattern transfer. The shorter etch times may cause transients in etching characteristics to become more pronounced. Electrical damage, as measured by Schottky diode ideality factor, has been reported to be constant during etch times of 2 min for GaAs [30] or to change modestly with long overetch times on an etch stop layer [31] in RIE systems. Optical damage time dependence has been reported for ion exposure times > 1 min as measured by PL [32] and cathodoluminescence [33]. Cathodoluminescent intensity was reported to decrease after an etch time of 7 min compared to a sample etched for 1 min.

The competition between creation and removal of damage by dry etching determines whether the devices will have low damage or not. Slow etching of Si has been shown to lead to accumulation of damage [10] while faster etch rates for GaAs have been shown to lead to a denser but shallower damage layer near the surface [34]. Etching damage extends deeper than the predicted

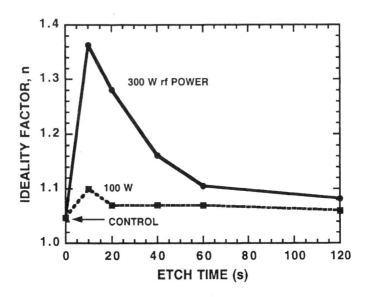

Fig. 8. Changes in GaAs Schottky diode ideality factor with etch time for different rf powers. The etch condition was 50 W microwave power, Cl_2/Ar flowing at 3/27 sccm, 1.5 mTorr chamber pressure and stage position of 12 cm.

ion stopping range due to both defect diffusion and ion channeling during etching [35, 36]. Also, InP based materials are typically more susceptible to damage from ion bombardment than GaAs-based materials as measured by Rutherford backscattering [37] and PL spectra taken from MQW material [34].

The time dependence of etch-induced damage from an ECR plasma source in GaAs and $In_{0.53}Ga_{0.47}As$ was compared. Variations in rf power, etch chemistry, and temperature were investigated to determine their respective effects on damage saturation time. The etch time dependence of n is shown in Figure 8. The etch condition was 50 W microwave power, Cl_2/Ar flowing at 3/27 sccm, 1.5 mTorr, and a source distance of 12 cm. The n initially increased after 10 s etch time and then decreased for longer etch time. However, n is still higher after etching for 120 s compared to an unetched sample. The diodes etched with 100 and 300 W rf power on the stage had n of 1.10 and 1.36 after 10 s but improved to 1.06 and 1.08, respectively, after 120 s. Likewise, the ϕ_b improved after an initial degradation at 10 s. After 10 s of etching, the ϕ_b of the diodes etched with 300 W rf power degraded from 0.75 to 0.59 eV but returned to 0.72 eV after 120 s. The diodes etched with 100 W rf power showed similar changes. Both the n and ϕ_b neared saturation after the 120 s. However, etch times > 120 s were not possible because the 1 μm thick epitaxial layer would have been etched through. With the 100 and 300 W rf power, the etch rates were constant with etch time at 110 and 204 nm/min, respectively. The $|V_{dc}|$ was 80 and 240 V for 100 and 300 W rf power, respectively. At these $|V_{dc}|$ and having a vacuum tight system with load lock, the native oxide and surface residue on GaAs can probably be removed instantaneously once dry etching starts. Therefore, no induction time was observed under these etch conditions. The changes in n and ϕ_b could be related to the diffusion of dry etching induced defects. As GaAs is etched, defects created at the surface could diffuse farther into the material due to defect enhanced diffusion, therefore lowering surface damage at longer etch times. On the other hand, the defects generated during the beginning of the etch could be etched away more easily at longer etch time due to disorder generated at the surface. Disorder at the surface caused by the etching may be responsible for the reduction in channeling [38].

The effects of the etch chemistry on the time dependent diode degradation were investigated. It was found that increasing the percentage of Cl_2 during the etch improved the diode characteristics. Figure 9 shows the changes in ϕ_b with etch time for different Cl_2 percentages in Ar. The etch condition used was similar to the one used in Figure 8 with the rf power fixed at 100 W. The diodes etched with Ar only showed a decreasing ϕ_b with etch time. The ϕ_b was 0.61 eV after 10 s and continued to degrade to 0.53 eV after 120 s. With the addition of only 5% Cl_2, the decrease in ϕ_b was less compared to Ar sputtering and the diode characteristics improved with etch time. At 10 s etch time, maximum degradation was observed and ϕ_b changed from 0.67 to 0.70 eV from 10 to 120 s. Less damage was induced with 10% Cl_2 in Ar and further increases in Cl_2 percentage did not affect the barrier height etch time dependence. When the samples were etched by Ar sputtering, the ϕ_b continued to degrade with etch time. However, the addition of 5% Cl_2 to Ar improved the diodes with etch time. The GaAs etch rate for 5% Cl_2 in Ar was 40 nm/min, substantially faster than the typical sputter rate of 5 nm/min when only Ar was used for etching. Thus, if the damaged layer was removed during the faster etching, the GaAs diodes tended to show better electrical characteristics [6]. This illustrates the competition between damage creation and damage removal that occurs during etching.

Capacitance-voltage measurements of diodes etched for different times did not show a time dependence. There was some carrier depletion of 24% near the intrinsic depletion region and it extended as deep as 400 nm below the etched surface. However, neither the degree or depth of depletion changed with etch time. This may be due to the large intrinsic depletion layer near the surface that makes C-V measurements insensitive to changes in defect density very close to the top surface. This indicates that the time dependent effects may be related to surface changes only.

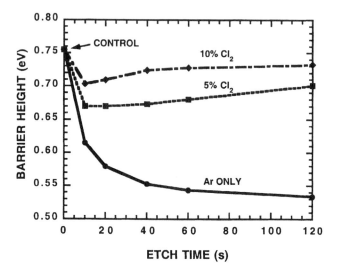

Fig. 9. Etch time dependence of barrier height for different Cl_2 in Ar percentages during etching. The etch condition was 50 W microwave power, 100 W rf power, total flow of 30 sccm, 1.5 mTorr and 12 cm below the source.

To determine whether the diffusion of defects could play a role in the etch time dependence of the damage to GaAs, the stage temperature in the etching chamber was controlled between −130 and 350°C. Increased temperature could cause defects to be more mobile and move away from the surface more quickly while reduced etch temperature could reduce the mobility of the defects and prevent them from moving away from the surface. As shown in Figure 10, the ϕ_b of diodes etched at an elevated stage temperature improved more quickly than those etched with the stage at room temperature. The ϕ_b improved from 0.66 to 0.75 eV after etching from 10 to 40 s. The diodes etched at room temperature only improved from 0.59 to 0.70 eV in the same time. Besides the diffusion effect, the reduction in defects could also be related to the annealing of defects by the higher stage temperature. The diodes etched at −130°C displayed much less etch time dependence as the ϕ_b changed from 0.62 eV after 10 s of etching to 0.64 eV after 1 min. This indicates that the etch time dependent changes in electrical characteristics are mostly related to the diffusion of defects away from the surface of the material.

Dry etching induced damage has been shown to cause changes in material that can degrade device characteristics. The demands for integration of electronic and optoelectronic devices on the same wafer have made it important to understand how the etch-induced damage can affect the electrical and optical properties of the material [20, 39]. Characterization of etch-induced

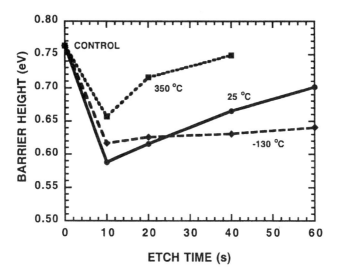

Fig. 10. GaAs Schottky diode barrier height dependence on etch time for stage temperatures of -130, 25 and 350°C. Etch condition was 50 W microwave power, 300 W rf power, Cl_2/Ar flowing at 3/27 sccm, 1.5 mTorr chamber pressure and stage position of 12 cm.

damage to the optical properties of AlGaAs/GaAs and InP/InGaAs have shown that defects can propagate deep into the material by a combination of ion channeling and diffusion [32, 34, 36]. Etch-induced damage has been shown to affect the electrical properties of materials as well by changes in mobility [30], sidewall damage depth [6, 40], and contact resistance and Schottky diode ideality factor [6, 9, 13, 18]. Previous work has also shown that the use of higher stage power during etching can affect the residual damage depth and distribution of the etch-induced defects [6, 10, 15]. Furthermore, optical characteristics of near surface quantum wells for epitaxial layers grown on InP substrates have been shown to be adversely affected by etch-induced damage to a greater extent than epitaxial layers grown on GaAs substrates [20, 41]. Therefore, the choice of material for device fabrication should be given careful consideration given the processing conditions.

In Figure 11, the sheet resistivity (ρ_s) of AlGaAs/InGaAs and AllnAs/InGaAs QW materials is shown to vary with the stage power used during etching. The etch condition was 100 W source power, Cl_2/Ar flowing at 2/28 sccm, a chamber pressure of 2 mTorr, and a source to sample distance of 12 cm. This etch condition was chosen to provide a slow, controllable etch rate and a higher concentration of Ar was used to generate more defects. The ρ_s for the AllnAs/InGaAs QW unetched sample was 5.5

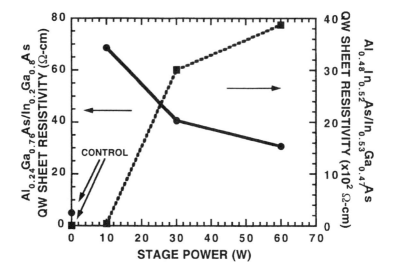

Fig. 11. Variations in sheet resistivity for AlGaAs/InGaAs and AlInAs/InGaAs QW materials etched with different stage powers. The etch condition was 100 W source power, Cl₂/Ar flowing at 2/28 sccm with a chamber pressure of 2 mTorr and the source to sample distance fixed at 12 cm. The samples were etched to a depth of 15 nm.

compared to 5.0 Ω-cm for the AlGaAs/InGaAs QW unetched sample. However, after etching 15 nm from the surface with 60 W stage power, the ρ_s extracted from the AlInAs/InGaAs QW structure was 3876 Ω-cm compared with 31 Ω-cm for the AlGaAs/InGaAs QW structure. Thus, the AlInAs/InGaAs QW structure exhibited a greater deviation from the control sample, indicating a greater effect from the etch-induced damage.

While the AlInAs/InGaAs QW structure showed greater degradation with increasing stage power, the AlGaAs/InGaAs QW sample had a higher ρ_s at 10 W stage power of 69 Ω-cm and it decreased to 31 Ω-cm as the stage power was increased to 60 W. Over this same range of stage powers, the $|V_{dc}|$ increased from 32 to 140 V. The changes in the ρ_s measured in the AlGaAs/InGaAs QW structure could be related to a different residual damage depth in AlGaAs/InGaAs QW structures for variations in ion energy. At lower ion energies, the defect density tends to be lower but the residual damage depth is deeper as compared to higher ion energies. This reflects the important balance between the damage generation and removal rates [6, 21, 39]. While a greater number of defects are generated with increased stage power and the associated higher ion energy, the etch rate also increases, thus removing some of the generated defects, resulting in a shallower residual damage layer. For the AlGaAs/InGaAs QW structures with a stage power of 10 W, the etch rate was 32 nm/min but with a stage power of 60 W, the etch rate increased to 56 nm/min. Thus the damage removal rate may have increased more rapidly than the generation rate, in effect leading to a lower number of residual defects to degrade the conduction in the well at the higher stage power.

InP-based materials have previously been shown to exhibit a different etch time dependence on the etch-induced damage compared to GaAs because of different diffusion, generation, or removal rates. Due to the involatility of the $InCl_x$ compounds, the etch rate for the AlInAs/InGaAs QW structure was slower than that of the AlGaAs/InGaAs QW materials. When the stage power was increased from 10 to 60 W, the etch rate increased from 4.5 to 12.0 nm/min in the AlInAs/InGaAs QW structure. Thus the differences in ρ_s extracted from AlInAs/InGaAs and AlGaAs/InGaAs QW structures could be related to the different damage removal rates. The etch rate of the AlInAs/InGaAs QW structure did not increase significantly with the increased stage power while for the same range of $|V_{dc}|$, the damage generation rate probably did. Thus, more concentrated damage in the well would lead to a higher ρ_s at higher stage power.

In Figure 12, the ρ_s extracted from AlInAs/InGaAs QW structures is shown to have a greater dependence on etch time than the AlGaAs/InGaAs QW structures. The AlInAs/InGaAs and AlGaAs/InGaAs QW materials were etched with similar condition as in Figure 11 with stage powers of 60 and 10 W, respectively. These stage powers were chosen because the structures showed the greatest effects of the etch-induced damage at these respective stage powers. It has been previously shown that changes in stage power do not affect the etch time dependence characteristic of etch-induced damage, only the magnitude of the damage [20]. The AlGaAs/InGaAs QW structure showed little deviation from the control sample after being etched for 20 s as the ρ_s increased from 5.0 to 5.5 Ω-cm. However, etching for 28 s increased the ρ_s to 53.0 Ω-cm. The AlInAs/InGaAs QW structure showed a marked increase from the control sample in ρ_s for all etch times. After 20 s of etching, the ρ_s increased from 5.5 to 16.9 Ω-cm and continued to increase to 427.0 Ω-cm after 60 s. While the AlInAs/InGaAs QW structure degraded more severely at shorter etch time, both material systems showed increased degradation, indicating that damage was accumulating in the well. These variations in the etch time dependence between AlGaAs/InGaAs and AlInAs/InGaAs QW structures show that there is probably a difference in how the damage is generated and propagates in the two material systems.

When no etching occurs, in the case of the sidewalls of conducting wires, the sidewall damage no longer depends on the etch rate since damage removal can be neglected. In Figure 13, the conductance of wires etched in AlGaAs/InGaAs and AlInAs/InGaAs QW structures shows that the extracted cutoff width for the AlInAs/InGaAs QW structure is greater

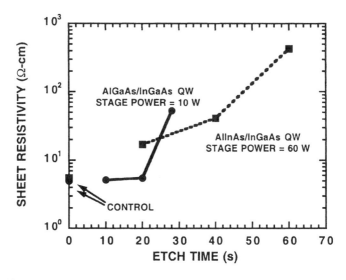

Fig. 12. Effect of changes in etch time on the sheet resistivity of the AlGaAs/InGaAs and AlInAs/InGaAs QW transmission lines. A similar etch condition to that of Figure 11 was used with the stage power set to 10 and 60 W for the AlGaAs/InGaAs and AlInAs/InGaAs QW conducting wires, respectively.

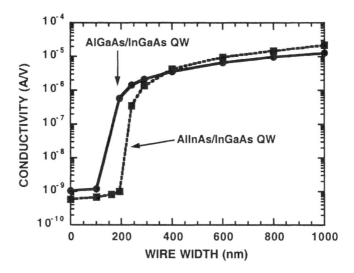

Fig. 13. Comparison of the etch induced sidewall damage in AlGaAs/InGaAs and AlInAs/InGaAs QW conducting wires. The etch condition was 500 W source power, 50 W stage power, Cl_2 flowing at 6 sccm with a chamber pressure of 0.15 mTorr and a source to sample distance of 12 cm.

than that for the AlGaAs/InGaAs QW structure. Wires were defined by electron beam lithography with widths of 100 nm to 1 μm and etched to a depth of 700 nm in an etch condition of 500 W source power, 50 W stage power, Cl_2 flowing at 6 sccm, a chamber pressure of 0.15 mTorr, and a source to sample distance of 12 cm. The $|V_{dc}|$ was 80 V and the etch rate for both materials was 280 nm/min. The conductance of the 1 μm wires was 2.2 $\times$ 10^{-5} and 1.3 $\times$ 10^{-5} Ω^{-1} for the AlInAs/InGaAs and AlGaAs/InGaAs QW structures, respectively. The AlInAs/InGaAs QW wires became non-conducting (9.9 $\times$ 10^{-9} Ω^{-1}) at a width of 193 nm while the AlGaAs/InGaAs QW structures were still conductive (5.7 $\times$ 10^{-7} Ω^{-1}). The electrical cutoff width for AlGaAs/InGaAs QW wires was 146 nm, while it was 243 nm for the AlInAs/InGaAs QW structures. Thus, when the effect of etch rate or damage removal rate can be neglected from the residual damage profile, the AlInAs/InGaAs QW structures still show a greater degradation from etching than the AlGaAs/InGaAs QW structures. Previously, it has been reported that the presence of a strained layer tends to getter defects [14] and cause higher degradation of strained quantum wells. In Figure 13, a lower degree of damage was observed for the strained AlGaAs/InGaAs QW than the unstrained AlInAs/InGaAs QW materials. The difference could be due to the small strain in the AlGaAs/InGaAs QW structures and the different barrier materials.

3.3.   Electrical Degradation of Heterostructure Bipolar Transistors and In-plane Gated Transistors

Although wet chemical etching is often used for self-aligned AlInAs/GaInAs HBTs, the sidewall etch rate is difficult to control and the dimensions of the emitter must be kept larger than what is desirable. A controllable sidewall profile can be obtained by dry etching. One disadvantage to dry etching is the potential for ion-induced damage to the etched surface. There are also several advantages to using a high ion energy. For example, a high ion energy has been found to minimize the roughness that would otherwise occur due to the low-volatility of the $InCl_x$ (x = 1 to 3) etch products [43]. High ion energy also assists in minimizing the etch initiation time by efficiently removing the surface oxide at the start of etching [44]. However, the use of high ion energy promotes damage to the etched surface.

The effects of ion energy on specific contact resistivity (ρ_c) for lightly doped n-type GaInAs has been studied [13]. The base layer is heavily doped p^+-GaInAs and so the effects of ion energy on the contact resistance of this material needs to be determined. Damage to GaInAs has been found to increase the sheet resistance (R_s) for $|V_{dc}|$ above 100 V when only Ar or H_2 was used to etch the material [45]. We have found that for n-type material, increasing ion energy decreases the ρ_c. However, the ρ_c increases with

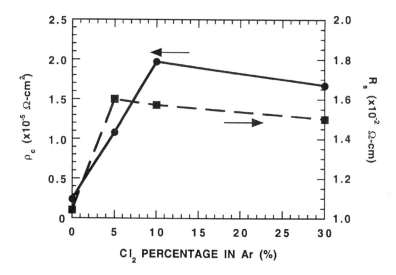

Fig. 14. Effect of microwave power on specific contact resistivity and sheet resistivity of GaInAs. The samples were etched with $|V_{dc}|$ of 130 V, 13 cm, Cl_2/Ar at 1/9 sccm, 1 mTorr and 30°C.

increasing ion energy for p-type material. Also, the amount of damage and the depth of the damage into the material increase as the ion energy is increased.

For samples etched with microwave power ranging from 50 to 150 W, ρ_c and R_s remain mostly constant as shown in Figure 14. As microwave power was varied from 0 to 150 W, the rf power was varied from 42 to 48 W to maintain constant $|V_{dc}|$ at 130 V. Meanwhile, the etch rate increased from 6 to 45 nm/min as the microwave power was increased from 0 to 150 W. For the wet etched GaInAs control sample, ρ_c is 7.5×10^{-5} Ωcm^2 and R_s is 1.45×10^{-2} Ω-cm. The results indicate that surface damage is independent of ion flux since ρ_c and R_s do not change with microwave power. Even though it is possible for more damage to be generated at higher ion flux, the faster etch rate at higher microwave power may remove the generated defects more efficiently and compensate the increased damage. However, ρ_c is only 2.1×10^{-5} Ω-cm^2 over this range of microwave power, approximately 70% lower than the control sample due to defect generation. When only rf power was applied and no microwave power was used, ρ_c decreased further to 1.1×10^{-5} Ω-cm^2. The increased surface damage when no microwave power was applied may be related to more damage accumulation since the etch rate is reduced. It has been previously shown that a balance exists between defect generation and removal, and the value of $|V_{dc}|$ determines whether increasing microwave power enhances or degrades device performance [6].

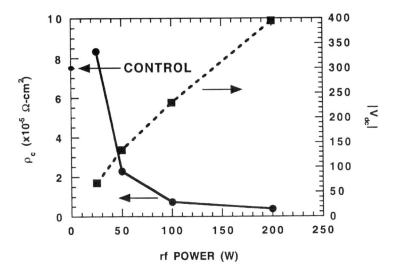

Fig. 15. Reduction of ρ_c and increase in dc bias with rf power. The etch condition was 50 W microwave power, 13 cm, Cl_2/Ar at 1/9 sccm and 30°C.

The changes in $|V_{dc}|$ due to the variations in etch condition have a strong influence on ρ_c. The dependence of ρ_c and $|V_{dc}|$ on rf power is shown in Figure 15. For an rf power of 25 W ($|V_{dc}| = 68$ V), the dry-etched sample showed similar ρ_c as the wet-etched sample. Using higher rf power is desirable since the etch rate increases from 14 to 120 nm/min as the rf power increases from 25 to 200 W. However, increasing the rf power to 200 W caused $|V_{dc}|$ to increase to 395 V and the higher energy ions created more damage. The specific contact resistivity was significantly lower when the rf power was increased from 25 to 200 W, decreasing by a factor of 21 from 8.3 to 0.4×10^{-5} Ω-cm^2. However, R_s was found to be independent of rf power; therefore, ρ_c is more sensitive to surface damage than R_s. The increased damage at higher rf power is caused by the increased ion energy since the ion flux is only weakly affected by changing rf power. These results indicate that low ion energy is essential to minimize etch-induced damage.

Figure 16 shows the forward I-V curves after etching with 100% Ar and 10 to 30% Cl_2 addition in Ar. Ar sputtering results in an ohmic-like contact, whereas the samples etched with 10 to 30% Cl_2 are very similar to the control sample. From the C-V measurements, the doping profiles after etching were the same as the control sample when 10% Cl_2 was added in Ar. The reduction of etch-induced damage by the addition of reactive gas can be related to the increase in etch rate or to passivation of the surface by reactive chlorine.

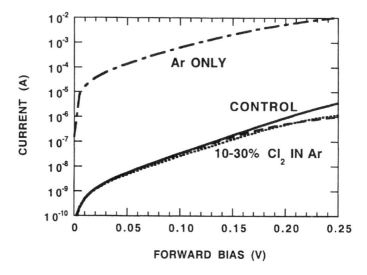

Fig. 16. Forward current for Schottky diodes after etching with 50 W microwave power, 50 W rf power, 1 mTorr, 12 cm and 30°C. Compared to Ar sputter etching (— — — — —), 10% (— — — —) and 30% (- - - - - - - -) Cl_2 addition in Ar improve diode performance substantially.

The extracted ϕ_B from the forward I-V curves and the VBR after etching also vary with different Cl_2 percentage in Ar. The wet etched sample had a ϕ_B of 0.63 eV and a V_{BR} of 21 V. The diode characteristics were severely degraded after Ar sputtering. The ϕ_B decreases from 0.62 to 0.35 eV when no Cl_2 is added in the discharge. Likewise, the V_{BR} decreases from 21 to 0 V. With 10% Cl_2 addition, both the ϕ_B and the V_{BR} were similar to the wet etched sample. The results indicate the importance of avoiding physical sputtering by introducing reactive gases to minimize surface damage.

The I-V curves of the dry-etched diodes were comparable to the control sample for most of the etch conditions except when pure Ar sputter etching was used. However, with 10% Cl_2 addition in the plasma, from transmission line measurements ϕ_c was only 25% of the control sample while from the diode measurements the n, ϕ_B, and V_{BR} were approximately the same as the control sample. This shows that the transmission line method is more sensitive to surface damage than Schottky diode measurements.

For the emitter etching of self-aligned HBTs, not only is low damage etching important, precise endpoint detection is also essential since the base layer can be only 60 nm thin. If the emitter layer is not completely removed, the HBTs will not be functional since ohmic contact to the base cannot be formed. On the other hand, too much over-etching of the emitter layer will decrease the base layer thickness and increase the base contact resistance.

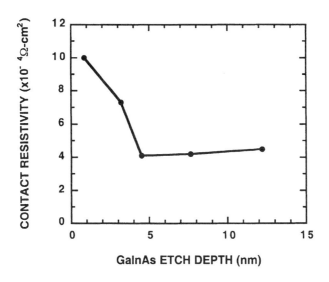

Fig. 17. Transmission line measurements of the contact resistivity of the etched surface as a function of GaInAs etch depth. The plasma conditions were 50 W microwave power, 100 W rf power, Cl_2/Ar flow at 3/27 sccm, 2 mTorr chamber pressure and 13 cm source-to-sample distance.

Therefore, a sensitive and non-invasive *in situ* diagnostic technique, such as optical emission spectroscopy (OES), is needed to provide monitoring and feedback to control the remaining base layer thickness after emitter etching. Using the Ga optical emission signal at 417.2 nm, precise endpoint of the AlInAs/GaInAs emitter-base heterostructure was achieved by stopping on the base layer within 5 nm [19].

Transmission line measurements were used to evaluate the base layer conductivity after the emitter was etched. Figure 17 shows the ρ_c at the AlInAs/GaInAs etched surface after 0 to 18 s of overetch time. The amount of overetching was determined by monitoring Ga emission signal using OES. For 1 and 3 nm of overetching, where Al could still be detected on the etched surface by X-ray photoelectron spectroscopy, ρ_c was 1.0×10^{-3} and 7.3×10^{-4} Ω-cm^2. After the AlInAs emitter layer was completely removed with 4.5 nm of the GaInAs base layer was removed, ρ_c decreased to 4.1×10^{-4} Ω-cm^2. The value of ρ_c is essentially constant for the 4.5, 7.5, and 12.5 nm of overetching into the GaInAs base layer. The reasons for the high value of ρ_c for the GaInAs layer are that the GaInAs layer was nominally undoped and the contacts were not alloyed. The decrease in ρ_c as the AlInAs layer is removed is expected since the energy bandgap of GaInAs is lower than that of AlInAs, resulting in lower contact resistance when the GaInAs base layer is reached. The AlInAs emitter layer was completely removed when etching was stopped 6 s after the increase in the Ga optical emission signal. In this case, only 4.5 nm of the base layer will be removed when the etching is stopped with 6 s overetch time.

Even though 6 s overetch time was sufficient to completely remove the AlInAs emitter layer in this case, the overetch time needed will vary and it depends on the layer structure of the samples, etch conditions, and plasma chamber conditions. Therefore, calibration runs are needed with the corresponding surface and electrical analysis to identify the exact overetch time needed for given devices and etch condition. However, once the correlation between the overetch time and the remaining GaInAs base layer thickness is made, the increase in the Ga optical emission signal can be used to endpoint the emitter etching precisely even in the presence of run-to-run variations in etch rate or layer thickness.

Lower dimensional semiconductor nanostructures are potential candidates for future electronic devices because of their more favorable density of states and electronic properties. One example is the novel IPG quantum wire transistor in which the width of the one-dimensional (1-D) channel can be modulated by the electric field from two in-plane side gates [46]. These IPG quantum wire transistors offer several advantages over the more conventional surface split-gate [47, 48] or grating-gate [49, 50] quantum transistors. The two-dimensional electron gas (2DEG) in the side gates and the channel of IPG transistors are isolated by air gaps, with the 2DEG in the side gates providing electrostatic modulation of the width of the channel. Since the electric field from the 2DEG in the side gate is parallel to the 2DEG in the channel, this provides a very efficient coupling of the electric field into the channel. The air gap between the gates and the channel results in a negligible gate leakage current, and the small dielectric constant of the air ($\varepsilon_{air} = 1$) provides a much smaller gate capacitance compared to the surface gate structures. A cutoff frequency as high as 10 THz has been predicted for the IPG transistors [51].

The isolation between the gates and the channel of IPG transistors formed using focused Ga^+ ion beam, low energy Ar ion exposure, wet etching, or dry etching have been reported [52–54]. In the case of dry etching, the bombardments from the high-energy ions could introduce defects into the semiconductors [6, 13]. It is crucial that these etch-induced damage be minimized since any defects introduced will degrade the device performance and wipe out the advantages of using these quantum effect devices. In-plane gated quantum wire transistors were fabricated using dry etching in a Cl_2/Ar plasma generated with an ECR source. The starting material for this device is a modulation-doped $GaAs/Al_{0.24}Ga_{0.76}As$ heterostructures with an 15-nm-thick $In_{0.15}Ga_{0.85}As$ undoped quantum well. A 2DEG with high mobility is formed at the $In_{0.15}Ga_{0.85}As$ layer. Using a 80 nm SiO_2 layer as the etch mask, the air gaps between the side gates were defined by electron

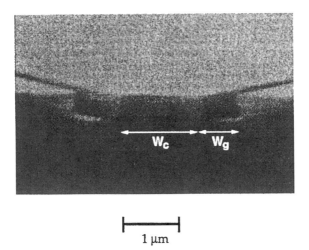

$$\vdash\!\!-\!\!-\!\!\dashv$$
1 µm

Fig. 18. Scanning electron micrograph showing the cross-section of the in-plane gates. The in-plane gates were etched by a Cl_2/Ar plasma generated with 20% Cl_2 using 50 W microwave power and 100 W rf power at 0.5 mTorr.

beam lithography at 50 keV. Dry etching in a Cl_2/Ar plasma generated with the ECR source was then carried out under different etch conditions with an etch depth of 350 nm to ensure good isolation of the two gate regions from the conducting channel. Figure 18 shows the cross-section of the in-plane gates after dry etching. A Cl_2/Ar plasma generated with 20% Cl_2, 50 W microwave power, and 100 W rf power at 0.5 mTorr was used for etching. It can be seen that vertical profile and smooth surface morphology were achieved on the in-plane gates.

Figure 19 shows the drain-source current (I_{DS}) vs. drain-source voltage (V_{DS}) at different gate-source voltage (V_{GS}) for IPG quantum wire transistors with channel width (W_c) of 440 nm and gate isolation width (W_g) of 400 nm. The in-plane gates were etched using a Cl_2/Ar plasma with 20% Cl_2 generated with 50 W microwave and 20 W rf power at 0.5 mTorr. Good field effect transistor (FET) characteristics was observed on the devices, and the in-plane gates are seen to be effective in modulating the channel width. The channel can be completely pinched off with a gate bias less than -1 V. The drain-source current levels were found to be higher for IPG transistors with larger W_c. The saturated drain-source current (I_{DSAT}) at $V_{GS} = 2$ V was 68 and 153 µA for IPG transistors with $W_c = 440$ and 800 nm, respectively. This is expected since the channel conductance increased with W_c. No current can be measured on devices with $W_c = 130$ nm. This cut-off width is much wider than what has been obtained before on GaAs conducting wires etched

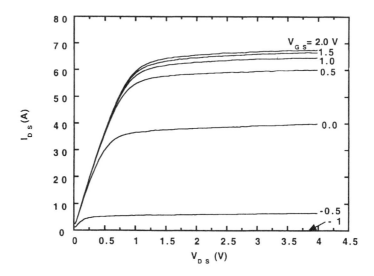

Fig. 19. I_{DS} vs. V_{DS} characteristics of IPG quantum wire transistors with $W_c = 440$ nm and $W_g = 400$ nm. The transistors were etched in a Cl_2/Ar plasma with 20% Cl_2 generated using 50 W microwave and 20 W rf power at 0.5 mTorr.

under similar conditions [6]. Current can be measured on GaAs conducting wires that are only 40 nm wide. This could be related to the higher sensitivity of the quasi 1-D channel to sidewall damage as compared to the conducting wire which is not confined in the vertical direction.

The drain current and gate leakage current of IPG transistors etched using rf power ranging from 50 to 250 W rf power were measured. The $|V_{dc}|$ was 85 and 320 V at 50 and 250 W rf power, respectively. The width of the gate isolation for these devices were 130 nm and $W_c = 370$ nm. As rf power was varied, the widths of the channel and gate isolation did not change and the etch profile remained vertical. It was found that the I_{DS} increased from 33 to 56 μA at $V_{GS} = 2$ V when the rf power was increased from 50 to 250 W. In Figure 20, the gate leakage current at $V_{GS} = 2$ V is shown to increase from 0.10 to 0.27 μA as the rf power increased from 50 to 250 W with $W_g = 130$ nm. For devices with $W_g = 410$ nm, a similar dependence of leakage current on rf power was observed but the amount of leakage current was lower due to the large gate-source separation. Therefore, the increase in the drain current could be related to the increase in the gate leakage current at higher rf power. Damage induced by dry etching at higher ion energy could introduce surface states into the device and result in the increased drain current and leakage current observed.

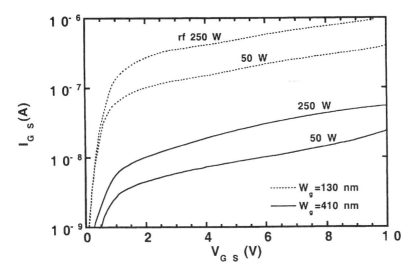

Fig. 20. Changes in the gate leakage current characteristics for in-plane gated quantum wire transistors etched at 50 W and 250 W rf power under similar condition as in Figure 19. The width of the channel was 370 nm and W_g was 130 and 410 nm.

3.4. Dependence of Optical Properties on Dry Etch Conditions

Etched mirrors are not only necessary for optoelectronic circuit integration, they are also essential components for fabricating surface-emitting laser arrays [55] and diode ring lasers [56]. Dry etching allows these mirrors to be fabricated with high packing density and vertical profiles. In order to achieve high reflectivity on these mirrors, the surfaces of the sidewalls should be smooth because surface roughness can cause light scattering and reduce reflectivity. In addition, etch-induced defects have to be minimized because these defects can act as non-radiative recombination centers. A vertical etch profile is also crucial to ensure that all the light will be reflected with the same direction. Dry etched mirrors fabricated on GaAs/AlGaAs or GaInAs/GaAs based materials have been reported [57–59].

Dry etching conditions can be optimized to fabricate mirrors with high reflectivity. Dry etched mirrors for the $In_{0.20}Ga_{0.80}As/GaAs$ based waveguides were fabricated using a Cl_2/Ar plasma generated with an ECR source. Figure 21(a) shows a typical $In_{0.20}Ga_{0.80}As/GaAs$ mirror etched down to 2.8 μm deep. The mirrors were etched in using 10% Cl_2 generated with 50 W microwave power and 70 W rf power at 0.5 mTorr. It can be seen that a vertical profile and smooth surface morphology have been obtained.

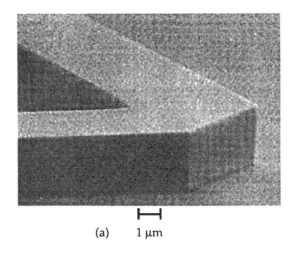

(a)   1 μm

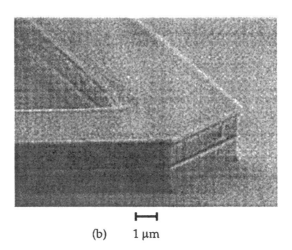

(b)   1 μm

Fig. 21. Scanning electron micrographs of the $In_{0.20}Ga_{0.80}As/GaAs$ total internal reflecting mirror. The mirror was etched in a Cl_2/Ar plasma using (a) 10% and (b) 80% Cl_2, 50 W microwave power and 70 W rf power at 0.5 mTorr, to an etch depth of 2.8 μm.

The etch rate was 0.11 μm/min. The selectivity of GaAs to Ni was 42 under these etch conditions, so the 125 nm thick Ni mask is sufficient for this etch depth. These highly controllable profile and surface morphology are critical for high reflectivity mirrors. Figure 21(b) shows the scanning electron micrograph for a mirror etched with 80% Cl_2 in a Cl_2/Ar plasma. A slightly undercut profile and rougher sidewalls were found for the mirrors etched with

80% Cl_2 as shown in Figure 21(b). While the $Al_{0.30}Ga_{0.70}As$ cladding layers still maintained nearly vertical profile, the $In_{0.20}Ga_{0.80}As$ /GaAs MQW in between the cladding layers were recessed. This excess etching for MQW is probably due to the presence of high reactive chlorine concentrations when 80% Cl_2 is used.

Reflectivity was measured on the totally reflecting mirrors at the corner of a 90° bent waveguide. The width of the waveguides was 5 μm. Since the corner mirror is inclined at 45° to the waveguides, total internal reflection occurs. The mirror reflectivity measurements were made by coupling light from a 1.15 μm HeNe laser into the cleaved facet of the L-shaped waveguides, and the light output from the other end was measured using a Ge detector. The reflectivity of the mirrors is given by the ratio of the light output from the L-shaped waveguide to that from a straight waveguide with equivalent length. The reflectivity measured for the mirrors etched with 10% and 80% Cl_2 were 84% and 47%, respectively. A 44% reduction in reflectivity is found on the mirrors with the slight undercut profile and rougher sidewalls. This could be caused by light being reflected at different angles from the undercut mirror sidewalls or by surface roughness. These results show that the reflectivity of the mirrors are highly sensitive to the etch profile and sidewall roughness. In order to produce vertical profile with smooth sidewalls for the mirrors, the dry etching technique should be optimized to generate directional reactive species at low pressure to reduce scattering. High selectivity to the Ni etch mask is also important to maintain smooth sidewalls and to avoid the formation of tapered profile due to mask erosion.

The effects of ion energy and ion flux used for drying the mirrors on the mirror reflectivity were measured and compared. Figure 22 shows the changes in the mirror reflectivity as a function of rf power. The sidewall damage width (W_s) extracted from GaAs conducting wires etched at different rf power is also shown for comparison [6]. The samples were etched with 10% Cl_2 in a Cl_2/Ar plasma generated at 0.5 mTorr with 50 W microwave power and rf power ranging from 70 to 200 W. Vertical profile and smooth sidewalls were achieved on all the mirrors etched within this rf power range. The etch depth was 2.8 μm. It can be seen that even though the W_s increases substantially from 9.9 to 20.4 nm with rf power, there is no significant change in the mirror reflectivity. The reflectivity is 84% at 70 W rf power and 93% at 200 W rf power. Since this is within the range of measurement uncertainty, reflectivity is mostly independent of the rf power used during etching. These results suggest that the additional electrical degradation on the sidewalls induced at higher rf power does not cause any reduction in the mirror reflectivity.

Another crucial parameter in dry etching is the ion flux. Figure 23 shows the reflectivity for mirrors etched with different microwave powers. The changes in the unalloyed contact resistance extracted from GaAs transmission lines are also plotted. The samples were etched with different microwave power at

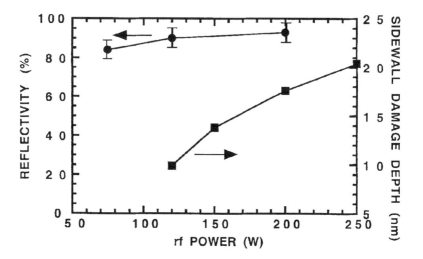

Fig. 22.   Reflectivity (•) and sidewall damage depth (■) as a function of rf power. The samples were etched under similar conditions shown in Figure 21(a) with rf power ranging from 70 to 200 W.

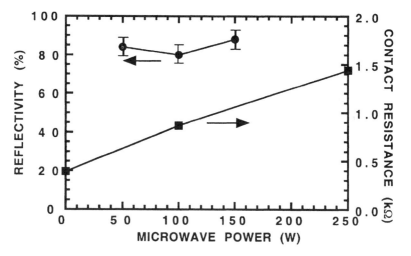

Fig. 23.   Effects of ion flux on the mirror reflectivity (•) and unalloyed contact resistance measured from GaAs transmission lines (■). The samples were etched under similar conditions as in Figure 21(a) with microwave power varying from 50 to 150 W at a fixed $|V_{dc}|$ of 120 V.

a fixed $|V_{dc}|$ so that the ion flux is varied as the microwave power is changed but the ion energy is nearly constant. The microwave power was varied from 50 to 150 W at 120 V. Vertical profiles and smooth surfaces were obtained on all the mirrors. Similar to the case of the ion energy, there is no significant change in the reflectivity with ion flux; the reflectivity remains at 90% in all cases studied. The increase in the unalloyed contact resistance measured from transmission lines etched at higher microwave power suggests that more surface defects are induced on samples etched at higher ion flux, but this does not seem to affect the reflectivity of the etched mirrors. These results show that the reflectivity of the mirrors are highly sensitive to the etch profile and sidewall roughness, but are not influenced significantly by the ion energy or the ion flux used for etching.

The MQW layers used for PL measurements consisted of 7 quantum wells of varying dimensions and different distances from the surface grown on top of a 1 μm thick GaAs buffer layer over a semi-insulating GaAs substrate. The alternating $Al_{0.30}Ga_{0.70}As$/GaAs MQW layers with well widths of 2, 3, 4, 6, 8, 10, and 15 nm were placed approximately 52, 75, 99, 125, 153, 183, and 448 nm below the surface. Prior to dry etching, the samples were wet etched to different depths so that after dry etching for different times, the 2 nm well would be the same distance from the surface for all samples. A solution of $H_2O{:}NH_4OH{:}H_2O_2$ at 130:3:1 was used which provided nearly constant etch rates of 306 nm/min for $Al_{0.30}Ga_{0.70}As$ and GaAs. Changes in PL intensity from a wet etched control sample can be related to etch-induced damage [32]. Samples were first wet etched to different depths and dry etched for different times so that the total material removed for all samples was the same (110 nm) and the 2 nm well was 52 nm from the surface for all samples. The intensity of the peaks was normalized to the intensity from the well 448 nm below the surface that was assumed to be undamaged for all etch conditions. Similar to the C-V measurements, the PL spectra of the samples did not exhibit changes in PL intensity or damage depth after the initial 10 s of etching as shown in Figure 24. Reduction in PL intensity was observed for quantum wells down to 153 nm below the surface. However, there was no variation with etch time after the initial 10 s of etching. Etching at 100 W rf power showed similar behavior but the damage was only 52 nm. The deep damage depths show that channeling of ions and the enhanced diffusion of defects into the material could occur.

Changes in the PL intensity caused by dry etching induced damage in semiconductors can be compared to conductance of quantum well structures. Gratings with linewidths down to 110 nm and an area of 500 μm $\times$ 500 μm were used for PL measurements. For conductance measurements, 160 μm long conducting wires with lateral dimensions down to 120 nm were used. These gratings and wires were defined by electron beam lithography and etched using an ICP source. The etch condition was 150 W source power and

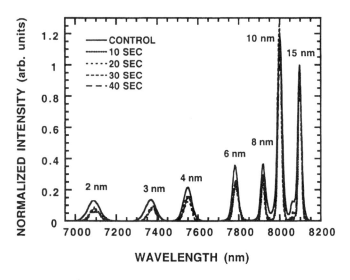

WAVELENGTH (nm)

Fig. 24. Photoluminescence spectra of multiple quantum well structure after etching for different times. Etch condition was 50 W microwave power, 300 W rf power, Cl_2/Ar flowing at 3/27 sccm, 1.5 mTorr chamber pressure and stage position of 12 cm.

200 W stage power with Cl_2 flowing at 6 sccm, chamber pressure at 0.15 mTorr, and the source to sample distance was 12 cm. The $|V_{dc}|$ was 270 V and an etch rate of 300 nm/min was achieved. At pressures below 1 mTorr with a Cl_2 plasma, vertical profiles, smooth surfaces, and high etch rates for gratings and wires with small dimensions were obtained.

The $In_{0.15}Ga_{0.85}As$ PL signals at 925 nm were normalized to the GaAs substrate signal at 820 nm and the intensities were all divided by a fill factor for each grating so that the active area was the same for all gratings. The PL signal, as a function of etch condition, was found to be independent of the illuminating laser intensity. In Figure 25, it is shown that both the PL intensity and conductance decreased with wire width. The conductance of the wires degrades more quickly for smaller wire width as compared to the PL intensity. The wires became effectively non-conducting when the wire width decreased to 120 nm. The conductance decreased from 4.3×10^{-6} to 8.3×10^{-9} Ω^{-1} as the wire dimension was reduced from 550 to 120 nm. However, a measurable PL signal was still detected from gratings with linewidth of 110 nm. The PL intensity decreased by 83% as the linewidth was varied from 530 nm to 110 nm, and the gratings were still producing an optical signal at these sizes. The extracted cutoff width for the optical signal was 33 nm while the cutoff width for the electrical signal was 136 nm. These results indicate a difference in how dry etching induced damage can affect the electrical and optical

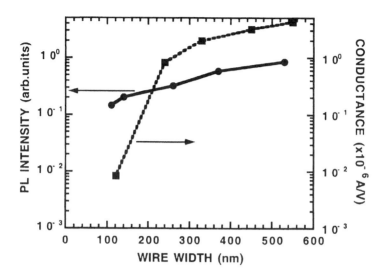

Fig. 25. Comparison of PL intensity and conductivity for wires of different widths showing different cutoff widths for optical and electrical signals. The etch condition was 150 W source power, 200 W stage power, Cl_2 flowing at 6 sccm, a chamber pressure of 0.15 mTorr and a source to sample distance of 12 cm.

properties of an InGaAs quantum well as evidenced by the different cutoff widths of etched wires and gratings. While the etch-induced damage may render the material non-conducting, the quantum well may still be sufficiently intact to allow for radiative recombination of the carriers generated by the incident laser in the PL measurement to allow the optical signal to be detected.

The effects of the stage power used during etching on the optical characteristics of etched gratings are shown in Figure 26. The etch condition was the same as that used in Figure 24 except the stage power was varied from 50 to 200 W. The PL intensity from the gratings etched with 200 W stage power were significantly lower than those etched with 50 W. The PL intensity from the gratings with 530 nm linewidth decreased by 70% and the 110 nm lines decreased by 22% as the stage power was increased from 50 to 200 W. The $|V_{dc}|$ for these conditions was 80 and 270 V, respectively. The degradations in the PL signal with increasing stage power were probably related to the increase in ion energy, which could increase the etch-induced damage along the sidewalls and reduce the luminescence from the quantum well.

Optical losses, light propagation, and electro-optical modulation of optoelectronic devices can be influenced by defects generated due to dry etching [60]. Figure 27 shows the optical losses and the full width half maximum

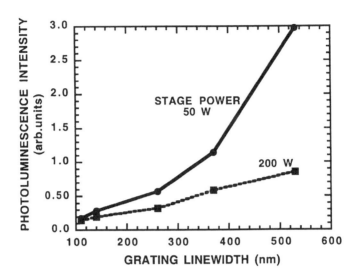

Fig. 26. Changes in stage power during etching cause a degradation in PL intensity from etched gratings. A similar etch condition to that of Figure 24 was used with stage power varied between 50 and 200 W.

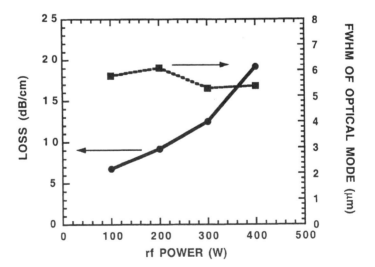

Fig. 27. Optical loss and the FWHM of the optical mode of GaInAs/InP waveguide as a function of rf power. The waveguides were etched using CH_4/H_2 with 150 W rf power at 15 mTorr.

(FWHM) of the optical mode for GaInAs/InP waveguides etched at various rf power. The samples were etched in a RIE system with 15 sccm CH_4 and 50 sccm H_2 at 15 mTorr. The waveguides were 5 μm wide and they were etched to a depth of 2.2 μm. The absorption and scattering losses were measured by monitoring the transmission, which is defined as the ratio between optical power coupled into the waveguide and optical power transmitted through the waveguide, for various waveguide lengths. The FWHM of the optical mode was determined by the light intensity profile from the output of the waveguides. As shown in Figure 27, the optical losses increase almost linearly with rf power, from 6.8 dB/cm at 100 W to 19.6 dB/cm at 400 W. The increase in optical loss with rf power is probably related to larger scattering or absorption losses due to increased surface roughness and/or disordering. On the other hand, the FWHM of the optical mode remained nearly constant between 5.2 and 6.2 μm. Similarly, the voltage required for π phase shift of the light propagating in 3 mm long waveguide was 3.6 to 3.8 V, independent of the rf power used for etching.

4. REMOVAL AND PASSIVATION OF DEFECTS

In dry etching, multiple requirements such as fast etch rate, vertical profile, high etch selectivity, smooth surface, and low damage need to be satisfied. Even though some of the etch conditions can provide low surface damage, it may not be feasible to apply such low damage etch conditions in order to meet the other requirements. Often, some surface damage may be induced after dry etching and it is important to develop techniques to remove the surface defects so that high device performance can be obtained. Surface damage can be reduced or removed by wet chemical etching, thermal anneal, and two-step etching. In addition, damage removal and passivation using low-energy chlorine species have been developed [61]. These techniques provide better reproducibility and controllability than the conventional wet etching or annealing, and is therefore much better suited for the fabrication of devices with small dimensions. Low-energy chlorine species [6, 17, 61] were generated with the ECR source without applying any rf power, and they can be used to physically remove the etch-induced damage *in situ* after the etching. For some device applications, however, precise control in etch depth and linewidth are necessary and it is more desirable to have the damage passivated without any additional etching. In the Cl_2 plasma passivation, a native oxide is allowed to form on the dry etched GaAs before exposure to the low-energy chlorine species so that no etching can occur [61]. Damage passivation using a Cl_2 plasma can be carried out at room temperature and it does not de-activate the donors. Plasma passivation and damage removal of dry etched GaAs have been studied. Electrical measurements using Schottky diodes, transmission lines, and conducting wires were used for evaluating the effectiveness of the damage removal, and comparisons of the various passivation and damage removal techniques were made.

4.1.   Wet Chemical Etching

Dry etching induced damage layer can be removed by wet chemical etching in steps. To determine the damage depth, samples were first dry etched, followed by removing 3, 6, 12, and 18 nm of the dry etched surface by wet etching as shown in Figure 28. Damage removal was carried out in wet chemical solution consisted of $H_2O:C_6H_8O_7$(citric acid)$:H_2O_2:H_3PO_4$ (220:55:5:1). This etch solution provides a slow etch rate of 0.7 nm/min and smooth surface morphology. A control sample was used for comparison and it had a ρ_c of 5.7×10^{-5} Ω-cm^2. After etching with 50 W microwave power, 50 W rf power ($|V_{dc}| = 130$ V), a Cl_2/Ar flow of 1/9 sccm, at 13 cm, and 1.0 mTorr, ρ_c decreased to 0.7×10^{-5} Ω-cm^2. Using wet etching to remove 6 nm of the dry etched surface, ρ_c increased to 6.4×10^{-5} Ω-cm^2. This result is comparable to a study of GaAs etched with an ECR source that shows a damage depth of 12 nm [62].

Damage removal for two additional etch conditions is also shown in Figure 28. The conditions were chosen because significant damage can be generated. Etch conditions were similar to those listed above for curve A except higher rf power at 200 W was used for curve B and only Ar was used as the etch gas for curve C. With higher rf power, a fast etch rate of 90 nm/min was obtained due to high ion energy ($|V_{dc}| = 395$ V). With pure Ar sputter etching, a slow etch rate of 14 nm/min was obtained and $|V_{dc}|$ was 145 V. Figure 28 shows that more damage is generated at higher rf power or with pure Ar sputter etching. In addition, the damage depth is greater under these conditions. The damage depth at 200 W rf power is 18 nm. This is probably due to the larger ion penetration depth at higher ion energy. The damage depth is only 6 nm with 10% Cl_2 in Ar, whereas the damage depth from Ar sputter etching is 12 nm. The $|V_{dc}|$ for conditions A and C are 130 V and 145 V, respectively. Since $|V_{dc}|$ is similar for these two etch conditions, it appears that the lower etch rate of Ar sputter etching prevents efficient damage removal and results in larger damage depth. In all these three different etch conditions, wet chemical etching was very effectively in removing dry etching induced damage and the ρ_c of the transmission lines returned to their original levels after the damaged layer was removed.

In order to gain insight into the etch time dependent damage of the GaAs Schottky diodes and $In_{0.53}Ga_{0.47}As$ transmission lines, the depth profile of the damage was studied by damage removal using wet etching. Wet etching solution similar to the one used above was used to etch GaAs and $In_{0.53}Ga_{0.47}As$ at rates of 27.6 and 30.0 nm/min. The results are shown in

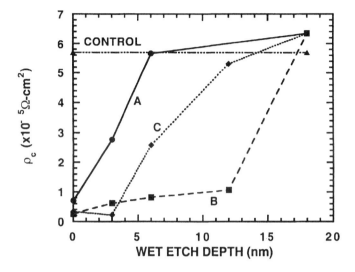

Fig. 28.   Damage removal using wet chemical etching as measured by changes in ρ_c. The etch condition for curve A was 50 W microwave power, 50 W rf power, 13 cm, Cl_2/Ar flow rate at 1/9 sccm, 1 mTorr and at 30°C. For curve B the rf power was increased to 200 W and for curve C only Ar was flowed at 10 sccm.

Figures 29 and 30. The dry etch condition was 50 W microwave power, 300 W rf power, Cl_2/Ar flowing at 3/27 sccm, 1.5 mTorr chamber pressure, and stage position of 12 cm. As mentioned earlier, the ϕ_b of GaAs diodes at the etched surface was lower for shorter etches. When the etch time was increased from 10 to 60 s, the degradation in ϕ_b was reduced. The results show that the damage creation was greatest near the top 5 to 10 nm of the etched surface for GaAs and the etch time dependent changes may be confined to this layer. After removing 15 nm from the dry etched surface, the diode characteristics were completely recovered to the original levels.

The depth profiling for $In_{0.53}Ga_{0.47}As$ obtained using wet chemical removal is shown in Figure 30. In contrast to the GaAs, the damage as measured by contact resistance was more intense for the longer etch. This shows that the defects in $In_{0.53}Ga_{0.47}As$ tend to stay near the top 10 nm of the etched surface. While the samples dry etched for 10 and 60 s showed large variation in contact resistance at the surface, after wet etching 5 nm, the difference in contact resistance was much less. In contrast to the GaAs, wet etching to 20 nm did not fully return the material to the control sample level. Typically, wet chemical etching is effective in removing dry etching induced damage. The damage depth, and hence the layer thickness to be removed by wet chemical etching, depends on the dry etch conditions.

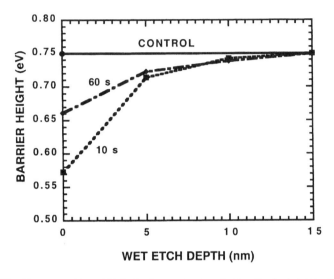

Fig. 29. Slow wet etching of GaAs for damage depth evaluation. Barrier height is plotted for different wet etch depths after dry etching for 10 and 60 s. Etch condition was 50 W microwave power, 300 W rf power, Cl_2/Ar flowing at 3/27 sccm, 1.5 mTorr chamber pressure and stage position of 12 cm.

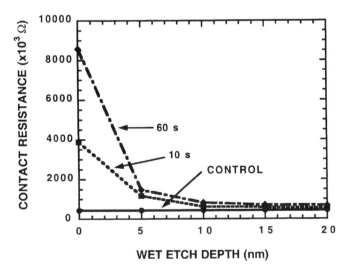

Fig. 30. p-$In_{0.53}Ga_{0.47}As$ damage depth profiling for 10 and 60 s etch times. Contact resistance was measured after wet etching to different depths. Etch condition was the same as in Figure 29.

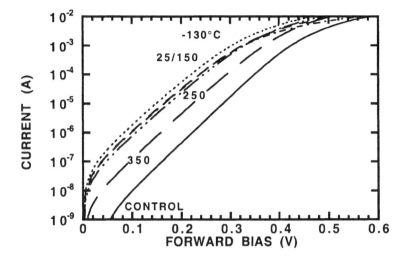

Fig. 31. Effects of etching temperature on the forward I–V characteristics of GaAs diodes. The samples were etched at 0.6 mTorr with 30% Cl_2 in a Cl_2/Ar plasma, 50 W microwave power and 200 W rf power.

The drawbacks of damage removal using wet chemical etching include difficulties in controlling etch rate, achieving smooth surface morphology, and avoiding undercut profile.

4.2.   Thermal Annealing

The I–V characteristics of GaAs Schottky diodes etched at temperatures between −130 and 350°C are shown in Figure 31. The samples were etched with a Cl_2/Ar mixture with 30% Cl_2, 50 W microwave power, 200 W rf power at 0.6 mTorr and 18 cm source to sample distance. Lower leakage current was observed for samples etched at higher temperatures. This is possibly caused by the more efficient removal of the surface damage due to the faster etch rate as well as the annealing effects at higher temperature. The GaAs etch rate increased from 77 to 398 nm/min as the temperature was increased from −130 to 350°C. Figure 32 shows the I–V characteristics of the GaAs diodes after dry etching and rapid thermal annealing (RTA). The samples were etched at 25°C under similar condition as described above. After etching, RTA was carried out at temperatures between 400 and 540°C in a N_2/H_2 ambient for 2 min. The ideality factor improves from 1.23 to 1.07 and barrier height increases from 0.63 to 0.70 eV after the etched sample were annealed at 480°. This suggests that etch-induced damage could be

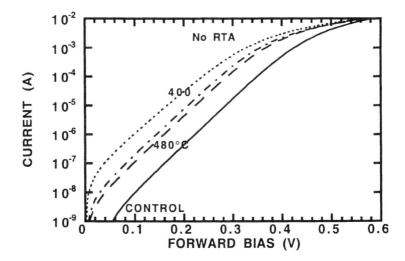

Fig. 32. Forward I–V characteristics of GaAs after dry etching and RTA. The samples were etched at $25°$C under similar condition as described in Figure 31. Rapid thermal annealing was carried out in a N_2/H_2 ambient for 2 min.

partially removed by annealing. No further recovery was observed by using annealing temperatures $> 480°$C or longer annealing time.

The unalloyed contact resistance of GaAs transmission lines etched at different temperatures is shown in Figure 33. The samples were etched with 20% Cl_2 in a Cl_2/Ar plasma using 50 W microwave power and 200 W rf power at 0.5 mTorr. The contact resistance decreased from 2.8 to 1.6 kΩ as the etch temperature was increased from 25 to $350°$C, indicating that less damage was induced on the surface at higher temperature. The improvement of the contact resistance may be related to the defect annealing effects and the faster GaAs removal rate at higher etch temperature. The etch rate increased from 221 to 292 nm/min as the temperature was increased from 25 to $350°$C under this etch condition. Figure 33 also shows the reduction of W_s measured from the conducting wires etched at higher etch temperature. The samples were etched under similar etch condition as the transmission lines except that 10% Cl_2 was used. As the etch temperature was increased from 25 to $350°$C, W_s decreased from 13.1 to 2.7 nm.

4.3. Two-Step Etching

The motivation for using two different etch conditions to remove the emitter layer was to reduce roughness and damage simultaneously. The roughness

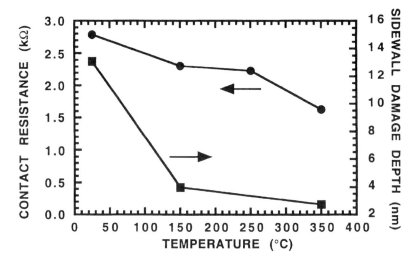

Fig. 33. Unalloyed contact resistance of transmission lines (•) and W_s of conducting wires (■) as a function of etch temperature. The samples were etched with 50 W microwave power and 200 W rf power at 0.5 mTorr in a Cl_2/Ar plasma.

measured by atomic force microscopy was shown to decrease with increasing rf power due to more efficient removal of the non-volatile $InCl_x$ etch products. Additionally, a high rf power at the beginning of the etch removes surface oxide more uniformly. This also improves the surface morphology. However, increasing rf power also increases the ion energy which generates more damage. Therefore, a two-step etching can be used to minimize surface damage while still maintaining smooth surface and vertical profile. The first portion of the emitter etch is done at high rf power to remove surface oxides, minimize roughness, and reduce etching time. The second portion of the etch is done at low rf power to remove the damage created in the first portion of the etch.

For the study of damage removal using two-step etching, a 600 nm thick n-type GaInAs layer (doping $= 10^{16}$ cm^{-3}) was etched for 150 nm. The samples were etched using either low rf power at 30 W ($|V_{dc}| = 50$ V), high rf power at 160 W ($|V_{dc}| = 225$ V), or a combination of the two. The amount that was etched at 160 W was varied from 0 to 150 nm and the rest of the 150 nm was etched at 30 W rf power, resulting in the percentage of the GaInAs layer thickness etched at 225 V to vary from 0% to 100%. The etching was carried out with 50 W of microwave power, 3/27 sccm Cl_2/Ar gas flow, 2 mTorr chamber pressure, and 13 cm source distance. Figure 34 shows how the contact resistivity decreased as the percentage of

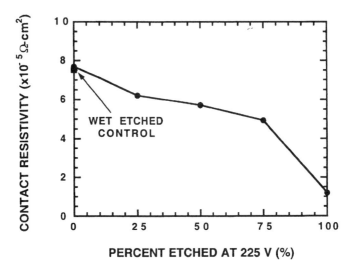

Fig. 34. Effect of two step etching as shown by decrease in contact resistivity of n-GaInAs from the control value with increase in percent etched at 225 V $|V_{dc}|$. The etch condition was 50 W of microwave power, Cl_2/Ar at 3/27 sccm, 2 mTorr pressure and 13 cm source distance.

the n-type GaInAs layer thickness etched at 225 V $|V_{dc}|$ increased from 0% to 100%. For etching the entire 150 nm with the low power, low damage condition, ρ_c was 7.7×10^{-5} Ω-cm^2. After wet etching 150 nm of GaInAs, ρ_c was 7.5×10^{-5} Ω-cm^2, very similar to the low damage dry etched case. The wet etch solution was similar to the one described above and the etch rate was 42 nm/min. When just high rf power was used, ρ_c decreased to 1.2×10^{-5} Ω-cm^2. This decrease could be due to the creation of defects at higher rf power which increases the leakage current, thereby reducing ρ_c.

Even though there is no damage detected by the transmission line measurement when etching with 30 W rf power, this would be impractical for etching the HBT emitter layer. First, the etch rate is only 14 nm/min such that 20 min would be needed to etch the emitter stack. Second, the surface morphology degrades due to inefficient removal of the $InCl_x$ etch products. Therefore, a compromise can be made with two-step etching where 50% of the emitter will be etched using 160 W rf power, and 50% can be etched using only 30 W rf power for damage removal. From Figure 34, it can be seen that a significant improvement in ρ_c of the n-type GaInAs layer occurs by including the 50% damage removal step, but full recovery of ρ_c does not occur even with 75% of the material etched with the low damage condition. The reason for this could be the enhanced diffusion of defects introduced by dry etching [12]. Since the damage depth could be quite extensive for the high power case, and the etch rate is low for the low damage case, it is conceivable that some defects generated in the first etch step diffuse into the material faster than they can be removed during the second etch step. The damage depth has been measured previously for similar conditions [13]. The damage depth was 6 nm at 130 V $|V_{dc}|$ and it increased to 18 nm for 395 V $|V_{dc}|$. It is these defects generated below the surface in the first etch step and not removed by the second low damage etch step that prevent complete damage removal.

The surface morphology of dry etched AlInAs/GaInAs layers can be better than the ones formed by wet chemical etching. Comparison of dry and wet etched HBT emitters is shown in Figure 35. The dry etch condition was 50 W of microwave power, 100 W of rf power, Cl_2/Ar gas flow at 3/27 sccm and 2 mTorr chamber pressure. For all of the dry etched samples, the root mean square (RMS) roughness was < 2.5 nm, while the RMS roughness of the wet etched sample was 4.4 nm. In general, a smooth dry etched surface is obtained with optimized dry etch condition and the stoichiometry of the dry etched surface is not changed after etching.

Two-step etching was also investigated using the HBT structure. First, the effect of rf power on ρ_c for p^+-GaInAs was studied as shown in Figure 36. The rf power was varied from 30 to 250 W and the Ga emission intensity was used to stop the etch on the p^+ GaInAs base layer. These were all single step etches where the rf power was held fixed for the entire etch. Before the transmission lines were annealed, increasing the rf power from 30 to 250 W caused ρ_c to increase over 20× from 3.3×10^{-5} to 6.7×10^{-4} Ω-cm^2. The wet etched sample had the lowest ρ_c for the emitter etches before annealing. It was 2.1×10^{-5} Ω-cm^2, slightly lower than the value for the sample etched at 30 W rf power. After annealing, ρ_c was lower but still showed the increase with rf power. In the case of etching p-type GaInAs, it appears that the defects generated act to compensate the p-type dopants and reduce the effective doping level, thereby increasing ρ_c.

The HBT emitter layers were also formed using a two step etch. The rf power was initially 150 W. When the Ga intensity decreased indicating that the 100 nm thick GaInAs cap layer was removed, the rf power was reduced to 30 W to facilitate damage removal. The ρ_c obtained for this two step etch was only 1.0×10^{-4} Ω-cm^2 before annealing. This is significantly lower than the value obtained when the rf power was kept at 150 W for the entire etch, and it is lower than the etch where the rf power was kept at 100 W. However, as was the case with n-type material, the two step etch was not as good at preventing damage as was a single low damage etch. In this case, ρ_c was not as low as the single etch using only 30 W rf power.

The contacts on the p^+-GaInAs base layer were annealed to lower ρ_c and possibly remove the etch-induced damage. The contact resistivity of the wet etched sample decreased slightly from 2.1×10^{-5} to 1.6×10^{-5} Ω-cm^2 after

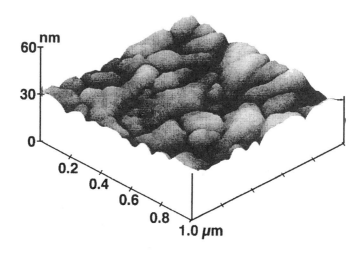

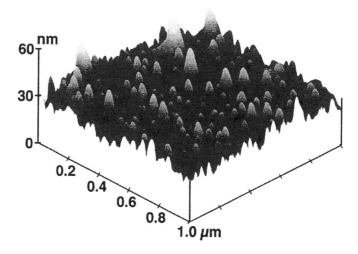

Fig. 35.   Comparison of dry etched (top) and wet etched (bottom) HBT emitter. The dry etch condition was 50 W of microwave power, 100 W of rf power, Cl_2/Ar gas flow at 3/27 sccm and 2 mTorr chamber pressure. The RMS roughness was 2.5 nm for the dry etched sample and it was 4.4 nm for the wet etched sample.

annealing at 300°C for 1 min. For all of the dry etched samples, the decrease in ρ_c after annealing was much more substantial. This suggests that not only was the damage partially removed by annealing, but the dry etched surface was more readily annealed. For the sample etched with 30 W of rf power, the

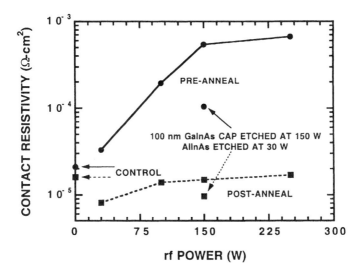

Fig. 36. The use of low rf power allows low ρ_c to be obtained for dry etched p^+-GaInAs. Also shown is ρ_c measured for a two step etch. The contact resistivity before (•) and after (•) annealing are shown. The etch condition was 50 W of microwave power, Cl_2/Ar at 3/27 sccm, 2 mTorr pressure and 13 cm source distance.

annealed ρ_c was reduced from 3.3×10^{-5} to 8.1×10^{-6} Ω-cm^2. The two step etch sample also had a lower ρ_c after annealing of 9.6×10^{-6} Ω-cm^2 than did the wet etched sample. One possible explanation could be the smooth surface morphology of the dry etched materials as shown in Figure 35. Using optimized dry etch condition, surface morphology can be improved and the stoichiometry of the dry etched surface is not changed after etching, which provides better electrical characteristics for HBTs.

4.4. Damage Removal by Low Energy Reactive Species

Damage removal using conventional wet etching technique suffers from the disadvantage of having to expose the dry etched sample to the ambient before the wet etching. The controllability of the removal step could be a problem because of the formation of native oxide or surface impurities on the samples. A more desirable technique is to remove the surface damage *in situ*, and this could be achieved by using low-energy chlorine species generated in the same plasma system right after the dry etching step. When surface oxide on GaAs is removed, low-energy chlorine species can be used to etch GaAs controllably as shown in the layer by layer etching technique [63]. Figure 37 shows the self-limited etch rate for GaAs as etching was cycled between

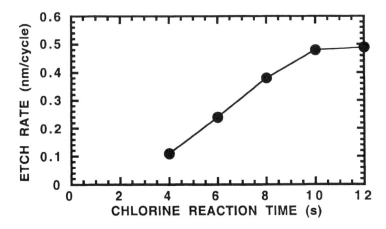

Fig. 37. GaAs etch rate as a function of chlorine reaction time. The samples were etched with 200 cycles at 25°C. Each cycle consisted of: (i) chlorine species generated with 35 W microwave power at 1 mTorr using 10% Cl_2 in a Cl_2/Ar mixture and (ii) Ar ions generated with 50 W microwave power and 6 W rf power at 1 mTorr for 5 s.

adsorption of low-energy chlorine species and etch product desorption by Ar ion bombardment. The samples were etched with 200 cycles at 25°C. Each cycle consisted of (i) reactive chlorine species generated using 35 W microwave power and 10% Cl_2 in a Cl_2/Ar mixture; (ii) removal of excess species by pumping to a base pressure of 5×10^{-5} Torr; and (iii) Ar ions generated using 50 W microwave power and 6 W rf power for 5 s with $|V_{dc}|$ of 38 V. The pressure was 1 mTorr and the samples were kept at 23 cm below the ECR source. The etch rate increased from 0.11 to 0.48 nm/cycle as the chlorine species reaction time was increased from 4 to 10 s. At reaction time $\geq$ 10 s, the etch rate remained approximately constant at 0.48 nm/cycle. At reaction time < 10 s, the etched surface was probably partially covered with chlorine species and therefore the etch rate increased with reaction time as more chlorine species adsorbed on the etched surface. Complete surface coverage by the chlorine species was probably achieved at reaction time $\geq$ 10 s, and the etch rate saturated. The saturated etch rate shows that in the absence of any ions, the chlorine species do not etch GaAs significantly. The I–V characteristics of the GaAs diodes etched with low-energy chlorine species were similar to that of the unetched sample and were independent of the chlorine species reaction time used. The negligible surface damage is probably related to the presence of reactive species during the adsorption step and the low energy Ar ions used during the desorption step. This damage free layer by layer etching technique could therefore be applied as an effectiveness *in situ* damage removal technique.

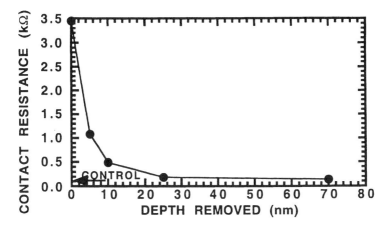

Fig. 38. Recovery of unalloyed contact resistance after damage removal using low-energy chlorine species. The samples were first etched at 0.5 mTorr, 100 W microwave power and 100 W rf power with 20% Cl_2 in Cl_2/Ar. Low-energy chlorine species for the damage removal were generated under similar conditions, except with 50 W microwave power and no rf power was applied.

Figure 38 shows the recovery of the unalloyed contact resistance extracted from dry etched GaAs transmission lines after removing various depths using low-energy reactive chlorine species. All the samples were first dry etched using 20% Cl_2 in a Cl_2/Ar plasma with 100 W microwave power and 100 W rf power at 0.5 mTorr. The low-energy reactive chlorine species were generated using the condition similar to the dry etching step except the microwave power was 50 W and no rf power was applied. Since no rf power was used, the ion energy of the reactive chlorine species was low and they can be used to remove the damage. The unalloyed contact resistance recovered from 3.5 kΩ without any damage removal to 0.1 kΩ after 25 nm or more of the dry etched surface was removed and is very close to the control sample. This shows that the low-energy reactive chlorine species is effective in removing the etch-induced damage.

This technique has also been used to remove the sidewall damage of the conducting wires. The conductance as a function of wire width for the conducting wires with and without damage removal is shown in Figure 39. Both samples were dry etched in a Cl_2/Ar plasma using 10% Cl_2, 50 W microwave power and 200 W rf power at 0.5 mTorr. One of the samples was then exposed to low-energy reactive chlorine species generated under similar conditions except without any rf power. After removing 10 nm with the low-energy reactive chlorine species, W_s decreased from 13.1 to 4.0 nm, which agrees with the measured removal depth. This shows that damage

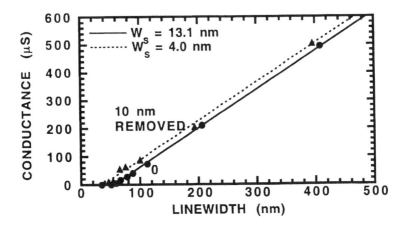

Fig. 39. Reduction of W_s of conducting wires after damage removal using low-energy chlorine species. The etch condition was 35 W microwave power and 200 W rf power at 0.5 mTorr using 10% Cl_2 in a Cl_2/Ar plasma. Low-energy chlorine species were generated under the same condition without any rf power.

removal using the low energy reactive chlorine species is mainly damage free and can be used as an effective and controllable *in situ* damage removal technique.

4.5. Surface Passivation and Coating

For some device applications, precise etch depth is needed and it is not desirable for the damage removal step to have significant etching. Therefore, plasma passivation of the etch-induced damage has been developed to reduce surface defects without actual removal of the semiconductor materials [61]. For plasma passivation, samples were first dry etched using high rf power, and a damaged layer was generated on the surface. The samples were then exposed to air for ∼ 2 hrs so that a native oxide can be formed on the etched surface. After that, the samples were re-loaded into plasma system, and the damaged layer was passivated by low-energy chlorine species generated using microwave power alone. Etching by the low-energy chlorine species was avoided by allowing the native oxide to form on the etched surface prior to the passivation step. Complete recovery in the electrical characteristics of dry etched Schottky diodes and unalloyed transmission lines were achieved with Cl_2 plasma passivation with no observable etching during the passivation step.

The changes in the forward I–V characteristics of the dry etched GaAs diode after Cl_2 passivation are shown in Figure 40. All samples were first

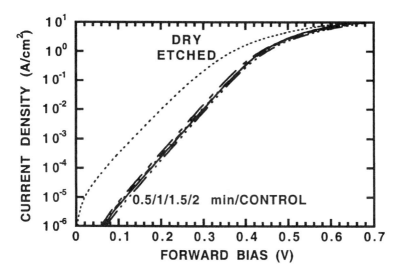

Fig. 40. Forward I–V characteristics showing complete recovery after the samples were etched and passivated with a Cl_2 plasma for time ranging from 0.5 to 2.0 min. Samples were first etched in a Cl_2/Ar plasma with 20% Cl_2, 50 W microwave power and 200 W rf power at 0.5 mTorr. The Cl_2 plasma for passivation was generated with 50 W microwave power at 2 mTorr and 25°C.

dry etched with a Cl_2/Ar plasma using 2/8 sccm Cl_2/Ar, 50 W microwave power, and 200 W rf power at 0.5 mTorr. High rf power was chosen intentionally so that more defects are generated and the changes in the electrical characteristics can be more easily observed. After the dry etching step, the samples were taken out from the plasma system and exposed to air for 3 hrs. This allows oxides to be formed on the surface. The dry etched surface was then passivated with a Cl_2 plasma generated with 50 W microwave power at 2 mTorr for time ranging from 0.5 to 2.0 min. Since oxides were present on the dry etched samples, there was no etching of GaAs by this Cl_2 plasma when no rf power was applied. It can be seen that the diodes have high leakage current after dry etching, but they recover completely to the level of the control sample with only 0.5 min of Cl_2 plasma passivation.

The recovery of the diode characteristics is not related to the removal of the damaged surface layer, since the etch depth during the Cl_2 plasma passivation was found to be negligible. This shows that the recovery is probably related to surface passivation. Since the Cl_2 plasma was generated with microwave power alone and no rf power was applied at the stage, the ion energy of the chlorine reactive species is very low (< 20 eV). These low-energy chlorine reactive species do not etch GaAs considerably when the native oxides have not been removed, but they may passivate the dangling bonds on the oxidized GaAs surface and/or form a stable surface layer that result in the excellent electrical characteristics observed. If the passivation was carried out immediately after the dry etching without exposure to air, considerable GaAs etch rates could be measured since no native oxides were present on the GaAs surface. Under this condition, it has been found that at least 25 nm needs to be removed from the dry etched surface for complete recovery. The Cl_2 plasma passivation shown in Figure 40, however, was done with native oxides present on the etched surface, and complete recovery was observed with no measurable etching.

Since Cl_2 dilution with Ar is commonly used to optimize the etch rate, profile, and surface morphology, the effects of Ar addition during Cl_2 plasma passivation were also investigated. Figure 41 shows the changes in n and ϕ_B extracted from the forward I–V measurements after dry etching and a 30 s passivation with Ar addition in the Cl_2 plasma. The plasma for passivation was generated under similar etch condition as shown in Figure 40 except with 0 to 60% Ar addition to Cl_2. The n and ϕ_B for the control sample were 1.03 and 0.76 eV, respectively. After dry etching, n and ϕ_B degrade to 1.32 and 0.63 eV, respectively, but they recover to the same level as the control sample after passivation with a Cl_2 plasma with no Ar addition, as shown in Figure 40. As more Ar was introduced in the Cl_2 plasma during passivation, however, the passivation becomes less effective. With 60% Ar addition, n and ϕ_B recover only to 1.26 and 0.64 eV, respectively. These results show that Ar addition in Cl_2 is not desirable for plasma passivation, and this may be related to the dilution effect. The additional Ar can also occupy some of the surface sites on GaAs, and less chlorine species can be accommodated on the GaAs surface to form a good passivating layer.

The pressure used for Cl_2 plasma passivation was also varied in order to find the optimal conditions for this passivation technique. Figure 42 shows the changes in the electrical characteristics of the unalloyed GaAs transmission lines after dry etching and passivated with Cl_2 plasma at different pressure has been studied. The samples were passivated under similar condition shown in Figure 40 for 30 s with pressure ranging from 2 to 8 mTorr. After dry etching, the unalloyed contact resistance increases from 0.1 to 3.8 kΩ, but it recovers back to the control level after 30 s Cl_2 plasma passivation at pressure ranging from 2 to 8 mTorr. This agrees with the results obtained above from the Schottky diode measurements and shows that Cl_2 plasma was indeed very effective in passivating the etch-induced damage on GaAs. The pressure used for Cl_2 plasma passivation has no significant effect on the electrical characteristics of GaAs diodes or transmission lines.

For device applications, it is also important that the passivated surface remains stable at high temperature. For this purpose, the thermal stability of the Cl_2 passivated surface has also been investigated. Samples of GaAs were first dry etched and then passivated for 30 s in a Cl_2 plasma under the

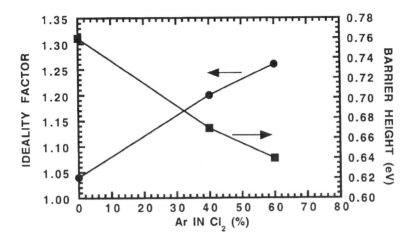

Fig. 41. Ideality factor and ϕ_B extracted from diodes etched and passivated in a Cl_2 plasma with Ar addition for 30 s. The plasma for passivation was generated under similar condition as in Figure 40 with 0 to 60% Ar.

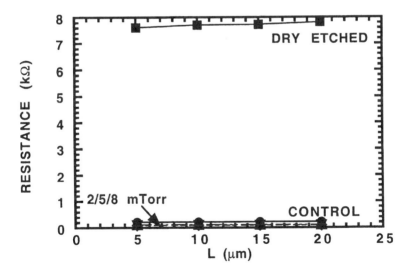

Fig. 42. Contact resistance of the unalloyed GaAs transmission lines that were etched and passivated with a Cl_2 plasma at different pressure for 30 s.

same conditions as shown in Figure 40. After passivation, the samples were subjected to temperatures between 200 and 450°C in a N_2 ambient for 3 min. Diode measurements on these samples showed that the Cl_2 passivation was stable and there was no change in electrical characteristics of the passivated samples at temperatures up to 450°C. This result suggests that the GaAs surface passivated with the Cl_2 plasma has good thermal stability and is suitable for device fabrication.

The electrical characteristics of the dry etched GaAs after N_2 or N_2/H_2 plasma passivation were also investigated. Table I shows the changes in the I–V characteristics of the dry etched GaAs diodes after being passivated in a N_2 plasma at different temperature and microwave power. The samples were first dry etched in a Cl_2/Ar plasma under similar conditions as before, and then passivated with a N_2 plasma generated at 5 mTorr for 2 min. Unlike the Cl_2 plasma passivation, passivation with N_2 plasma is not effective at 25°C or with 50 W microwave power. However, with 500 W microwave power, the I–V characteristics improve with higher passivating temperatures, and become similar to the control sample at 350°C. Compared to Cl_2 plasma, N_2 plasma passivation of dry etched induced damage requires much higher temperature and microwave power. Annealing the dry etched samples at 350°C without the N_2 plasma only results in partial recovery of the diode characteristics. Therefore, the improvements shown in Table I are not just an annealing effect, but rather the reactive species in the N_2 plasma are also responsible for the complete recovery observed. The recovery of the electrical characteristics in a N_2 plasma could be related to the formation of a nitride layer which passivates the dangling bonds on the surface. It probably requires high temperature to promote the nitride formation and/or partially anneal out the etch-induced damage, it also needs high microwave power to provide high concentrations of reactive species for the nitridation.

Since H_2 could remove the excess As_2O_3 and As that are responsible for the formation of the AsGa antisite defects [64], the effects of adding H_2 to the N_2 plasma were also investigated. GaAs samples were first dry

TABLE I

Changes in the diode characteristics as a function of temperature and microwave power during N_2 plasma passivation. The diodes were passivated for 2 min in a N_2 plasma generated at 5 mTorr.

Temperature	Microwave Power (W)	n	ϕ_B (eV)	$|V_{BR}|$ (V)
Control	–	1.03	0.76	20.2
Dry Etched	–	1.32	0.63	12.5
250°C	500	1.16	0.69	16.7
350°C	500	1.05	0.76	20.0
350°C	50	1.19	0.67	18.2

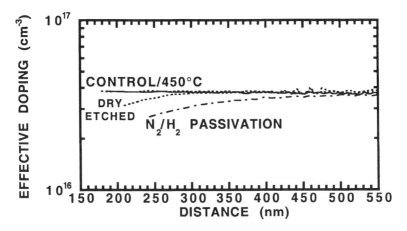

Fig. 43. Doping profiles extracted from C-V measurements showing re-activation of the dopants after annealing in N_2 for 3 min. The diodes were first dry etched and passivated with a N_2/H_2 plasma, generated with 50% H_2 and 500 W microwave power at 2 mTorr for 2 min.

etched under the same condition as before, and then passivated for 2 min using a N_2/H_2 plasma with $H_2\%$ ranging from 0 to 100%. The plasma was generated with 500 W microwave power at 2 mTorr. The addition of H_2 has no effect on the I–V characteristics and the results are similar to passivation by N_2 plasma. However, the C-V measurements reveal significant dopant passivation when H_2 is added, as shown in Figure 43. After passivation with 50% H_2 in the N_2 plasma, the dopants were depleted from the surface up to a depth of 400 nm. This dopant passivation effect, however, can be annealed out. Figure 43 shows that the doping concentrations of the etched and N_2/H_2 passivated diodes recover after being annealed at 450°C in a N_2 ambient for 3 min. The hydrogen in the passivated samples are probably being driven out at high temperature and this re-activates the Si-dopants, resulting in the recovery observed.

Passivation of dry etched GaAs using a H_2S plasma was studied. Figure 44 shows the changes in $|V_{BR}|$ and ϕ_B for the dry etched GaAs diodes after passivation in H_2S plasma and then annealed for 2 min at different temperatures. The samples were first dry etched in a Cl_2/Ar plasma under similar condition as before, and then passivated in a H_2S plasma generated with 100 W rf power at 200 mTorr for 5 min. It can be seen that annealing at 300°C is needed after H_2S plasma passivation, and even then the recovery is not complete. Measurements on transmission lines showed a decrease of the unalloyed contact resistance from 3.8 to 2.8 kΩ after being etched and passivated with the H_2S plasma. Similar to the diode results in Figure 44,

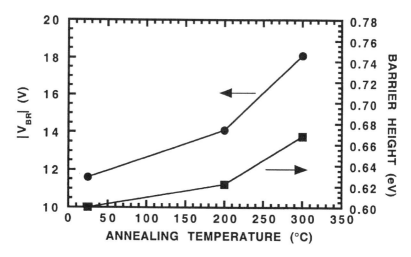

Fig. 44. Changes in $|V_{BR}|$ and ϕ_B with annealing temperature on diodes etched and passivated with a H_2S plasma generated using 100 W rf power at 200 mTorr for 5 min. Annealing was carried out *in situ* after the passivation for 2 min.

the recovery is not complete since the unalloyed contact resistance for the control was only 0.1 kΩ. The improvement in the electrical characteristics after H_2S passivation could be related to the formation of a thin protective S layer on the GaAs surface. It has been reported that the Ga-S bonds can reduce surface state density on GaAs [65]. However, high temperature annealing is needed to desorb the excess S on the GaAs surfaces [66, 67]. Similar to N_2/H_2 plasma passivation, a large decrease in dopant concentration near the surface is observed from the C-V measurements for the H_2S passivated samples due to the presence of H_2 in H_2S. However, this effect can be alleviated by using lower rf power to generate the H_2S plasma for passivation.

The effectiveness of Cl_2 plasma passivation on dry etched IPG transistors were demonstrated [68]. The IPG transistors were first dry etched with 20% Cl_2 generated using 50 W microwave and 200 W rf power at 0.5 mTorr. After the dry etching, the samples were exposed to air for ~ 2 hrs which allows native oxide to form on the surface. The dry etched IPG transistors were then passivated with a Cl_2 plasma generated with 50 W microwave power at 2 mTorr for time ranging from 1 to 2 min. The gate leakage current as a function of V_{GS} for these transistors were shown in Figure 45. For the IPG transistors with $W_g = 400$ nm, the gate leakage current at $V_{GS} = 2$ V was 40 nA, and it decreased to 4.4 and 3.5 nA after being passivated with the low-energy chlorine species for 1 and 2 min, respectively. Since no rf

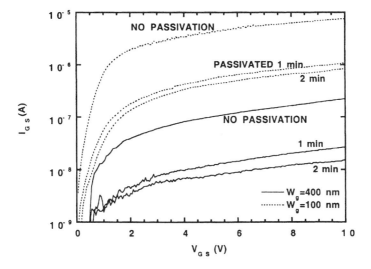

Fig. 45. Reduction of the gate leakage current for dry etched IPG transistors after being passivated with low energy chlorine species for different time. All transistor were first dry etched with 20% Cl_2 using 50 W microwave power and 200 W rf power at 0.5 mTorr. The Cl_2 plasma for passivation was generated using 50 W microwave power at 2 mTorr.

power was used for generating the plasma for passivation, the ion energy of the chlorine species is very low, and no etching can be measured during the passivation step due to the presence of the native oxide on the surface. Therefore, the decrease in the gate leakage current shown in Figure 45 is not caused by the removal of the surface damaged layer or changes in the etch depth, but is probably related to passivation. It is possible that the etch-induced defects that caused the gate leakage current were passivated with the low-energy chlorine species.

5. SUMMARY

Dry etching induced surface damage has been characterized using various electrical, surface analysis, and optical techniques. The surface damage depends mainly on the ion energy, ion flux, concentration of reactive species, etch temperature, etch time, and etched materials. High ion energy or sputtering with inert gases results in high defect density and the formation of a non-conducting damaged layer on the surface. It is found that higher ion energy, higher ion flux, or lower etch temperature causes a larger W_s, and these are similar to the results observed on the Schottky diodes and unalloyed transmission lines. Enhanced defect diffusion has been observed from TEM for GaAs etched at higher temperature. However, the electrical characteristics of Schottky diodes, unalloyed transmission lines, and conducting wires actually improve due to lower defect density by annealing. Degradations could be observed after dry etching HBTs or IPG quantum wire transistors. However, the device degradation can be minimized by using optimized etch conditions. Dry etched mirrors have high reflectivity up to 93%. The reflectivity is independent of the ion energy or the ion flux used as long as smooth sidewalls and vertical profile were maintained.

Surface damage can be removed by wet chemical etching or thermal annealing. However, two-step etching provides better control of etch profile and surface morphology while maintaining low surface damage. Damage removal using low-energy chlorine species has been demonstrated to lower the contact resistance of the transmission lines as well as reduce W_s of the conducting wires. Complete recovery to damage free surface can be obtained for both dry etched surface and sidewalls. The low-energy chlorine species can also passivate the surface damage with minimal etching if native oxide is allowed to form on the etched surface prior to passivation. With Cl_2 plasma passivation, no further etching is needed to reduce surface damage. This provides more precise control of the dimensions and profile of nanostructures. Typically, dry etching induced surface damage depends on the balance between defect generation rate and defect removal rate (etch rate). Surface damage can be minimized by etching with low ion energy, low ion flux, high etch temperature, low pressure, high Cl_2 concentration, fast etch rate, and damage removal or passivation using low-energy chlorine species.

Acknowledgements

The author would like to thank her students K.T. Sung, K.K. Ko, S. Thomas III, and E.W. Berg for their significant contributions to the work presented in this chapter.

References

1. S.W. Pang, G.A. Lincoln, R.W. McClelland, P.D. Degraff, M.W. Geis and W.J. Piancentini, *J. Vac. Sci. Technol.*, **B1**, 1334 (1983).
2. S.W. Pang, M.W. Geis, N.N. Efremow and G.A. Lincoln, *J. Vac. Sci. Technol.*, **B3**, 398 (1985).
3. S.W. Pang, "Radiation damage in dry etching", *Microelectronic Engineering*, **5**, 351 (1986).
4. H.F. Wong, D.L. Green, T.Y. Liu, D.G. Lishn, M. Bellis, E.L. Hu, P.M. Petroff, P.O. Holtz and J.L. Merz, *J. Vac. Sci. Technol.*, **B6**, 1906 (1988).
5. R. Cheung, S. Thoms, M. Watt, M.A. Foad, C.M. Sotomayor-Torres, C.D. W. Wilkinson, U.J. Cox, R.A. Cowley, C. Dunscombe and R.H. Willams, *Semicond. Sci. Technol.*, **7**, 1189 (1992).

6. K.K. Ko, S.W. Pang, T. Brock, M.W. Cole and L.M. Casas, *J. Vac. Sci. Technol.*, **B12**, 3382 (1994).
7. K.T. Sung and S.W. Pang, *J. Vac. Sci. Technol.*, **A12**, 1346 (1994).
8. E.L. Hu, C.H. Chen and D.L. Green, *J. Vac. Sci. Technol.*, **B14**, 3632 (1996).
9. K.K. Ko and S.W. Pang, *J. Electrochem. Soc.*, **141**, 255 (1994).
10. K.T. Sung, S.W. Pang, M.W. Cole and N. Pearce, *J. Electrochem. Soc.*, **142**, 206 (1995).
11. S. Yapsir, G. Fortuno-Wiltshire, J.P. Gambino, R.H. Kastl and C.C. Parks, *J. Vac. Sci. Technol.*, **A8**, 2939 (1990).
12. N.G. Stoffel, *J. Vac. Sci. Technol.*, **B10**, 651 (1992).
13. S. Thomas III and S.W. Pang, *J. Vac. Sci. Technol.*, **B12**, 2941 (1994).
14. H. Nakanishi, K. Wada and W. Walukiewicz, *J. Appl. Phys.*, **78**, 5103 (1995).
15. M.W. Cole, K.K. Ko and S.W. Pang, *J. Appl. Phys.*, **78**, 2712 (1995).
16. M. Hafizi, W.E. Stanchina, R.A. Metzger, J.F. Jensen and F. Williams, *IEEE Trans. Elec. Dev.*, **40**, 2178 (1993).
17. K.K. Ko and S.W. Pang, *J. Electrochem. Soc.*, **141**, 255 (1994).
18. S. Thomas III, E.W. Berg and S.W. Pang, *J. Vac. Sci. Technol.*, **B14**, 1807 (1996).
19. S. Thomas III, H.H. Chen and S.W. Pang, *J. Vac. Sci. Technol.*, **B15**, 681 (1997).
20. E.W. Berg and S.W. Pang, *J. Vac. Sci. Technol.*, **B15**, 2643 (1997).
21. E.W. Berg and S.W. Pang, *J. Vac. Sci. Technol.*, **B16**, 3359 (1998).
22. J.W. Coburn, *Plasma Etching and Reactive Ion Etching*, (American Vacuum Society Monograph Series, New York, 1982).
23. G.S. Oehrlein, *Handbook of Plasma Processing*, ed. by S.M. Rossnagel, J.J. Cuomo and W.D. Westwood, (Noyes Publications, New Jersey, 1990).
24. J. Asmussen, R. Mallavarpu, J.R. Hamman and H.C. Park, *Proc. IEEE*, **62**, 109 (1974).
25. J. Hopwood, D.K. Reinhard and J. Asmussen, *J. Vac. Sci. Technol.*, **B6**, 268 (1988).
26. J. Hopwood, *Plasma Sources Sci. Technol.*, **1**, 109 (1992).
27. J.H. Keller, *Plasma Sources Sci. Technol.*, **5**, 166 (1996).
28. P. Collot, C. Gaonach and N. Proust, *Mat. Res, Soc. Symp. Proc.*, **144**, 507 (1989).
29. M. Knoedler, L. Osterling and H. Shtrikman, *J. Vac. Sci. Technol.*, **B6**, 1573 (1988).
30. S. Agarwala and I. Adesida, *Sixth International Conf. on Indium Phosphide and Related Materials*, 391 (1994).
31. Forchel, B.E. Maile, H. Leier and R. Germann, *Physics and Technology of Submicron Structures: Proceedings of the Fifth International Winter School*, 26 (1988).
32. C.-H. Chen, D.L. Green and E.L. Hu, *J. Vac. Sci. Technol.*, **B13**, 2355 (1995).
33. M. Rahman, N.P. Johnson, M.A. Foad, A.R. Long, M.C. Holland and C.D.W. Wilkinson, *Appl. Phys. Lett.*, **61**, 2335 (1992).
34. E.L. Hu, C.-H. Chen and D.L. Green, *J. Vac. Sci. Technol.*, **B14**, 3632 (1996).
35. S. Murad, M. Rahman, N. Johnson, S. Thoms, S.P. Beaumont and C.D.W. Wilkinson, *J. Vac. Sci. Technol.*, **B14**, 3658 (1996).
36. G. Yu, C.-H. Chen, A.L. Holmes, Jr., S.P. DenBaars and E.L. Hu, *J. Vac. Sci. Technol.*, **B15**, 2672 (1997).
37. R.S. Williams, *Solid State Comm.*, **41**, 153 (1982).
38. N.G. Stoffel, *J. Vac. Sci. Technol.*, **B10**, 651 (1992).
39. H. Linke, I. Maximov, D. Hessman, P. Emanuelsson, W. Qin, L. Samuelson, P. Omling and B.K. Meyer, *Appl. Phys. Lett.*, **66**, 1403 (1995).
40. M. Rahman, N.P. Johnson, M.A. Foad, A.R. Long, M.C. Holland and C.D.W. Wilkinson, *Appl. Phys. Lett.*, **61**, 2335 (1992).
41. H. Chen, D.G. Yu, E.L. Hu and P.M. Petroff, *J. Vac. Sci. Technol.*, **B14**, 3684 (1996).
42. Blakeslee, Chemistry and Defects in Semiconductor Heterostructures, *Mater. Res. Soc. Sym. Proc.*, **148**, 217 (1989).
43. S. Thomas III and S.W. Pang, *J. Vac. Sci. Technol.*, **B13**, 2350 (1995).
44. D.J. Kahaian, S. Thomas III and S.W. Pang, *J. Vac. Sci. Technol.*, **B13**, 253 (1995).
45. Ren, T.R. Fullowan, S.J. Pearton, J.R. Lothian, R. Esagui, C.R. Abernathy and W.S. Hobson, *J. Vac. Sci. Technol.*, **A11**, 1768 (1993).
46. Y. Hirayama, A.D. Wieck and K. Ploog, *J. Appl. Phys.*, **72**, 3022 (1992).
47. J. van Wees, H. van Houten, C.W.J. Beenakker, J.G. Williamson, L.P. Kouwenhoven, D. can der Marel and C.T. Foxon, *Phys. Rev. Lett.*, **60**, 848 (1988).

48. Y. Ochiai, T. Onishi, J.P. Bird, M. Kawabe, K. Ishibash, Y. Aoyaki and S. Namba, *Jpn. J. Appl. Phys.*, **30**, 3859 (1991).

49. K. Ismail, W. Chu, D.A. Antoniadis and H.I. Smith, *Appl. Phys. Lett.*, **52**, 1071 (1988).

50. K. Ismail, D.A. Antoniadis and H.I. Smith, *Appl. Phys. Lett.*, **54**, 1130 (1989).

51. J. Nieder, A.D. Wieck, P. Grambow, H. Lage, D. Heitmann, K.V. Klitzing and K. Ploog, *Appl. Phys. Lett.*, **57**, 2695 (1990).

52. Y. Takagaki, F. Wakaya, S. Takaoka, K. Gamo, K. Murase and S. Namba, *Jpn. J. Appl. Phys.*, **28**, 2188 (1989).

53. D. Wieck and K. Ploog, *Appl. Phys. Lett.*, **56**, 928 (1990).

54. Y. Feng, T.J. Thornton, M. Green and J.J. Harris, *Superlattices and Microstructures*, **11**, 281 (1992).

55. J.J. Yang, M. Sergant, M. Jansen, S.S. Ou, L. Eaton and W.W. Simmons, *Appl. Phys. Lett.*, **49**, 1139 (1986).

56. J.J. Liang and J.M. Ballantyne, *J. Vac. Sci. Technol.*, **B12**, 2929 (1994).

57. P. Buchmann, H.P. Dietrich, G. Sasso and P. Vettiger, *Microelectron. Engin.*, **9**, 485 (1989).

58. N.C. Frateschi, M.Y. Jow, P.D. Dapkus and A.F.J. Levi, *Appl. Phys. Lett.*, **65**, 1748 (1994).

59. K.K. Ko, K. Kamath, O. Zia, E. Berg, S.W. Pang and P. Bhattacharya, *J. Vac. Sci. Technol.*, **B13**, 2709 (1995).

60. Thirstrup, S.W. Pang, O. Albrektsen and J. Hanberg, *J. Vac. Sci. Technol.*, **B11**, 1214 (1993).

61. K.K. Ko and S.W. Pang, *J. Vac. Sci. Technol.*, **B13**, 2376 (1995).

62. M.A. Foad, S. Thoms and C.D.W. Wilkinson, *J. Vac. Sci. Technol.*, **B11**, 20 (1993).

63. K.K. Ko and S.W. Pang, *J. Vac. Sci. Technol.*, **B11**, 2275 (1993).

64. E.S. Aydil and R.A. Gottscho, *Mat. Sci. Forum*, **148–149**, 159 (1994).

65. T. Ohno, *Phys. Rev.*, **B44**, 6306 (1991).

66. Y. Gu, E.A. Ogryzlo, P.C. Wong, M.Y. Zhou and K.A.R. Mitchell, *J. Appl. Phys.*, **72**, 762 (1992).

67. J.S. Herman and F.L. Terry, Jr., *J. Vac. Sci. Technol.*, **A11**, 1094 (1993).

68. K.K. Ko, E. Berg and S.W. Pang, *J. Vac. Sci. Technol.*, **B14**, 3663 (1996).

CHAPTER 6

Defects induced by Metal-Semiconductor Contacts Formation

TSUGUNORI OKUMURA

Department of Electrical Engineering, Tokyo Metropolitan University, 1-1 Minami-ohsawa, Hachioji, Tokyo 192-0397, Japan

1. IMPORTANCE OF SCHOTTKY-BARRIER CONTROL

1.1.   Introduction

In order to transfer signal and/or power into and out of semiconductor devices, virtually all kinds of devices must have an "interface". The interface is a plane where two different materials are brought into contact with each other. Whenever two different materials are brought into contact, a potential barrier is established at the interface. Such a potential barrier gives rise to asymmetry in an electrical characteristic of the interfaces. The larger the potential barrier, the more remarkable asymmetry becomes. At the interface between a metal and a semiconductor, this energy barrier is usually called a Schottky barrier, which is characterized by its value, Schottky barrier height (SBH). From a device point of view, metal-semiconductor interfaces are divided into two categories: a low-resistance ohmic contact and a rectifying Schottky contact. For both contacts, SBH is one of the most important parameters to be controlled as described below.

Over the past decade, there has been significant progress in understanding mechanisms of Schottky barrier formation owing to the development of various analytical techniques for clean and atomically-controllable surfaces in ultrahigh vacuum condition. Such studies also lead to various new techniques to control the properties of metal-semiconductor contacts at the atomic scale. However, practical contacts for device application are not devoid of imperfections introduced during device fabrication processes. The interfacial structure is generally modified by reaction and diffusion upon heat treatment during processing, and such modification might rarely occur homogeneously. From the practical point of view, we should understand the details of such device processing-related issues in the Schottky contacts. This chapter deals with the effects of inhomogeneity in the metal-semiconductor interface as well as imperfections in the sub-surface region of semiconductors on the characteristics of the Schottky contacts to compound semiconductors, mainly GaAs.

1.2.   Ohmic Contacts in Compound Semiconductor Devices

An ideal ohmic contact can flow current without any voltage drop through it, i.e., zero contact resistance. In optoelectronic devices, realization of low-resistance ohmic contacts is a critical issue for low-voltage as well as high-efficiency operation. If the contact resistance is high, for instance, in laser diodes, an actual applied voltage becomes large due to a voltage loss at the contacts, which might heat up the devices and hence degrade the device performance and reliability. In representative electronic devices with compound semiconductors, such as metal-semiconductor field effect transistor (MESFET) and high electron-mobility transistor (HEMT), source and drain contacts must be ohmic. If their contact resistance is high, a parasitic series resistance decreases the effective transconductance [128].

Now let us discuss how to decrease the contact resistance at the metal-semiconductor interface. From the current-voltage characteristic of the metal-semiconductor junction the ohmic contact resistance is given by,

$$r_c = \left. \frac{dV}{dJ} \right|_{V=0} \cong 0, \tag{1}$$

because only small voltage should be applied over a low-resistance contact. Here, r_c is usually referred as the specific contact resistivity. When the doped-impurity concentration is rather high in semiconductors, the current flowing through the contact is determined predominantly by the thermionic-field emission (TFE) mechanism [122]. Under certain conditions specified by Padovani and Stratton, the specific contact resistivity in the TFE regime is approximately given by [115],

$$r_c = \frac{k_B}{q A^{**} T} \exp\left(\frac{q \phi_B}{E_{00}} \right), \tag{2}$$

$$E_{00} = \frac{h}{4\pi} \left(\frac{N_I}{m^* \varepsilon_S} \right)^{1/2}, \tag{3}$$

where k_B is the Boltzmann constant, q the electronic charge, A^{**} the effective Richardson constant, T the absolute temperature, $q \phi_B$ the Schottky barrier height, h Plank's constant, N_I the impurity concentration, m^* the effective mass, and ε_S the static dielectric constant of the semiconductor.

This equation indicates that a good ohmic contact is realized by either lowering the Schottky barrier height or increasing the doping concentration. The specific contact resistivity is inversely proportional to exponential of the doped-impurity density, $\exp(N_I)^{1/2}$. Actually, this relationship was experimentally confirmed for GaAs ohmic contacts by Braslau [15]. However, it is known that the maximum doping concentration in compound semiconductors depends strongly on the conductivity type [154]. The highest doping densities are in the range of 10^{19} cm^{-3} for n-GaAs smaller than for p-GaAs, and for both n- and p-Si. If we assume a typical SBH value of 0.8 eV for n-GaAs, the doping density should be as high as 10^{20} cm^{-3} in order to realize $r_c = 10^{-6}$ Ωcm^2. On the other hand, p-GaAs can be doped to a carbon acceptor as high as the order of 10^{21} cm^{-3} by using metal-organic sources [179]. Since the SBH is, furthermore, rather low (typically 0.6 eV) for p-GaAs, realization of good ohmic contacts is much easier than for n-GaAs. The situation is definitely inverse for InP. The doping limit for InP is much lower for p-type than for n-type. The SBH is much higher for p-InP than for n-InP [55]. Therefore, further reduction of SBH is still challenging for ohmic contacts of some devices.

1.3. Schottky Contacts in Compound Semiconductor Devices

As for a rectifying contact, a Schottky contact to compound semiconductors
is indispensable for gate electrodes in MESFET as well as HEMT devices
to which an input signal is applied [23]. The Schottky-gate contact is
a key element of such gated-devices. An ideal gate electrode controls
the current flow in an underlying semiconductor channel without drawing
current through the gate electrode. For Schottky contacts, the current-voltage
characteristic (J-V) is empirically expressed as [122],

$$J = J_S \exp\left(\frac{qV}{nk_BT}\right)\left\{1 - \exp\left(-\frac{qV}{k_BT}\right)\right\}, \qquad (4)$$

where n is the ideality factor which can be close to unity if the current through
the barrier consists solely of the thermionic emission current. In general, n
becomes greater than unity due to a variety of reasons as will be discussed
later. J_S is the saturation current density under reverse bias voltage and given
by,

$$J_S = A^{**}T^2 \exp\left(-\frac{q\phi_B}{k_BT}\right). \qquad (5)$$

Hence, the SBH used in the FET gates will be expected to be high to reduce
the saturation current density J_S. Since the drift mobility is much higher for
electrons than for holes in most semiconductors, n-channel devices are more
attractive for high-speed-circuit application. Most metals give rather high
SBH of 0.8–0.9 eV to n-GaAs. Unfortunately, a SBH greater than 0.4 eV
is, on the other hand, rather often difficult to obtain on n-InP [99], and as
a result, several techniques have been reported for increasing the effective
SBH on n-InP [127, 147, 166].

The Schottky barrier height is also important in MESFET as well as
HEMT, since it is directly related to the threshold voltage of the FET's.
The width of a current channel is controlled by changing the depletion-layer
width expanding from the Schottky contact. The depletion-layer width W is
determined by the SBH ϕ_B, the applied bias V and the doping density N_D
in the channel. If the impurity is uniformly doped through the channel, W
becomes,

$$W = \sqrt{\frac{2\varepsilon_S\left(V_{bi} - V - \frac{k_BT}{q}\right)}{qN_D}}, \qquad (6)$$

$$\phi_B = V_{bi} - \frac{k_BT}{q}\ln\left(\frac{N_C}{N_D}\right), \qquad (7)$$

where V_{bi} is the built-in-potential, ε_S the dielectric constant of the semiconductor, and N_C the effective density-of-states of the conduction band. For an MESFET, the threshold voltage can be obtained with equation (6) when the depletion-layer width is equal to the channel thickness. The threshold voltage V_{TH} of n-channel HEMT devices is given by, [83]

$$V_{TH} = \phi_B - \frac{qN_D d^2}{2\varepsilon_S} - \frac{\Delta E_C}{q}. \tag{8}$$

Here, N_D and d are the doping density in the carrier-supplying layer and its layer-thickness, respectively, and ΔE_C the discontinuity of the conduction band. Equation (8) tells us that uniformity of SBH in addition to doping, the layer thickness, and the hetero-interface should be a key issue for VLSI application of HEMT's as well as MESFET's.

As described above, Schottky barrier height (SBH) is the most important parameter at metal-semiconductor interfaces regardless of a "Schottky" gate contact or an ohmic contact. In addition to realize a proper SBH value for each device application, uniform and thermally stable materials are required to realize reliable compound semiconductor devices.

2. MODELS OF SCHOTTKY BARRIER FORMATION

2.1. Schottky and Bardeen Limits

From a historical consideration by Schottky [124], the potential barrier at metal-semiconductor interfaces is determined by the difference between the metal work function $q\phi_M$ and the electron affinity of a semiconductor $q\chi_S$ as,

$$q\phi_{Bn} = q(\phi_M - \chi_S) \qquad \text{for } n\text{-type semiconductors}, \tag{9}$$
$$q\phi_{Bp} = E_G - q(\phi_M - \chi_S) \qquad \text{for } p\text{-type semiconductors}, \tag{10}$$

where E_G is the energy band gap for semiconductors. Such relationship was also derived by Mott independently [90]. Schottky's prediction indicates that the barrier height can be controlled using metals with different work functions, yet the following experimental results have shown work-function independence of the barrier height for most of covalent semiconductors, including Si, GaAs, InP, etc. [66].

This phenomenon is usually refered to as "Fermi-level pinning" at the metal-semiconductor interface. High-density of surface states, whether they are intrinsic or extrinsic, are induced in the vicinity of the metal-semiconductor interface, and hence the Fermi-level at the semiconductor surface is pinned at a certain position which is determined by the nature of the states. In the very early stage of the development of field-effect transistors, Bardeen postulated continuum states at the semiconductor surface before metal deposition [7]. If the Fermi level moves across a certain energy position within the gap upward or downward, the surface states are charged negatively (acceptor-like) or positively (donor-like), respectively. This energy position is called as the charge-neutrality level (CNL). In the extreme of high density of states, the surface states screen the effect of a metal overlayer and the Fermi level at the semiconductor surface is stabilized (pinned) at the CNL. In this Bardeen limit, the barrier height is independent of metals.

2.2.   Fermi-level Pinning Strength

In actual metal-semiconductor interfaces, SBH dependence on the metal work-function lies in between two limits described in the above section. The strength of the Fermi-level pinning is given by the index of interface behavior (or called as the slope factor) S_ϕ defined by,

$$S_\phi = \frac{\partial \phi_{Bn}}{\partial \phi_M}. \tag{11}$$

$S_\phi = 0$ and 1 correspond to the Bardeen and Schottky limits, respectively. In the 1970's, the relationship between semiconductor ionicity and the index S_ϕ was investigated with clean semiconductor surfaces. For most covalent semiconductors, such as Si, GaAs and InP, S_ϕ is around 0.1 being insensitive to different metals, in other words the Fermi level is rather tightly pinned [66].

Cowley and Sze [25] and Rhoderick et al. [122] generalized Bardeen's model. Consider an n-type semiconductor brought into contact with a metal across a very thin insulating layer with thickness of δ and permitivity of ε_i. The density of surface states is assumed to be constant D_s near the CNL which is located at $q\phi_{CNL}$ above the top of the valence band. If the insulating layer is so thin that electrons in the metal can easily communicate with the surface states, the electron occupation at the states is governed by the Fermi-level in the metal. For such a contact, the SBH is given by,

$$q\phi_{Bn} = S_\phi(q\phi_M - q\chi_S) + (1 - S_\phi)(E_G - q\phi_{CNL}), \tag{12}$$

$$S_\phi = \left\{ 1 + \frac{q^2 \delta D_S}{\varepsilon_i \varepsilon_0} \right\}^{-1}. \tag{13}$$

When $\delta D_S \to 0$ and $\delta D_S \to \infty$, then $S_\phi \to 1$ (Schottky limit) and $S_\phi \to 0$ (Bardeen limit), respectively.

The Bardeen and its extended model assume a uniform distribution of the surface states throughout the energy gap in semiconductors. The surface states can be sometimes discrete rather than continuous, such as the deep levels in the unified defect model [142] as described later. In this case, the surface Fermi level will be pinned near either the donor or acceptor states depending on the conductivity type of the semiconductor and/or the metal work function [122].

Now, the question arises what is the origin of the surface states. A number of mechanisms for the surface Fermi-level pinning have been proposed [17]. The proposed models for the Schottky barrier formation for compounds semiconductors, such as GaAs and InP, can be divided into two categories: intrinsic and extrinsic. Here, intrinsic surface (interface) states are defined as localized states created near the metal-semiconductor interface without a change in chemical or electrical properties of either medium.

2.3. Metal-Induced Gap State (MIGS) Model

The metal-induced gap state (MIGS) model relies on the intrinsic electronic states in the semiconductor energy gap induced by the tails of the electron wave functions in the metal. After the first argument by Hein, Louie and Cohen [75, 76] carried out detailed theoretical studies of MIGS at the interface between aluminum and various semiconductors. They showed that continuous electronic states are created in the semiconductor gap and that the penetration depth of charge into the semiconductor becomes small with increasing the band gap (GaAs→ZnSe→ZnS). Tejedor *et al.* [150] provided a clear one-dimensional model in which gap states due to the metal wave functions are evolved with compensation of conduction- and valence-band density-of-states of the semiconductor. In this sense, the MIGS is a metal-perturbed electronic states in the semiconductor. Separation between the bonding (valence band) and antibonding (conduction band) states becomes incomplete due to the penetration of electrons in the metal. Since the total number of the electronic states should be conserved, the smaller the surface density-of-states D_S, the wider the band gap energy. From equation (13), the slope factor S_ϕ increases with decrease in D_S, and in fact S_ϕ tends to large for wide gap semiconductor.

The MIGS tends to pin the Fermi level at the CNL of the interface. Further work done by Tersoff [151, 152] showed a simple estimate of the pinning strength in terms of the electronic susceptibility of the semiconductors. For 20 different semiconductors and insulators, Mönch [93] demonstrated an empirical (semi-theoretical) trend of the interface index (S_ϕ) as a function of the electronic susceptibility at optical frequencies ε_∞ as,

$$\frac{\alpha}{S_\phi} = 0.1(\varepsilon_\infty - 1)^2. \tag{14}$$

where α is approximately unity. δD_S in equation (13) can be related to dielectric properties of the semiconductor [92, 125].

2.4.   Extrinsic Models For Fermi-Level Pinning

The extrinsic mechanisms include defects, impurities, lattice disorder in the sub-surface region of semiconductors and/or the interfacial layer between the metal and the semiconductor. Actually, formation of such imperfections near the metal-semiconductor interface is greatly dependent on the process of metal deposition. The relationship between the Schottky-contact behavior and particular process will be discussed in Sections 3 and 4. In this section, unified models accounting for insensitivity of the SBH for compound semiconductors are briefly reviewed. It should be noted that in some cases more than one mechanisms can be used to account the experimental results in addition to process-related issues.

In 1970's, Spicer's group made photoemission measurements with synchrotron radiation for a wide variety of metals and oxygen on n-type and p-type GaAs(110), GaSb(110) and InP(110) [142, 143, 144]. Although a well-prepared (110) cleaved surface is unpinned (no intrinsic surface states) for most compound semiconductors due to lattice relaxation [21, 116, 141], the final position of the surface Fermi-level after metal deposition shows relatively small variation with metal overlayers. Fermi-level stabilization starts even at a submonolayer coverage of metal adatoms, and this means that other mechanism than MIGS, if it coexists, plays an important role in formation of the surface states. They concluded that at least two defect levels at or near the semiconductor surface are necessary to fit the experimental results; one must be a donor and the other an acceptor. For n-type and p-type GaAs, the pinning positions are 0.75 and 0.5 eV, respectively, above the top of the valence band. Their unified defect model (UDM), which was later refined as the Advanced Unified Defect Model (AUDM) for GaAs,[145] was also applied to account for the interfacial properties of III-V MIS (metal-insulator-semiconductor) structures.

A model for the creation of defects is based on the local energy released by adsorption of depositing atoms onto the semiconductor surface. An adsorbed atom gives up the heat of condensation, and this energy, for a typical value 60 kcal/mol (2.5 eV), is of the order of or larger than the semiconductor bond energy. In the 1979 paper, they thought that the 0.75 eV acceptor and the 0.5 eV donor at the GaAs surface were attributed to missing As and Ga (not a simple vacancy but maybe more complicated), respectively. Weber, in 1982 [174], suggested that two sets of energy levels of the arsenic antisite defect As_{Ga} in GaAs (D^+/D^0: $E_V + 0.75$ eV, D^{++}/D^+ : $E_V + 0.52$ eV) are surprisingly in good agreement with the above pinning levels. Since the As_{Ga} antisite is a double donor, a lower lying acceptor level is needed to explain pinning at midgap in n-Gaas. Hence, the gallium antisite defect Ga_{As} (acceptor) [35] was included in the model.

It is interesting that in this defect model the position of the surface Fermi-level slightly depends on interface chemistry and particularly off-stoichiometric composition at the interface [145]. For noble metals with a large electronegativity, such as Au, the electropositive component (Ga) in GaAs diffuses preferentially into the Au layer and forms alloy with Au [137], and hence the ratio of $[As_{Ga}]/[Ga_{As}]$ increases in the near interface. Accordingly the Fermi-level slightly moves upward to the conduction band minimum due to the increase of the donor concentration $[As_{Ga}]$. For less electronegative metals (Ti and Al), on the other hand, the strong bonding between the metal and As atoms takes place (chemical trapping of anions) [14, 16] resulting in the increase of Ga concentration and its outdiffusion. In this case, the surface Fermi-level tends to go down to the valence-band maximum according to the $[Ga_{As}]$ increase.

Hasegawa and his group pointed out that in various III–V compounds there exists a strong correlation among the Fermi-level position of metal-semiconductor, insulator-semiconductor and semiconductor-semiconductor interfaces [45, 54]. The interface states measured at the insulator-semi-conductor interfaces have a U-shape density-of-states whose magnitude strongly depends on processing for the interface formation. The Fermi-level at the interface is pinned at a minimum point of U-shaped density-of-states, i.e., the charge-neutrality level, which is close to an empirical hybrid orbital energy E_{HO} approximately 5 eV below the vacuum level. In an actual interface between two different materials, some extent of lattice disorder arises due to fluctuation in angle and length of atomic bonds, the presence of mismatch dislocation and preferential chemical reaction. In fact, such lattice disorder in the vicinity of the interfaces was observed by transmission electron microscopy (TEM) and the Ratherford backscattering measurement [46, 48]. Consequently, separation between the bonding and antibonding states becomes insufficient, just like the MIGS model, so that a continuum of the disorder-induced gap states (DIGS) emerges into the semiconductor energy gap.

In most metal-semiconductor contacts, the metallurgical interaction occurs resulting in formation of one or more new phases. The effective work function model, proposed by Freeouf and Woodall [39], suggests that the Fermi level at the interface is not determined by surface states but rather is related to the work function of microclusters formed upon chemical interaction between semiconductor and metal. In the effective work function model, the Schottky "ideal" model described in equation (9) is modified as follows:

$$q\phi_{Bn} = q(\phi_{eff} - \chi_S),$$ (15)

where ϕ_{eff} is a weighted average of the work functions of the different interface phases. For most of III–V compound semiconductors, metallization and/or oxidation results in the presence of excess anion species segregated to the intimate. For a typical example, heating a gold film deposited on GaAs results in release of free arsenic and formation of Au-Ga compounds, such as Au_7Ga_2, Au_2Ga, and a hexagonal gold-gallium phases [70]. In an ordinary device processing, the substrate surface is covered with native oxides. Thurmond *et al.* determined the Ga-As-O phase diagram showing that there is no tie-line between arsenic oxides (As_2O_3 and As_2O_5) [153]. The arsenic oxides cannot be, therefore, stable in contact with GaAs, and metallic As will be segregated as,

$$3 \cdot As_xO_y + 2y \cdot GaAs \rightarrow y \cdot Ga_2O_3 + (3x + 2y) \cdot As. \qquad (16)$$

The condition for driving this reaction to the right is that the Gibbs free energy is negative. The free energies of As_2O_3, As_2O_5 and Ga_2O_3 are -138, -184 and -237 Kcal/mol, respectively. Therefore, the surface of GaAs exposed to air is brought in contact with Ga_2O_3 and free As at thermal equilibrium. Such oxide reactions were examined for GaAs, InAs and InSb [32, 126].

3. INHOMOGENEOUS REACTION AND CONTACT PROPERTIES

3.1.   Inhomogeneous Metal-Semiconductor Reaction

The control of Schottky barrier heights at metal-semiconductor interfaces is critical to the design of high performance electronic and optoelectronic devices. Good contacts for device application require well-controlled barrier height proper to either ohmic or rectifying contacts, thermal stability and uniformity. In the LSI application of compound semiconductors, particularly, both metallurgical and electrical uniformity are significantly important, not only among contacts but also within an individual contact. Therefore, it is still challenging to establish the process technology of routinely making a large number of homogeneous contacts. The interfacial structure is generally modified by reaction and diffusion upon heat treatment used during device fabrication processes [137]. Such modification might occur inhomogeneously.

In general, the deposited metal-film consists of polycrystal grains rather than a single crystal. In the polycrystalline case, fast diffusion of atoms through grain boundary plays a very important role in the intermixing which takes place on heating. In the Al-GaAs system fabricated in a usual manner, gallium can migrate through Al polycrystalline films at 250°C [19] and accordingly the electrical characteristics will change. On the other hand, Missous *et al.* [86] showed that an *in situ* Al-GaAs contact prepared molecular beam epitaxy survived even at 500°C. The Al films formed in this manner were almost perfect single crystals with the (100) Al plane parallel to the (100) GaAs plane, and were therefore almost free from grain boundaries, which are responsible for inhomogeneous reaction at low temperatures. Miyazaki and Okumura [88] showed direct evidence of inhomogeneity of Schottky barrier heights (SBH's) at the Al/(111)Si interface due to inhomogeneous reaction through grain boundaries. Depending on the deposition condition, the Al/(111)Si interface consists either exclusively of the so-called type B (single grain) or mainly of type A with partial type B grains (mixed grains). The interface consisting of a single grain was thermally stable and the SBH's determined by the I-V and C-V methods did not change up to 550°C (slightly below the eutectic point of Al-Si, 577°C) for such single domain interfaces [87]. The SBH mapping can be obtained by scanning internal photoemission microscopy (IPEM) described below.

Practical contacts are composed of multiple materials, which are either in the form of alloy or multi-layered structure. Upon heat treatment, various compounds might be phase-segregated between the contact material and the semiconductor. These phases are randomly distributed in the vicinity of the interface, and hence the interface becomes rough and inhomogeneous in composition. For example, Au-Ge-Ni, one of the most practical ohmic contact materials to n-GaAs, reacts with GaAs resulting in formation of Ni_xGaAs, Ni_3Ge, β-AuGa, NiAs(Ge) and NiGe, whose composition and distribution depend on the annealing temperature [97]. Braslau pointed out that the contact resistance at alloyed Au-Ge-Ni/n-GaAs contact is determined by the protrusion of NiAs(Ge) nonuniformly formed [15]. Kuan et al. [67] and Shih et al. [131] showed the detailed interfacial structure studied by cross-sectional TEM and a good correlation between the contact resistance and the total area of the NiAs(Ge) patches. On the other hand, components with high surface tension, such as indium, tend to be phase segregated inhomogeneously upon reaction, sometimes balling up [56, 73]. Native oxides on the semiconductor substrate promote the inhomogeneous interfacial reaction. The semiconductor surface is exposed to atmosphere during fabrication process of devices and hence covered with native oxides, which are not, of course, guaranteed to be uniform in thickness as well as composition, sometimes are in the form of patches. Therefore, the insertion of thin near-noble metal (Ni, Pd, Pt) layers can be effective to suppress the interface irregularity due to native oxides [89, 137].

3.2. I-V and C-V Characteristics of Inhomogeneous Contacts

Over several decades, the relationship between SBH and microscopic structure of metal-semiconductor interfaces has been of fundamental as well as practical interest in the semiconductor science and technology.

Now, importance of the studies in this field has been increasing, since minuatualization and integration of semiconductor devices still proceed in high pace. Cross-sectional images of transmission electron microscopy (TEM) are very powerful in revealing metallurgical inhomogeneity, i.e., interfacial phases, bonding structures (epitaxial relationship) or the existence of thin intervening layers [74]. For example, the identification of the phases formed at the Au-Ge-Ni/GaAs interface and their correlation to the specific contact resistance were done by TEM as pointed out in the previous Section 3.1. Tung showed that the Schottky barrier height at $NiSi_2/Si(111)$ depends on the microscopic bonding structure at the interface; the type A and type B interfaces, where the $NiSi_2$ film shares the surface normal axis [111] with the Si substrate and rotated 0 and 180°, respectively, around this axis with respect to the Si substrate, have different Schottky barrier heights (0.65 and 0.79 eV, respectively) [155, 156]. Other analytical techniques, such as x-ray standing wave [162], medium energy ion scattering (MEIS) [77] and x-ray interference measurements [123], also reveled atomic information on the detailed structure of such epitaxial interfaces.

While structural as well as compositional analyses of the semiconductor interfaces are now available in a microscopic or even in an atomic scale, the SBH is usually characterized by using "macroscopic techniques", such as current-voltage (I-V) and capacitance-voltage (C-V) methods. Ohdomari *et al.* [103] pointed out that a transition metal silicide forms in patches on Si during the early stage of reaction, resulting in formation of a parallel contact (diode). Figure 1 schematically shows the parallel Schottky contact. For the parallel contact, the SBH's determined by the I-V and C-V methods consequently become apparent. If the size of mixed patches is not as small as either the Debye length or the depletion layer width of the barrier region, a simple model based on linear combination of two different interfacial phases can be applied to evaluation of the apparent barrier height [104]. In the I-V measurements, the apparent I-V barrier height ϕ_{app}^{IV} becomes,

$$\phi_{app}^{IV} = \frac{k_B T}{q} \ln \left[\left(\frac{S_l}{S_l + S_h} \right) \exp \left(-\frac{q\phi_l}{k_B T} \right) + \left(\frac{S_h}{S_l + S_h} \right) \exp \left(-\frac{q\phi_h}{k_B T} \right) \right],$$
(17)

when the thermionic emission mechanism alone is a current flowing over the barrier. Here, S_l or S_h is the area and ϕ_l or ϕ_h the barrier height, respectively, where the subscript l and h correspond the low- and high-barrier phases, respectively. This equation suggests that a small size patch does significantly affect on the lowering of apparent Schottky barrier height due to the nature of the exponential function. When the barrier height varies spatially on a scale less than, or comparable to, the depletion layer width, Tung gave analytical expressions for the potential and the current transport at inhomogeneous metal-semiconductor interfaces [157, 158]. The potential

(a) schematic picture of parallel Schottky contact

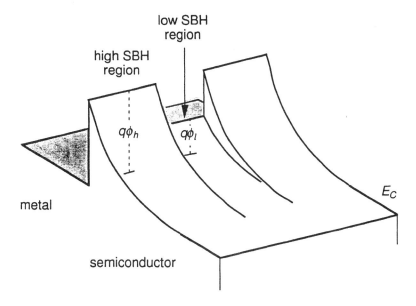

(b) Screening effect on the potential profile at the M/S interface

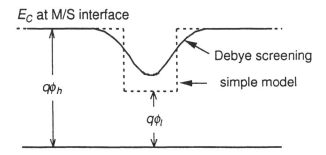

Fig. 1. Schematic picture of the parallel Schottky contact. (a) perspective view of the conduction band minimum (CBM) of the semiconductor. (b) If the size of the low SBH region is small compared with the Debye length, the CBM potential in the low SBH patch is pinched off.

for a small patch of low SBH, surrounded by the high SBH regions is "pinched off" (see Figure 1(b)) and increased towards a saddle point. This effect was experimentally verified by Olbrich *et al.* [114]. He showed that the presence of inhomogeneities in the Schottky barrier height leads to some anomalies in the I-V characteristics of the Schottky diodes, such as leakages, greater-than-unity ideality factors, soft reverse characteristics, and the temperature dependence of the ideality factor.

On the other hand, the apparent built-in potential V_{app} measured by the C-V method is given by [104],

$$\left(\frac{1}{V_{app}}\right)^{1/2} = \left(\frac{S_l}{S_l + S_h}\right)\left(\frac{1}{V_{bl}}\right)^{1/2} + \left(\frac{S_h}{S_l + S_h}\right)\left(\frac{1}{V_{bh}}\right)^{1/2}, \quad (18)$$

and C-V barrier height ϕ_{app}^{CV} becomes,

$$\phi_{app}^{CV} = V_{app} - \frac{k_B T}{q}\ln\left(\frac{N_C}{N_D}\right), \quad (19)$$

where V_{bl} and V_{bh} are the built-in potentials for the low-barrier and high-barrier regions, respectively. In contrast to I-V measurements, the apparent barrier height is not very sensitive the existence of the low-barrier regions, but it is dominated by the barrier height of the phase which occupies a large contact area in C-V measurements. It is often observed that the SBH's determined by different experimental methods, generally I-V and C-V measurements, are not always in good agreement with each other [38, 87, 103, 106]. Equations (17) and (18) suggest that inhomogeneity at the real metal-semiconductor interfaces is one reason for the above discrepancy among the measured SBH's. Some data showing $\phi_{app}^{IV} < \phi_{app}^{CV}$ can be attributed to such an inhomogeneous Schottky contact [87, 88].

3.3. Internal Photoemission Spectroscopy and Its Microscopy

As mentioned above, the most standard I-V and C-V techniques for characterizing electrical properties of the Schottky contacts are very macroscopic and just give one average (apparent) datum for one sample. A gap of spatial resolution between structural and electrical analyses is too far away at present. The techniques bridging the gap are ballistic electron emission microscopy (BEEM) and scanning internal photoemission microscopy (IPEM).

Ballistic electron emission microscopy (BEEM), firstly developed by Kaiser and Bell, [11, 12, 60] is based on scanning tunneling microscopy (STM). In BEEM applied to the metal-semiconductor interface, the current injected through the metal and into the semiconductor by the tunneling tip is monitored as a function of the bias voltage applied between the tip and the metal overlayer. When the applied bias voltage exceeds the Schottky barrier height and the metal-film thickness is sufficiently thin (less than about 10 nm) so that ballistic transport is dominant entirely through the metal overlayer, current injection into the semiconductor is possible. Since BEEM is based on STM, it is capable of characterizing Schottky contacts in a nanometer scale and visualizing the interface inhomogeneity. Although ballistic electron emission microscopy (BEEM) has been successfully applied to study internal electronic structure of buried semiconductor interfaces, its application has been restricted to sophisticated samples [11, 12, 49, 60, 118]. The metal thickness required for BEEM measurements should be too thin compared with practical semiconductor contacts. Furthermore, it was pointed out that the Debye screening length might limit the spatial resolution of BEEM. Hasegawa *et al.* [49] characterized the NiSi$_2$/Si(111) interface with mixed domains (type A and type B) by using STM/BEEM. Although the STM image could distinguish between type A and B domains in atomic resolution, the BEEM image showed very little difference between two. In order to see domains as small as 10 nm by using BEEM, the samples should be cooled down to or below 77 K and doped to higher level. It was, furthermore, pointed that an apparent inhomogeneity in the BEEM images often stems from the thickness variation of the metal overlayer [34] or carrier scattering at the interface [139], [140], and that the attenuation length of hot electrons in metals restricts the thickness of the metal overlayer (several tens of nanometers). Therefore, BEEM is not practical to characterize the actual metal-semiconductor contacts.

Internal photoemission measurements have been used to characterize various semiconductor interfaces. This is usually called as a "photoresponse" technique. Barrier height determination from photoemission threshold measurements has been considered the most accurate and reliable method because it is much less sensitive to the various problems of leakage, tunneling and interface layers than are the I-V and C-V methods.

When ultraviolet (UV) light or x-ray is incident upon a metal, the excited electrons in the metal can emit into vacuum. This effect is well-known as the photoemission effect. A similar effect takes place at interfaces between two solids. This is the so-called internal photoemission effect. The basic model for the internal photoemission process is shown in Figure 2, where we show a metal layer deposited on an n-type semiconductor. When a light with a photon energy of $h\nu > q\phi_{Bn}$ is incident upon a metal, the excited electrons in the metal can surmount the Schottky barrier and then photocurrent is generated due to the built-in field in the depletion region of the semiconductor. To determine the SBH from the internal photoemission measurements, we usually apply an approximate Fowler's formula [36] as,

$$Y \cong B \cdot (h\nu - q\phi_{Bn})^2, \tag{20}$$

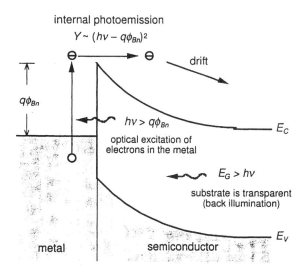

Fig. 2. Internal photoemission process at a metal-semiconductor interface. Monochromatic light with a photon energy below the energy gap of the semiconductor is not absorbed in the semiconductor region, and hence the probe light is able to impinge thorough the semiconductor substrate directly on the metal-semiconductor interface.

where Y is the photocurrent yield, $h\nu$ photon energy of the illuminated light and B the constant. Thus, plotting the square root of the photoyield as a function of photon energy, the Schottky barrier height is obtained by the so-called Fowler plot.

Consider a Schottky contact consisting two different phases with low and high barrier heights, $q\phi_{Bl}$ in area of S_l and $q\phi_{Bh}$ in area of S_h. The inset of Figure 3 illustrates such a contact; in this particular case the lower barrier phase might be a kind of intervening layer. If the constant B in equation (20) is assumed to be independent of each portion of the parallel contact, the averaged photoyield becomes,

$$Y \propto S_l \cdot (h\nu - q\phi_{Bl})^2 + S_h \cdot (h\nu - q\phi_{Bh})^2, \quad \text{for } h\nu \geq q\phi_{Bh}, \quad (21)$$

and

$$Y \propto S_l \cdot (h\nu - q\phi_{Bl})^2, \quad \text{for } q\phi_{Bh} > h\nu \geq q\phi_{Bl}. \quad (22)$$

For the parallel contact, the simple Fowler plot of the measured photoyield does not give a straight line at all. Okumura and Tu [105] showed that plotting the first derivative of the photoyield, $dY/d(h\nu)$, as a function of $h\nu$ is also capable of determining the barrier height in place of the square-root plot, and,

(a) uniform contact

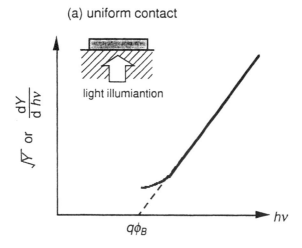

(b) parallel contact

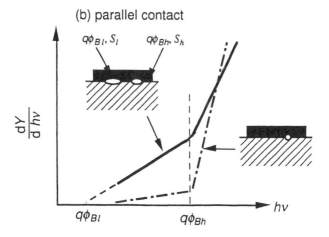

Fig. 3. Schematic illustration of the internal photoemission (IPE) spectrum for (a) a uniform Schottky contact and (b) parallel Schottky contact. For the parallel contact, the first derivative plot of the photoemission yield consists of two straight portions. If the area of the low-barrier phase decreases, the slope ratio will be changed as shown in the figure.

furthermore it is effective for quantitative evaluation for parallel contacts. The first derivative of equations (21) and (22) are,

$$\frac{dY}{d(h\nu)} \propto S_l \cdot (h\nu - q\phi_{Bl}) + S_h \cdot (h\nu - q\phi_{Bh}), \quad \text{for } h\nu \geq q\phi_{Bh}, \quad (23)$$

and

$$\frac{dY}{d(h\nu)} \propto S_l \cdot (h\nu - q\phi_{Bl}), \quad \text{for } q\phi_{Bh} > h\nu \geq q\phi_{Bl}. \quad (24)$$

Therefore, the first derivative plot of the photoresponse shows two linear portions as a consequence of equations (23) and (24). Figure 3 schematically shows the resultant spectra. The extrapolated energy of the lower linear portion gives the Schottky barrier height for the lower phase. The higher barrier phase starts to contribute to the photoresponse signal at photon energy of $q\phi_{Bh}$. In other words, the Schottky barrier height for the higher phase is determined by the photon energy at the kink. The slope ratio of the two linear portions in the plot corresponds to the area fraction, $S_l/(S_l + S_h)$, as expressed in equation (23). If the area of the higher energy phase increases accompanying with the dissociation of the lower barrier phase, the higher energy portion of the spectrum becomes steeper (see the dashed curve in Figure 3). Thus, the slope change in the spectrum reflects the fractional change in the parallel contact.

In general, the value of SBH's is smaller than that of semiconductor energy gaps E_G. Therefore, we can excite electrons in a metal layer from the semiconductor side, even when the metal thickness is extremely large. This configuration is the so-called "back illumination" [148]. This is the reason why the internal photoemission effect is applied to a nondestructive evaluation for "buried interfaces", and hence the IPE technique has the advantage over BEEM in characterization of the actual contacts for device application. Two dimensional mappings of photoemission current are obtained by translating (scanning) a specimen with respect to the axis of the focused laser beam, or vice versa (see Figure 4). Thus, the technique is a kind of scanning microscopy. Okumura et al. [108, 111, 132] developed the scanning internal-photoemission microscope (IPEM) by using laser beams from semiconductor laser diodes. It should be noted that this current is due to majority carriers in contrast to minority carriers in the optical-beam induced current (OBIC) [69] as well as the electron-beam induced current (EBIC) [175] techniques. Therefore, the diffusion length of minority carriers does not limit the spatial resolution in IPEM.

Equation (20) suggests that at least two values of photoyield at different photon energies can give the SBH. Okumura et al. [111] used 1.3 and 1.55 μm semiconductor lasers as a light source to build a microscope; the corresponding photon energies from the laser diodes are 0.95 and 0.80 eV,

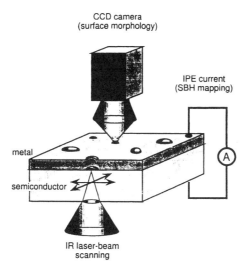

Fig. 4. Concept of scanning internal photoemission microscopy (IPEM). Two dimensional mappings of internal photoemission current are obtained by translating a specimen with respect to the axis of the focused laser beam, or vice versa. The corresponding surface morphology can be simultaneously observed by using a CCD camera.

respectively, and less than the energy gaps of typical semiconductors such as Si (1.12 eV), GaAs (1.42 eV) and InP (1.35 eV). Therefore, the laser beam is capable of impinging through substrates directly on metal-semiconductor interfaces. A schematic illustration of the IPEM apparatus is shown in Figure 5 [112]. A beam diameter of less than 2 μm was achieved in free space by using "pigtail-type" laser diodes (LD's) [132]. Two laser beams are merged and alternated with a fiber coupler in order to measure the photoyield values (Y) at two different photon energies. The energy accuracy of SBH's depends on the stability of the total system as well as quality of a sample Schottky diode. For a typical value of the signal-to-noise ratio for $\Delta Y/Y = 2 \times 10^{-3}$, the resultant accuracy's are estimated as 4 and 15 meV for the SBH's of 0.8 and 0.7 eV, respectively.

3.4. Case Study I: Al-GaAs Interfaces Characterized by IPE Spectroscopy

The thermal stability of metal-semiconductor interfaces is practically of great importance in device technology, since elemental device structures are suffered from a number of heat cycles during processing. Schottky diodes are also used for the characterization of shallow and deep impurities [78].

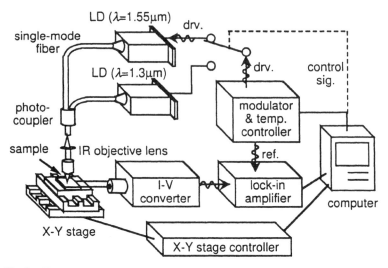

Fig. 5. The apparatus of scanning internal photoemission microscope (IPEM). Two laser beams from LD's (1.3 and 1.55 μm) are merged with a fiber coupler to determine the SBH with Fowler's plot.

In such cases, again, the Schottky barrier height, metallurgical stability and uniformity affect reliability in the experimental results and thus are relevant to the interpretation derived. For example, the controversy on the midgap electron-trap in GaAs, referred to as EL2, in relation to Al and Au Schottky metals was attributed to the Schottky barrier height and/or the reaction between metal and GaAs during DLTS measurements [68].

Aluminum is one of the most extensively-used metals for a Schottky contact which is used for characterization of impurities and deep levels in GaAs. Sinha and Poate [137] classified aluminum to a group of metals which exhibit very stable electrical and metallurgical properties with GaAs upon thermal annealing. In fact, the Al-GaAs Schottky contacts prepared by molecular-beam epitaxy showed remarkable stable electrical characteristics even after annealing for 1 hour at 500°C [86]. However, the barrier height of Al-Schottky contacts formed on the chemically-etched n-GaAs surface by a conventional evaporation technique is very often found low and varies upon annealing at relatively low temperatures (100–400°C) [47, 107].

Okumura and Shibata [107] analyzed inhomogeneity of the Schottky barrier height on the Al-GaAs interfaces, which were prepared by standard etching and evaporation process, and its change upon low-temperature annealing by internal photoemission spectroscopy. The barrier height determined by the I-V and C-V measurements as well as the diode ideality factor are plotted as a function of the annealing temperature as shown in Figure 6.

(a) Change in Schottky barrier height

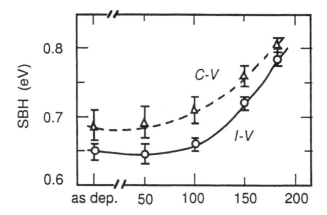

(b) Change in ideality factor (*n*-value)

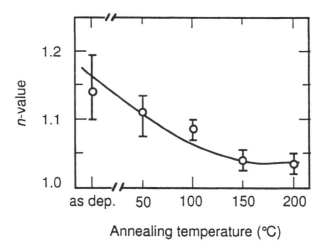

Annealing temperature (°C)

Fig. 6. Change in SBH's of Al/*n*-GaAs upon annealing. The GaAs surface should be covered with native oxides in the as prepared state, because the substrate was cleaned and etched in a standard procedure followed by evaporation of aluminum.

The barrier height for the as-deposited diode was around 0.65 eV, which
was low compared with the Au Schottky contacts fabricated on the same
substrate: 0.82–0.88 eV. Such a low barrier height for Al/n-GaAs contacts
has usually been observed and discussed in relation to native oxide(s) at the
interface [47]. A poor ideality factor of 1.15 in the as-deposited state can
be interpreted by a thin insulating layer at the interface [122] as well as
inhomogeneity at the interface [158]. The annealing at 50°C for 90 min did
not change the apparent barrier height very much, but improved the ideality
factor slightly. The increase in barrier heights started around 100°C. As the
annealing temperature increased, the diode characteristics improved. The
barrier height for the diode annealed at 200°C became comparable to, but
slightly lower than, that for Au/n-GaAs contacts. The ideality factor was
around 1.03 for the diodes annealed at 200°C. The barrier height determined
by the C-V measurements showed the same trend as that deduced from the
forward I-V curves, while the former was slightly higher than the latter.

The first derivative spectra of the photoresponse are shown in Figure 7 for
the Al/n-GaAs contacts in the as-deposited state as well as after annealing at
four different temperatures. The spectra are aligned with the lowest portion of
each curve in order to place emphasis on the kink around 0.9 eV. The lowest
threshold energy was determined to be 0.7 eV by extrapolating the linear
portion of the spectrum for the as-deposited contact which shows no obvious

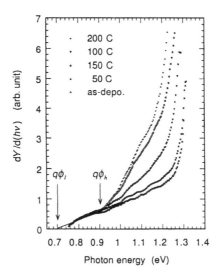

Fig. 7. Internal photoemission spectra of Al/n-GaAs in the as-deposited state and after
annealing at various temperatures. The first derivative plot of the internal photoemission yield
with respect to photon energy clearly shows that the Al/GaAs interface fabricated by a standard
process consists of two phases with different SBH's, 0.7 and 0.9 eV.

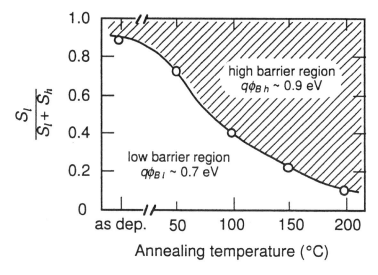

Fig. 8. Change of the low-SBH phase fraction in the Al/n-GaAs interface as a function of annealing temperature.

kink. As illustrated in Figure 3, the kink in the spectrum as well as the change in slope by annealing give evidence of the parallel contact in these Al/GaAs Schottky contacts; the spectrum with two linear portions means a two-phase mixture in the contact. The lowest threshold energy, 0.7 eV, corresponds to the barrier height for the lower-barrier phase. The energy at the kink, 0.9 eV, corresponds to the barrier height for the higher-barrier phase. The increase in the slope above 0.9 eV with increased annealing temperature indicates the dissociation of the lower barrier portion or the conversion to the higher phase. The fraction of the lower barrier portion, $S_l/(S_l + S_h)$, was calculated by applying equation (23) to the data shown in Figure 7. The result is shown in Figure 8. The area fraction of the lower-barrier phase decreased from about 0.9 in the as-deposited state to 0.1 after annealing at 200°C.

One can estimate the apparent barrier height in the I-V and C-V measurements by using equations (17) and (18) with the results shown in Figure 8. Table I summarizes these values, ϕ_{app}^{IV} and ϕ_{app}^{CV}, as well as the corresponding experimental ones. All apparent barrier heights with the C-V technique are larger than those with the I-V technique in any annealing stage, which agrees with discussion in Section 3.2. Relatively low Schottky barrier is often observed in Al diodes deposited on intentionally oxidized GaAs [47]. Christou and Day studied interdiffusion between Al thin films and GaAs by using Auger electron spectroscopy [19]. Their sputter-depth

TABLE I
Summary of the low-barrier phase fraction in Al/n-GaAs Schottky contacts. The experimental and theoretical values of SBH's determined by the I-V and C-V techniques are also included. Theoretical values are calculated by using equations (17)–(19).

Sample	$S_l/(S_l + S_h)$	ϕ_{app}^{IV}, exp	ϕ_{app}^{IV}, th	ϕ_{app}^{CV}, exp	ϕ_{app}^{CV}, the
as. depo.	0.88	0.70	0.65	0.74	0.69
50°C	0.72	0.71	0.65	0.77	0.69
100°C	0.40	0.72	0.66	0.83	0.71
150°C	0.22	0.74	0.72	0.87	0.76
200°C	0.12	0.76	0.78	0.88	0.81

unit: eV except for $S_l/(S_l + S_h)$

profiles for Al-GaAs prepared by standard etching and evaporation process showed the oxygen layer at the interface associated with the presence of native oxide for the as-deposited sample. On the other hand, the oxygen was completely transported with the Ga toward the aluminum surface after annealing at 250°C. The effect of annealing was to increase both the I-V and C-V barrier heights [86]. The enhancement of the Schottky barrier height was due to the formation of the (Ga,Al)As phase formed at the interface [24]. The native oxide layer might be, in this context, responsible for the lower barrier phase. Probably the increase in the barrier height by annealing corresponds to the exchange reaction between Al and Ga [65] or dissociation of the native oxide by the reaction [19].

3.5.   Case Study II: Thermal Degradation of Au/Pt/Ti-GaAs Schottky Contacts Characterized by IPEM

A three-layer structure of Ti/Pt/Au is one of the most important materials for semiconductor metallization to Si as well as GaAs. This material is also utilized for a gate contact of GaAs metal-Schottky field effect transistors (MESFET's). The first Ti layer and the top gold layer provide low resistance electrical paths and a strong bond to GaAs [165], respectively. Gold cannot be placed directly on GaAs, because it does not give a strong bond to GaAs and because it reacts at relatively low temperatures with GaAs forming Au-Ga intermetallic phases, which degrade electrical properties of the Schottky contact. However, a metallization consisting of gold on top of titanium would not work because gold diffuses into titanium at about 300°C [5]. Since the platinum midlayer plays an important role in providing a diffusion barrier between gold and titanium, such a three-layer structure gives a good adhesion, a stable Schottky barrier height, resistance to interdiffusion and good electrical conduction. Nevertheless, it was reported that the characteristics of Ti/Pt/Au to GaAs Schottky contacts changed upon annealing around 500°C [136]. Shiojima and Okumura demonstrated by scanning internal-photoemission microscopy (IPEM) that the inhomogeneous degradation of Ti/Pt/Au Schottky contacts to n-GaAs proceeds upon annealing above 450°C [133]. A good correlation was found between the local formation of a low-SBH phase and the degradation of macroscopic I-V characteristics.

Figure 9 shows (a) a plot of Y vs. annealing temperature and (b) annealing temperature dependence of SBH and diode ideality factor (n-value). The substrate used for the experiments was n-GaAs(100) with a carrier density of 1×10^{17} cm^{-3}. Electron beam evaporation was used to deposit 50 nm of Ti, then 50 nm of Pt, and finally 200 nm of Au. Standard photolithographic and lift-off techniques were used to define an array of rectangular diodes. Isochronal and isothermal annealing were conducted at temperatures between 200 and 500°C. For the diode in the initial state, the SBH had a relatively low value of about 0.68 eV, and the n-value was relatively high about 1.4, which are attributed to process damage introduced during the electron-beam evaporation. Such damage can be annealed out at temperatures as low as 200°C.

The diodes annealed at temperatures below 400°C gave a uniform internal-photoemission yield, within 2%, across the area of a diode. Upon annealing at 400°C, all diodes had SBH of about 0.72 eV and exhibited good I-V characteristics with an n-value of 1.2. The work of Sinha et al. shows that the improvement in the diode characteristics may be attributed to the formation of platinum arsenide at the interface [136]. Namely, the intimate phase onto GaAs changes from TiAs to PtAs$_2$.

For the diodes annealed above 450°C, the internal-photoemission yield, however, varied significantly across the area of the diode, and rectifying characteristics in I-V measurements disappeared. The data of isothermal annealing at temperatures at 480°C are given in Figure 10, which shows that the SBH decreased and the n-value increased as the annealing time increased. This trend was particularly noticeable for annealing time between 15 and 20 min. The data also showed that there was considerable variation in values measured for different diodes. In the I-V curves for the diodes with a high n-value, an excess current was superimposed on the thermionic emission current. The excess current was approximately proportional to the applied voltage, i.e., quasi-ohmic.

The IPEM measurements revealed that the regions of high photoemission yield, which appear bright in the maps, were formed for the sample annealed at 480°C longer than 15 min, where the I-V characteristics were degraded. Frequently, the bright regions started in the vicinity of the contact periphery and moved in toward the center of the electrode pattern. Figure 11(a) shows representative maps of the high internal-photoemission yield (high-Y) region

(a) Photoyield change upon isochronal annealing

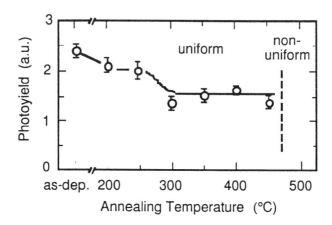

(b) SBH and *n*-value changes upon isochronal annealing

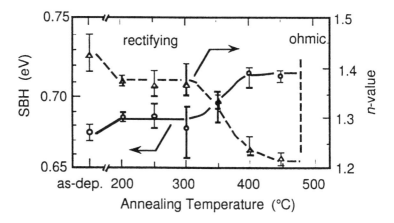

Fig. 9. Temperature variations of (a) the internal-photoemission yield and (b) the diode I-V characteristics (SBH and n-value) upon isochronal annealing for 20 min of Ti/Pt/Au contacts to n-GaAs.

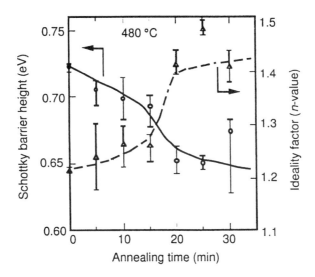

Fig. 10. Annealing-time variation of the SBH and the *n*-value upon isothermal annealing at 480°C of Ti/Pt/Au contacts to *n*-GaAs. Prior to the annealing at 480°C, the samples were annealed for 2 hours at 400°C.

detected in the sample annealed for 40 min at 480°C. From equation (20), one can expect that the lower SBH region gives higher photoemission yield at a constant photon energy. In fact, the measured SBH at the center was lower than that for the entire interface by 0.06 eV; the SBH in normal regions was around 0.82 eV determined with wavelength dependence of photoyield in IPEM measurements. Figure 11 also shows scanning images with (b) secondary electron, (c) Ga-LMM and (d) Au-MNN Auger signals by using samples whose GaAs-substrate was selectively removes. Tests showed that the total area of the high-Y regions tended to increase as annealing time increased. Figure 12 shows the excess current measured at a forward bias of 0.1 V and the number of the high-Y pixels were correlated with each other. It is, therefore, postulated that the degradation in rectifying characteristics observed after annealing above 450°C arises because the regions of gold-gallium intermetallic phases with quasi-ohmic I-V characteristics shorted out the regions that were still rectifying.

   From the above results, the thermal degradation of the Ti/Pt/Au-GaAs contacts can be attributed to the local interaction of gold and GaAs to form the gold-gallium intermetallic phase [10]. Gold probably reached the interface by migration around the edges of the contact or by diffusion along grain boundaries in the platinum film. The edge defined by the lift-off

(a) IPEM image

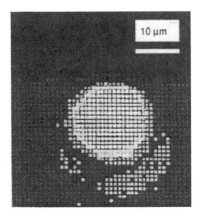

(b) SEM image

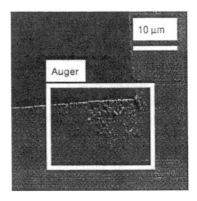

(c) Ga-LMM Auger image   (d) Au-MNN Auger image

Fig. 11. Interface images of Ti/Pt/Au-GaAs annealed at 480°C taken by (a) scanning internal photoemission microscopy (IPEM) and scanning Auger microscopy with (b) secondary electron (SEM), (c) Ga-LMM and (d) Au-MNN signals. SEM and Auger signals were taken in a different but equivalent region from the area as shown in (a) after removal of the GaAs substrate.

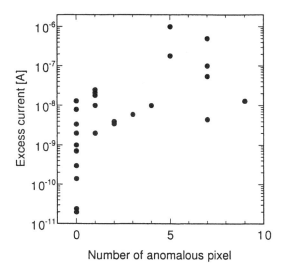

Fig. 12. The relationship between the number of the high IPEM signal (see Figure 11(a)) spots and the excess diode current of the thermally degraded Ti/Pt/Au-GaAs contacts measured at a forward bias of 0.1 V.

technique may be tapered off and the resistivity of platinum against the gold diffusion into the layers underneath is weakened there. It has been reported that the characteristics of gold-to-GaAs contacts change from rectifying to quasi-ohmic after annealing at temperatures above 400°C [20, 94]. Leung *et al.* [71] investigated the composition dependence of electrical properties, structure and phase morphology of gold-gallium alloy films codeposited on GaAs(100). The Schottky barrier height decreases from 0.95 to 0.64 eV with increasing composition of gallium from 0% (pure Au) to 90%. They showed that annealing of pure gold promotes formation of various gold-gallium compounds, the most prominent being Au_2Ga and pyramidal pits, and that the SBH value is reduced by the presence of Au_2Ga and localized current or field concentrations at the tip of the pyramidal pits. Coulman and his coworkers [182] showed the gold contact to the cleaved n-GaAs(110) surface becomes ohmic upon annealing due to very thin, long, predominantly Au crystallites formed on the GaAs surface at the periphery of the contact. These crystallites are crystallographically oriented along the ⟨110⟩ GaAs direction, and hence the findings are slightly different from those by the IPEM study with Ti/Pt/Au-GaAs(100).

4. EFFECT OF PROCESS-RELATED DEFECTS ON CONTACT PROPERTIES

4.1.   GaAs Schottky Contacts Fabricated in a Standard Device Process

Actual metal-semiconductor interfaces are not devoid of imperfections introduced during the contact formation, while most of past barrier height data have been taken with the contacts fabricated on clean and well-controlled surfaces of compound semiconductors. Some are non-polar (110) surface cleaved in an ultrahigh vacuum [141] and others as-grown (100) surface fabricated by molecular beam epitaxy [163]. Electrochemical etching of GaAs [62, 109] as well as InP [50, 178] can also provide oxide-free surfaces in electrolytic solutions for metal plating. On such well-controlled surfaces, Schottky metals are *in situ* deposited in the same vacuum chamber or in the same electrolytic solutions. A single mechanism, some of which have been discussed in Section 2, can sometimes account for the metal-work-function dependence of the measured SBH's.

On the other hand, semiconductor contacts for actual device application are fabricated on chemically etched or dry-cleaned semiconductor surfaces where some extrinsic layer, such as thin native oxides, process induced defects, and adsorbed impurity atoms, may exist. Therefore, the Schottky barrier height of the practical contacts is considered to be determined by multiple factors. In this section, we will discuss the relationship between the measured Schottky-diode characteristics and the process-related defects induced in the subsurface region of semiconductors.

Figure 13 shows Schottky barrier heights (SBH's) reported for (a) n-GaAs and (b) p-GaAs as a function of the metal work-function. All SBH data in the figure were determined by either the I-V or IPE methods. The reported values determined by the C-V measurements, and by using highly doped-substrate (> upper 10^{17} cm^{-3}) or the annealed samples with reacted interfacial layers are not included. The data are significantly scattered even for a certain metal. For example, the SBH value for Pt/n-GaAs contacts ranges from 0.65 eV [38] to 1.07 eV [177]. The contact with the highest SBH was fabricated by electrochemical process, which is free from electron- or ion-bombardment effects, while for the lowest one the Pt film was deposited by electron-beam (EB) deposition. The W/n-GaAs fabricated by CVD gave a rather high SBH of 0.81 eV [8], while other W contacts with 0.65–0.74 eV were fabricated by either EB deposition or sputtering [81, 135, 168]. Thus, the behavior of metal-semiconductor contacts depends on the details of their preparation and processing.

The bold straight lines in Figure 13 denote,

$$q\phi_{Bn} = 0.15 \cdot q\phi_M + 0.26 \text{ [eV]}, \tag{25}$$

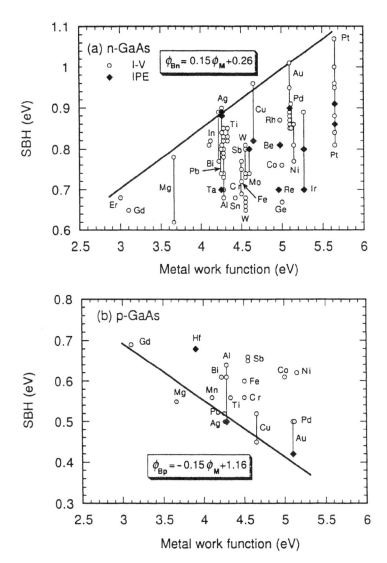

Fig. 13. The SBH's reported for (a) n-GaAs and (b) p-GaAs as a function of the metal work-function. All data in the figure were determined by either the I-V or IPE measurements. The reported values determined by the C-V method, and by using highly doped-substrate or the annealed sample with definite reaction products are not included. The references are as follows: Ag [79, 82, 117, 169, 170], Al [31, 58, 63, 82, 84, 159, 164, 169, 170], Au [43, 63, 71, 80, 82, 109, 135, 159, 169, 170], Be [82], Bi [159, 169, 170], Co [169, 170], Cr [13, 169, 170], Cu [82, 169, 170], Er [176], Fe [82, 169, 170], Ge [42], Gd [106], Hf [59], In [159], Ir [3, 167], Mg [159, 169, 170], Mn [169, 170], Mo [72, 169], Ni [63, 109, 169, 170], Pb [109, 169, 170], Pd [3, 169, 170], Pt [3, 38, 63, 82, 95, 177, 135], Re [167], Rh [3], Rh [3], Sb [22, 84, 159], Sn [109], Ta [18, 167], Ti [136, 169, 170], W [8, 81, 135, 168].

for (a) n-type and

$$q\phi_{Bp} = -0.15 \cdot q\phi_M + 1.16 \, [\text{eV}], \tag{26}$$

for (b) p-type, respectively. The sum of equations (25) and (26) becomes
1.42 eV, which is the bandgap energy of GaAs. It is noted that most of
data points for n-type and p-type scatter below equation (25) and above
equation (26), respectively. For n-GaAs, a similar trend was already pointed
out by Mönch, [91, 92] who discussed that the upper limit corresponds to the
SBH variation based on charge transfer due to electronegativity difference
between the metal and GaAs in the absence of process-induced defects (MIGS
limit). He also pointed out that the deviations in SBH from the MIGS limit
indicate high density of a single defect-level at $E_c - 0.65$ eV, which changes
the Fermi-level position at the interface. However, the level should posses an
amphoteric nature, if the same defect plays a dominant role in the Fermi-level
stabilization. As will be discussed in Section 4.3, the data deviated from
the MIGS line in both n- and p-type graphs can be explained by taking
into account the donor-type defects introduced in the semiconductor surface
during metal-deposition process.

4.2. Effect of Defects on the SBH Measurements

Schottky barrier height is usually determined by either the current-voltage
(I-V), or capacitance-voltage (C-V), or internal-photoemission (IPE) measurements. In the I-V measurement, the saturation current, which gives
the barrier height by using equations (4) and (5), is determined by using a
semi-logarithmic plot of the forward characteristic. Equations (4) and (5) are
based on the thermionic-emission theory, and therefore it should be noted that
the resultant value becomes just "apparent" when other transport mechanisms
are superimposed and/or some intervening layer exists at the interface. The
fact that the ideality factor (n-value) is close to unity is an important criterion
whether the measured value corresponds to a real Schottky barrier height or
not. In the C-V measurement, we extrapolate the data of C^{-2} vs. V onto the
voltage axis. The built-in-potential V_{bi} is given by using Equation (6) and the
relationship $C = \varepsilon_S S / W$ (S: the area of the contact). However, the model
assumes a homogeneous and uniform charge profile in the semiconductor
up to the interface. In the internal photoemission measurement, the Fowler
plot described in Section 3.3 gives the barrier height, which is usually most
reliable and less subject to errors.

Figure 14 shows the relationship between the values of SBH in the literature
determined through the I-V (SBH$_{I-V}$) and C-V (SBH$_{C-V}$) methods. In the
figure, the numbers in the bracket indicate the n-value of the diodes, and EB
and SP mean that the metals were deposited by electron-beam evaporation

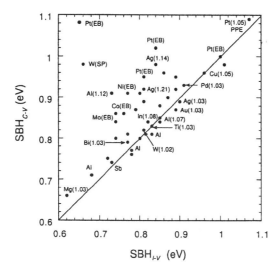

Fig. 14. The relationship between the reported values of SBH determined by the I-V and C-V methods. The numbers in the bracket indicate the n-value of the diodes. "EB", "SP" and "PPE" in the brackets mean that the metals were deposited by electron beam evaporation, sputtering and pulsed-photoelectrochemical process, respectively. The references are the same as those in Figure 13.

and sputtering, respectively. The relationship of $SBH_{C-V} > SBH_{I-V}$ is observed for many data, but the opposite for very few. For the diodes with an ideality factor close to unity, coincidence between the values determined by means of the two independent techniques is fairly good. On the other hand, the diodes with a large n-value or fabricated by EB or SP processes tend to give $SBH_{C-V} > SBH_{I-V}$. Amith and Mark [2] investigated the properties of Au-Schottky barriers on an n-GaAs (110) surface subjected to ion bombardment, and they showed that the damaged surface gives $SBH_{C-V} > SBH_{IPE} > SBH_{I-V}$. Okumura and his coworkers [38, 106] also showed the barrier heights of EB-deposited metals to n-GaAs(100) determined by I-V, C-V and IPE techniques were in the same order ($SBH_{C-V} > SBH_{IPE} > SBH_{I-V}$) and discussed in relation to the damage introduced by the metal deposition process. Many other reports suggested that there is discrepancy between SBH_{C-V} and SBH_{I-V} in the Schottky diodes with surface defects on Si [53, 96], GaAs [29, 30, 171, 172], and InP [161]. Then, we will discuss the effect of the near-surface defects on the Schottky diode characteristics.

Consider the following five defects or perturbations which modify the surface properties of semiconductors: (a) a surface defect with the same-type charge as the dopant, e.g., a donor-type defect in an n-type semiconductor, (b) a surface defect with the opposite-type charge to the dopant, e.g. a donor-type defect in a p-type semiconductor, (c) a defect in the sub-surface region which compensates the dopant, (d) a defect or an impurity which deactivates the dopant, like hydrogen, and (e) an amorphous surface-layer or a heavily disordered layer with high resistivity. Table II summarizes the effects of these surface modification induced during device fabrication processing on the I-V (log I vs V), C-V (C^{-2} vs V) and IPE ($Y^{1/2}$ vs $h\nu$) plots. Here, we assume that the Fermi-level position at the metal-semiconductor interface is the same before and after modification.

4.3. Surface Defects

First, surface defects with the same-type charge as the dopant will be discussed (see Table II(a)). Here, we call the defects which locate within about 1–10 nm below the semiconductor surface by a surface defect. The top-left drawing in Figure 15(a) schematically shows the space charge profile in the case that donor-type surface defects are introduced in an n-type substrate. Due to the high concentration of positive charges in the depletion layer, the electric field will increase near the metal-semiconductor interface (see Figure 15(b)). Therefore, the image-force lowering contributes the barrier-height reduction [129]. For the notations in Figure 15(a), the lowering is approximately given by,

$$\Delta\phi_{Bn} \cong \frac{q}{\varepsilon_S}\sqrt{\frac{N_{\text{defect}}d}{4\pi}}. \tag{27}$$

As the defect concentration further increases, the tunneling effect might become significant. The fact that the tunneling current superimposes on the thermionic-emission current means that the SBH value simply derived by equations (4) and (5) becomes lower than the real one and just apparent. In the C-V measurement, the n^+-layer near the surface will shrink the extent of the space charge region and cause an increase in the capacitance with the concomitant changes in the intercept. Based on the energy band diagram as shown in Figure 15(b), the C-V measurement gives the built-in-voltage which is an extrapolation of the parabolic potential in the depletion-layer edge. Therefore, V_{bi} will be apparently decreased. The Fowler plot in the IPE measurement will slightly shift toward lower energy due to the image-force lowering, and tailing of the spectrum will become significant due to the tunneling effect.

Second, surface defects with the opposite-type charge to the dopant will be discussed (see Table II(b)). This situation is the same as the counter doping to the substrate, for example the base-diffusion into the prefabricated collector region of a Si bipolar transistor. We will consider the case that donor-type surface defects are introduced in a p-type substrate (see the top-right drawing

TABLE II
Summary of the effects of imperfections in the surface or sub-surface region of
semiconductors on the electrical characteristics of Schottky-barrier diodes.
The table includes the most standard electrical techniques of current-voltage (O-V),
capacitance-voltage (C-V) and internal photoemission (IPE) measurements.

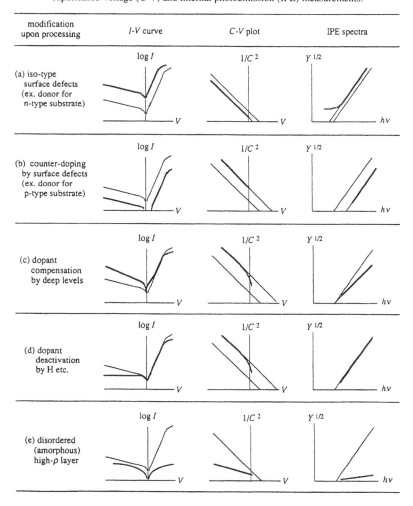

(a) space charge density profile (e-beam deposition case)

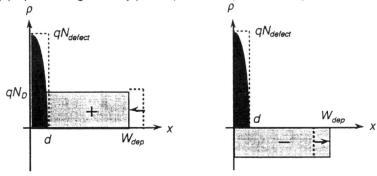

(b) energy band diagram and effective SBH change

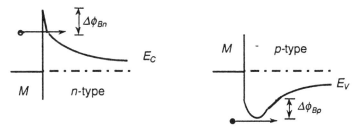

(c) space charge density profile (plasma-damage case)

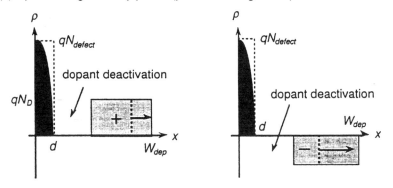

Fig. 15. The effect of the surface damage introduced during metal deposition or plasma processing. (a) Space charge distribution of the case that the donor-type surface defect is introduced and (b) the corresponding potential profiles. For an n-type substrate, the enhanced electric filed lowers the effective barrier height, while formation of n/p junction enhances the energy barrier for a p-type substrate. In plasma processing, (c) the effect of another defect or hydrogen atom deactivating the dopant might be superimposed to the surface defect.

in Figure 15(a)). If the defect concentration exceeds the original acceptor density, the near-surface region is converted to an n-type, and hence an n/p junction will be formed. This junction is not a standard pn-junction with a depletion layer sandwiched by two neutral regions, but a "partial junction" without an n-type neutral region. The n-type depletion region is directly brought into contact with a metal. As the positive space-charge due to the surface defect, $N_{defect}d$, increases, the downward band-bending in the surface region will be weakened and finally the energy band profile become concave as shown in Figure 15(b). Therefore, the energy barrier for a hole transport increases as the defect density increases. If $N_{\text{defect}} \gg N_A$ and $N_{\text{defect}}d \gg N_A W_{dep}$, the energy barrier for holes is raised approximately by [130],

$$\Delta \phi_{Bp} \cong \frac{q N_{\text{defect}} d^2}{2 \varepsilon_S}. \tag{28}$$

This enhancement of the energy barrier reduces the diode current in both forward and reverse directions, and hence the SBH determined by the I-V method becomes apparent and high. The internal photoemission spectrum shifts toward higher energy. On the other hand, the n^+ layer near the surface will expand the depletion-layer width and cause a decrease in the capacitance with an upward shift of the C^{-2}-V curve. Therefore, the extrapolated built-in-potential becomes higher than the real SBH.

Figure 16 shows typical I-V characteristics of the Schottky barrier diodes fabricated on (a) n-type and (b) p-type GaAs before and after an Ar plasma exposure at room temperature [113, 134]. The effective Schottky barrier height (SBH) was lowered for the n-type substrates (0.74 → 0.57 eV) and vice versa for the p-type substrates (0.61 → 0.88 eV). Bombardment of ions in plasma onto the GaAs surface is considered to be responsible for producing deficiency of arsenic atoms in the surface layer of GaAs [138, 171]. During sputtering, the semiconductor surface is also exposed to impinging ions, and a number of papers have reported the formation of donor-type defects in GaAs [2, 30, 172], InP [161] and Si [53, 96]. Therefore, the I-V curves as shown in Figure 16 can be attributed to the donor-type defect formation. We can estimate the depth (d) and the concentration (N_{defect}) of the donor-type defect in the subsurface region based on the energy band diagram as shown in Figure 15. Their typical values derived form the experimental results as shown in Figure 16 are 1–5 nm and 10^{20}–10^{21} cm^{-3}, respectively. Mullins and Brunnschweiler also estimated the similar quantities for the surface donor introduced in Mo/Si diodes fabricated by 1 kV sputtering, and obtained $d = 1 - 10$ nm and $N_{defect} = 10^{20} - 10^{21}$ cm^{-3} [96].

Similar changes in the I-V characteristics of GaAs Schottky diodes were reported and discussed in relation to electron-beam deposition [106]. An usual electron beam evaporator contains a hot filament emitting thermal

(a) *I-V* characteristics of plasma-irradiated n-GaAs

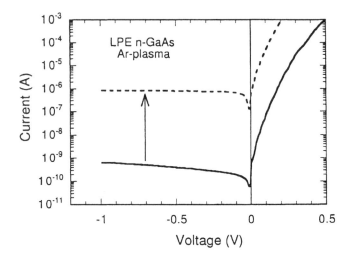

(b) *I-V* characteristics of plasma-irradiated p-GaAs

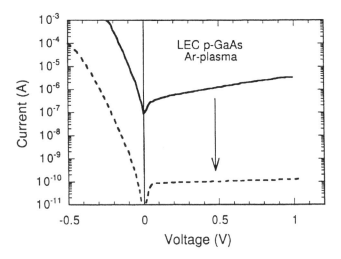

Fig. 16. Current-voltage (I-V) characteristics of the Al-Schottky diodes fabricated on (a) n-type and (b) p-type GaAs before (solid lines) and after (dashed lines) Ar-plasma exposure. These changes upon plasma exposure are attributed to introduction of donor-type surface defects: case (a) and (b) in Table II, respectively, for (a) n-type and (b) p-type in this figure. Improvement of the reverse characteristic after plasma exposure can be attributed to the effect of dopant deactivation classified to (d) in Table II.

electrons, which are accelerated through 3–10 keV. During electron-beam heating, optical radiation, x-ray emission, and stray or backscattered electrons can come from the E-gun source. Kleinhenz *et al.* [64] and Auret and his coworkers [98] observed electron traps in *n*-GaAs after EB deposition of various metals. They suggested that the EB-induced traps were not the consequence of x-rays as speculated previously, [40, 44, 101] but were caused by stray or backscattered electrons from the E-gun during deposition. However, the minimum energy of the electron beam for producing the primary defects in GaAs by elastic scattering was estimated to be 220–270 keV [27, 121]. It is therefore suggested that the defect species introduced in the subthreshold case are not the same as those in the case of above threshold energy electrons [6, 100]. Frenkel pair formation is not the mechanism and still controversial for defect creation in the subthreshold case. The excitations of L- or K-shell excitation are speculated to be responsible for the defect creation [173, 181].

4.4. Defects or Impurities in Sub-μm Region From the Surface

Some defects or impurities have an energy level deep in the energy gap. If their charge state is opposite to the dopant, they act as a charge compensator, which decreases the carrier density in a bulk region and the space charge density in a depletion region. In the I-V measurement of the Schottky diodes with such deep levels, the thermionic emission component does not change very much (see Table II(c)). In general, the region in which the dopant is compensated shows high resistivity. Due to the region of increased resistivity, the effect of a series resistance appears at high current level. The defects introduced in the subsurface region of the semiconductor give rise to the generation current component in a reverse-bias region of the Schottky diode. When the Schottky barrier height is high enough ($q\phi_B > E_G/2$) for a sufficient amount of minority carriers at the metal-semiconductor interface, the minority carrier injection occurs even in the Schottky contact [52]. Therefore, the recombination current component might be visible in the forward I-V curve of the contacts with high SBH. Generally, an efficient recombination center sits near the middle of the bandgap, and pulls the Fermi level toward itself. If such a compensated and high-resistive region exists between the metal and depletion region, the measured capacitance will be decreased with concomitant raise of the C^{-2}-V curve. Consequently, the estimated built-in-voltage becomes larger than the real one. In a lower bias-voltage region of the C^{-2}-V curve, the effect of the compensation might be seen. In the IPE spectrum, the slope, i.e., the photoemission yield, might be decreased.

The reverse I-V curves shown in Figure 16 are improved after plasma exposure from the viewpoint of a saturation characteristic. The deep levels discussed above cannot account for such a good saturation characteristic. Then, defects or impurities deactivating dopants in Table II(d) will be considered next. During dry etching and plasma-assisted deposition of dielectric's, hydrogen is well known as an impurity inevitably incorporated into semiconductors [120]. Hydrogen deactivates or passivates the activity of dopants in semiconductors. Due to deactivation, the carrier density and the space charge density in a depletion region is decreased, apparently just like the compensation case; so is the C^{-2}-V curve. In the I-V characteristics, the effect of the dopant deactivation on the thermionic-emission current is quite similar to (c). However, hydrogen is also known as a deactivator of deep-level defects [120]. If the generation-recombination centers in the original crystal are deactivated by hydrogen, the reverse characteristic of the diode might be improved. In fact, the saturation characteristic of the reverse current in Si-Schottky diodes can be improved by introduction of low-energy atomic hydrogen [4].

4.5.   Amorphous Surface Layer

Sputtering is a well-known technique for the deposition of good adherence metal layers, particularly refractory metals (W, Mo, Ti etc.) and their silicide or nitride (WSi, WN, WSiN, TiN) on compound semiconductors. An amorphous surface layer [2, 102] or a highly non-stoichiometric layer [138, 171], were identified on GaAs suffered from ion bombardment. At an ion energy of 100–200 eV and a total dose around 10^{15} ion/cm^2, an amorphous layer 2 to 3 nm deep is produced on GaAs surfaces [41]. Disordering of the subsurface layer is suggested to have a significant effect on the carrier transport through the metal-semiconductor interface. Some experiments tend to show that radiation damage produces semi-insulating layers [37, 160]. Loss of exponential relationships in the I-V curves were reported in the Schottky diodes with ion-sputtered GaAs surfaces (see Table II(e)) [2, 180]. Trap-assisted tunneling, hopping or space-charge limited currents are considered in order to account a relationship such as $I \propto V^m$ [149, 180]. As the thickness or the resistance of the disordered subsurface layer increases, the contact will become just like an MIS structure. Therefore, this resistance absorbs an appreciable amount of the voltage drop between the diode terminals, and the slope of the C^{-2}-V curve will be decreased; the resultant impurity concentration will apparently increase.

4.6.   Electrochemical Fabrication Process of Schottky Contacts Free From Process Damage

Formation of such imperfections near the metal-semiconductor interface is greatly dependent on the process of metal deposition. On the other hand, various efforts to suppress the formation of extrinsic defects on the compound semiconductor surface have been made, because the mechanism of Schottky barrier formation has been of fundamental interest in semiconductor physics [9]. Low temperature (LT) deposition of metals in an ultrahigh vacuum is one of attractive approaches to distinguish between intrinsic and extrinsic mechanisms, and a number of interesting results have been reported [146, 163]. On the other hand, electrochemical deposition of metals, namely electroplating, is an alternative process to the LT deposition in order to form good Schottky contacts to compound semiconductors. Heat of condensation [144] and electron or ion bombardment effects [2, 106], which are regarded as serious problem during vacuum deposition, are not significant during electroplating. Allongue et al. demonstrated that good Schottky contacts to n-GaAs have been fabricated by electroplating [1]. However, it was not obvious in their experiments whether the GaAs substrate surface was free from oxides before metallization.

Okumura and his coworkers demonstrated that in situ anodic etching was effectively used in the electrolytic solutions to remove surface oxides on GaAs just prior to the electrochemical deposition of metals [61, 109, 110]. The resultant Schottky diodes showed nearly ideal characteristics in current-voltage curves. They determined the conditions in which the GaAs surface becomes free from oxides in electrolytic solutions for electroplating. The oxide thickness on the GaAs surface is precisely controlled by photo-electrochemical process. The results indicate that the anodic potential for the in situ anodic etching is critical for complete removal of the oxides. The photocurrent transient measurements can be helpful in determining the applied potentials for photoanodic etching. Figure 17 shows surface images and tunnel spectra taken by electrochemical scanning tunneling microscopy (EC-STM) for the GaAs surfaces (a) with and (b) without residual oxides. In an acidic solution of about pH=2, the applied potential to n-GaAs of +0 V (vs. Ag/AgCl reference) provides a smooth and oxide free (no gap-state) surface, while +0.5 V gives a rough surface covered with anodic oxides.

Figure 18 shows the I-V plot for the Ni Schottky contact metallized on n-GaAs surface just after anodic etching at +0.1 V. The plot is also shown in the manner proposed by Missous and Rhoderick (see equation (4)) [85]. Linearity in a semilogalithmic plot holds over 25 orders of magnitude. The determined n-value was 1.00 ± 0.004, which is expected for lightly doped GaAs ($n = 1 \times 10^{16}$ cm^{-3}) used here. Electrical abruptness of this interface can be verified by coincidence in the barrier heights determined through three independent methods (I-V, C-V and J_S-T): 0.80 ± 0.013 eV, 0.81 ± 0.015 eV, and 0.80 ± 0.015 eV, respectively. The J_S-T measurement gives the value of $A^{**} = 7.03 - 8.65$ Acm^{-2}K^{-2}, which is very close to the theoretical value of 8.16 Acm^{-2}K^{-2} [122]. This fact also suggests that the fabricated metal-semiconductor interface consisted of very thin, if any, intervening layer.

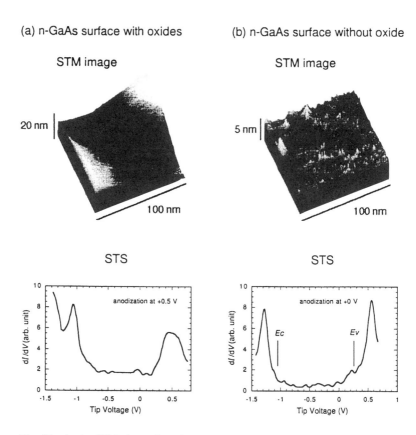

Fig. 17. *In situ* STM observation of anodized *n*-GaAs surfaces in Ni-salt solutions with pH=2.0. (a) If anodic reaction is carried out at a GaAs potential of +0.5 V (vs. Ag/AgCl), the resultant GaAs surface becomes rough and covered with native oxides. (b) When choosing a proper GaAs potential, such as +0.1 V (vs. Ag/AgCl), STS shows no gap state on the GaAs surface. The Schottky contact can be formed *in situ* on such an electrochemically etched GaAs surface, which might be free from process-related damage and native oxides.

The same photoelectrochemical process was applied to fabrication of Sn, Pb and Au Schottky contacts to *n*-GaAs. The electrical abruptness for these contacts were also verified in terms of coincidence in the Schottky barrier heights determined through *I-V* and *C-V* measurements. The Schottky barrier heights were 0.68 eV, 0.74 eV and 0.88 eV, respectively. In the analysis of chemical trends of Schottky barrier height, the slope $S_\phi = \partial\phi_{Bn}/\partial\phi_M$, becomes about 0.1, which is close to the value of 0.102 calculated by using equation (14) and not inconsistent with the MIGS model. However, the values of SBH's are by about 0.1 eV lower than Mönch's prediction [91, 92].

(a) *I-V* curve

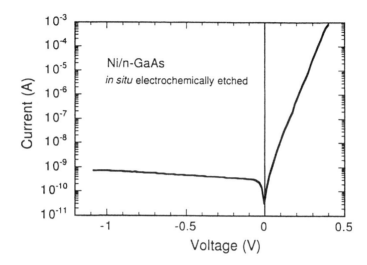

(b) Rhoderick plot of *I-V* curve

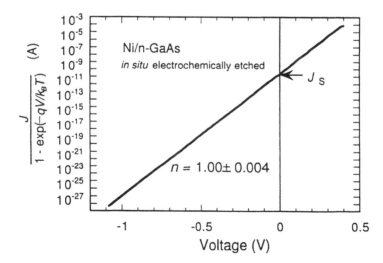

Fig. 18. Current-voltage characteristics of Ni/n-GaAs fabricated by *in situ* photoelectrochemical etching and succeeded electrodeposition technique. (a) the conventional log(J) vs. V plot and modified plot based on Equation (4) in the text. The resultant ideality factor (n-value) is very close to unity (the substrate donor concentration is as low as 1×10^{16} cm^{-3}).

Wu *et al.* tried to apply a similar process to fabrication of GaAs Schottky diodes [177]. They obtained work-function independence of the SBH (about 1.0 eV) for Ag, Cu, Co, Pd and Pt. The discrepancy between two results is still open for question; strain at the interface and/or hydrogen incorporation into GaAs from electrolytic solution might be the cause of the difference.

5. SUMMARY

In conclusion, we have examined the effects of inhomogeneity in the metal-semiconductor interfaces as well as imperfections in the sub-surface region of semiconductors on the characteristics of the Schottky contacts to compound semiconductors. For practical contacts for device application, processing issues are responsible for the electrical performance of the contact rather than the mechanism of the Fermi-level pinning at the interface. In addition to finding new contact materials which form a stable and uniform interface with compound semiconductors, it is still challenging to develop advanced metallization processing technology free from introducing damage. Electrical inhomogeneity in the Schottky contact with a thick metal can be evaluated in the order of μm scale by using IPEM, yet development of experimental techniques with a nm resolution and applicable to practical contacts is expected. In understanding the characteristics of practical semiconductor contacts, the experimentally observed deviations from the ideal characteristic should be interpreted by taking multiple effects relating to imperfections at or near the interface into account. Table II is indicative that application of two or three electrical characterization techniques, although they are neither modern nor surface sensitive tools, to the same sample can ascertain the imperfections introduced in the semiconductor surface, if any, during processing.

References

1. P. Allongue and E. Souteyrand, *J. Vac. Sci. & Technol.*, **B5**, 1644 (1987).
2. A. Amith and P. Mark, *J. Vac. Sci. Technol.*, **15**, 1344 (1978).
3. D.E. Aspnes and A. Heller, *J. Vac. Sci. Technol.*, **B1**, 602 (1983).
4. S. Ashok and K. Giewont, *Jpn. J. Appl. Phys.*, **24**, L533 (1985).
5. R. Audino, G. Destefanis, F. Gorgellino, E. Pollino and S. Tamagno, *Thin Solid Films*, **36**, 343 (1976).
6. F.D. Auret, L.J. Bredell, G. Myburg and W.O. Barnard, *Jpn. J. Appl. Phys.*, **30**, 80 (1991).
7. J. Bardeen, *Phys. Rev.*, **71**, 717 (1947).
8. P.M. Batev, M.D. Ivanovitch, E.I. Kafediiska and S.S. Simeonov, *Phys. Status Solidi a*, **45**, 671 (1978).

9. I.P. Batra (ed.), Metallization and Metal-Semiconductor Interfaces, *NATO ASI Ser. Ser. B*, **195**, Plenum Press, New York (1988).
10. E. Beam, III and D.D.L. Chung, *Thin Solid Films*, **128**, 321 (1985).
11. L.D. Bell and W.J. Kaiser, *Phys. Rev. Lett.*, **61**, 2368 (1988).
12. L.D. Bell, M.H. Hecht, W.J. Kaiser and L.C. Davis, *Phys. Rev. Lett.*, **64**, 2679 (1990).
13. O.Y. Borkovskaya, N.L. Dmitruk, R.V. Konakova and M.Y. Filatov, *Electron. Lett.*, **14**, 700 (1978).
14. L.J. Brillson, G. Margaritondo and N.G. Stoffel, *Phys. Rev. Lett.*, **44**, 667 (1980).
15. N. Braslau, *J. Vac. Sci. Technol.*, **19**, 803 (1981).
16. L.J. Brillson, *Thin Solid Films*, **89**, 461 (1982).
17. L.J. Brillson, *Contacts to Semiconductors: Fundamentals and Technology*, ed. by L.J. Brillson, p.333 (Noyes Pub., New Jersey, 1993).
18. J.A. Calviello and J.L. Wallace, *IEEE Trans. Electron Devices*, **ED-24**, 698 (1977).
19. A. Christou and H.M. Day, *Appl. Phys. Lett.*, **47**, 4217 (1976).
20. C.C. Chang and G. Quitana, *Thin Solid Films*, **31**, 265 (1976).
21. D.J. Chidi, *Surf. Sci.*, **99**, 1 (1980).
22. H. Cheng, X.-J. Zhang and A.G. Milnes, *Solid-State Electron.*, **27**, 1117 (1984).
23. C.Y. Chang and F. Kai (eds.), *GaAs High-Speed Devices* (John Wiley & Sons, New York, 1994).
24. C.-P. Chen, Y.A. Chang, J.-W. Huang and T.F. Kuech, *Appl. Phys. Lett.*, **64**, 1413 (1994).
25. A.M. Cowley and S.M. Sze, *J. Appl. Phys.*, **36**, 3212 (1965).
26. J. Comas and C.B. Cooper, *J. Appl. Phys.*, **37**, 2820 (1966).
27. J.W. Corbett and J.C. Bourgoin, *Defect creation in semiconductors, in Point defects in solids*, **2** (eds. J.H. Crawford, Jr. and L.M. Slifkin), (Plenum, New York, 1975).
28. E.D. Cole, S. Sen and L.C. Burton, *J. Electron. Mater.*, **18**, 527 (1989).
29. W.J. Devlin, C.E.C. Wood, R. Stall and L.F. Eastman, *Solid-State Electron.*, **23**, 823 (1980).
30. M.E. Edweeb, E.J. Charlson and E.M. Charlson, *Appl. Phys. Lett.*, **43**, 563 (1983).
31. S.J. Eglash, M.D. Williams, P.H. Mahowald, N. Newman, I. Lindau and W.E. Spicer, *J. Vac. Sci. Technol.*, **B2**, 481 (1984).
32. R.L. Farrow, R.K. Chang, S. Mcroczkowski and F.H. Pollak, *Appl. Phys. Lett.*, **31**, 768 (1977).
33. C. Feldman, *Phys. Rev.*, **117**, 455 (1960).
34. A. Fernandez, H.D. Hallen, T. Huang, R.A. Buhrman and J. Silcox, *J. Vac. Sci. Technol.*, **B9**, 590 (1991).
35. T. Figielski, *Appl. Phys.*, **A35**, 255 (1984).
36. R.F. Fowler, *Phys. Rev.*, **38**, 45 (1931).
37. A. Foyt, W. Lindley, J. Donnelly and C. Wolfe, *Solid-State Electron.*, **12**, 209 (1969).
38. C. Fontaine, T. Okumura and K.N. Tu, *J. Appl. Phys.*, **54**, 1404 (1983).
39. J.L. Freeouf and J.M. Woodall, *Appl. Phys. Lett.*, **39**, 727 (1981).
40. K.F. Galloway and S. Mayyo, *Solid State Technol.*, **22**, 96 (1979).
41. P. Guétin and G. Schréder, *J. Appl. Phys.*, **43**, 549 (1972).
42. H.R. Grinolds and G.Y. Robinson, *Appl. Phys. Lett.*, **34**, 575 (1979).
43. S. Guha, B.M. Arora and V.P. Salvi, *Solid-State Electron.*, **20**, 431 (1977).
44. M. Hamasaki, *Solid-State Electron.*, **26**, 299 (1983).
45. H. Hasegawa and H. Ohno, *J. Vac. Sci. Technol.*, **B4**, 1130 (1986).
46. H. Hasegawa, Li He, H. Ohno, T. Sawada, T. Haga, Y. Abe and H. Takahashi, *J. Vac. Sci. Technol.*, **B5**, 1097 (1987).
47. F. Hasegawa, M. Onomura, C. Mogi and Y. Nannichi, *Solid-State Electron.*, **31**, 223 (1988).
48. H. Hasegawa, H. Ohno, H. Ishii, T. Haga, Y. Abe and H. Takahashi, *Appl. Surf. Sci.*, **41/42**, 372 (1989).
49. Y. Hasegawa, K. Kuk, R.T. Tung, P.J. Silverman and T. Sakurai, *J. Vac. Sci. Technol.*, **B9**, 578 (1991).
50. H. Hasegawa, T. Sato and T. Hashizume, *J. Vac. Sci. Technol.*, **B15**, 1227 (1997).
51. V. Heine, *Phys. Rev.*, **A138**, 1689 (1965).
52. H.K. Henish, *Semiconductor contacts: An approach to ideas and models*, p.146 (Clarendon, Oxford, 1984).

53. G.J.A. Hellings, A. Straayer and A.H.M. Kipperman, *J. Appl. Phys.*, **57**, 2067 (1985).
54. E. Ikeda, H. Hasegawa, S. Ohtsuka and H. Ohno, *Jpn. J. Appl. Phys.*, **27**, 180 (1988).
55. A. Ismail, A. Brahim, J.M. Palau and L. Lasabatere, *Surf. Sci.*, **168**, 409 (1986).
56. A. Ismail, A. Brahim, J.M. Palau and L. Lasabatere, *Vacuum*, **36**, 217 (1986).
57. R.L. Jacobson and G.K. Wehner, *J. Appl. Phys.*, **36**, 2674 (1965).
58. N.M. Johnson, T.J. Magee and J. Peng, *J. Vac. Sci. Technol.*, **13**, 838 (1976).
59. K. Kajiyama, S. Sakata and O. Ochi, *J. Appl. Phys.*, **46**, 3221 (1975).
60. W.J. Kaiser and L.D. Bell, *Phys. Rev. Lett.*, **60**, 1406 (1988).
61. C. Kaneshiro, M. Shimura and T. Okumura, *Control of Semiconductor Interfaces*, p.181 (Elsevier, Amsterdam, 1994).
62. C. Kaneshiro and T. Okumura, *J. Vac. Sci. Technol.*, **B15**, 1595 (1997).
63. H.B. Kim and G.G. Sweeney, Gallium Arsenide and Related Compounds 1974, *Inst. Phys. Conf. Ser.*, **24**, p.307 (1975).
64. R. Kleinhenz, P.M. Mooney, C.P. Schneider and O. Paz, *13th International Conference on Defects in Semiconductors*, ed. L.C. Kimerling and J. Parsey, p.627 (Coronado, CA, 1984).
65. S.P. Kowalczyk, J.R. Waldrop and R.W. Grant, *Appl. Phys. Lett.*, **38**, 167 (1981).
66. S. Kurtin, T.C. McGill and C.A. Mead, *Phys. Rev. Lett.*, **22**, 1433 (1969).
67. T.S. Kuan, P.E. Batson, T.N. Jackson, H. Rupprecht and W.L. Wilkie, *J. Appl. Phys.*, **54**, 6952 (1983).
68. H. Kukimoto and S. Miyazawa (eds.), *Semi-Insulating III–V Materials*, Hakone, 1986, pp.403–420 (Ohm & North-Holland, Tokyo, 1986).
69. H. Leamy, *J. Appl. Phys.*, **53**, R251 (1982).
70. S. Leung, L.K. Wong, C.L. Bauer and A.G. Milnes, *J. Electrochem. Soc.*, **132**, 898 (1983).
71. S. Leung, T. Yoshiie, D.D.L. Chung and A.G. Milnes, *J. Electrochem. Soc.*, **130**, 462 (1985).
72. M.S. Lin, W.H. Su, J.C. Lou and T.F. Lei, *Proceedings of the 14th Conference on Solid State Device*, Tokyo, 1982, p.397 (1983).
73. T.-C. Lin, H.T. Kaibe, C. Kaneshiro, S. Miyazaki and T. Okumura, *Proc. InP and Related Materials (IPRM5)*, Paris, p.691 (IEEE, New Jersey, 1992).
74. Z. Lilliental-Weber, E.R. Weber and N. Newman, *Contacts to Semiconductors: Fundamentals and Technology*, ed. by L.J. Brillson, p.416 (Noyes Pub., New Jersey, 1993).
75. S.G. Louie and M.L. Cohen, *Phys. Rev.*, **B13**, 2461 (1976).
76. S.G. Louie, J.R. Chelikowsky and M.L. Cohen, *Phys. Rev.*, **B15**, 2154 (1977).
77. E.J. van Loenen, J.W.M. Frenken, J.F. van der Veen and S. Valeri, *Phys. Rev. Lett.*, **54**, 827 (1985).
78. D.C. Look, *Electrical Characterization of GaAs Materials and Devices* (Wiley. Chichester, 1989).
79. R. Ludeke, T.-C. Chiang and T. Miller, *J. Vac. Sci. Technol.*, **B1**, 581 (1983).
80. C.J. Madams, D.V. Morgan and M.J. Howes, *Electron. Lett.*, **11**, 574 (1975).
81. K. Matsumoto, N. Hashizume, H. Tanoue and T. Kanayama, *Jpn. J. Appl. Phys.*, **21**, L393 (1982).
82. C.A. Mead and W.G. Spitzer, *Phys. Rev.*, **134**, A713 (1964).
83. T. Mimura, K. Joshin and S. Kuroda, *FUJITSU Scientific Tech. J.*, **19**, 243 (1983).
84. M. Missous, E.H. Rhoderick and K.E. Singer, *Electron. Lett.*, **22**, 241 (1986).
85. M. Missous and E.H. Rhoderick, *Electron. Lett.*, **22**, 477 (1986).
86. M. Missous, E.H. Rhoderick and K.E. Singer, *J. Appl. Phys.*, **59**, 3189 (1986).
87. Y. Miura, K. Hirose, K. Aizawa, N. Ikarashi and H. Okabayashi, *Appl. Phys. Lett.*, **61**, 1057 (1992).
88. S. Miyazaki, T. Okumura, Y. Miura and K. Hirose, *Inst. Phys. Conf. Ser.* No. 149, p.307 (IOP, Bristol, 1996).
89. S. Miyazaki, T.-C. Lin, C. Nishida, H.T. Kaibe and T. Okumura, *J. Electronic Mater.*, **25**, 577 (1996).
90. N.F. Mott, *Proc. Cambr. Phil. Soc.*, **34**, 568 (1938).
91. W. Mönch, *Phys. Rev.*, **B37**, 7129 (1988).
92. W. Mönch, Metallization and Metal-Semiconductor Interfaces (NATO ASI Series B; Physics, **195**), ed. I.P. Batra, p.11 (Plenuim, New York, 1989).

93. W. Mönch, *Control of Semiconductor Interfaces*, eds. I. Ohdomari, M. Oshima and A. Hiraki, p.169 (Elsevier, Amsterdam, 1994).
94. S.P. Murarka, *Solid-State Electron.*, **17**, 869 (1974).
95. S.P. Murarka, *Solid-State Electron.*, **17**, 985 (1974).
96. F.H. Mullins and A. Brunnschweiler, *Solid-State Electron.*, **19**, 47 (1976).
97. M. Murakami, K.D. Childs, J.M. Baker and A. Callegari, *J. Vac. Sci. Technol.*, **B4**, 903 (1986).
98. G. Myburug and F.D. Auret, *J. Appl. Phys.*, **71**, 6172 (1992).
99. N. Newman, T. Kendelewicz, L. Bowman and W.E. Spice, *Appl. Phys. Lett.*, **46**, 1176 (1982), 922 (1983).
100. M. Nel and F.D. Auret, *Jpn. J. Appl. Phys.*, **28**, 2430 (1989).
101. T.H. Ning, *J. Appl. Phys.*, **49**, 4077 (1978).
102. O. Oelhafen, *J. Vac. Sci. Technol.*, **B1**, 787 (1983).
103. I. Ohdomari, T.S. Kuan and K.N. Tu, *J. Appl. Phys.*, **50**, 7020 (1979).
104. I. Ohdomari and K.N. Tu, *J. Appl. Phys.*, **51**, 3735 (1980).
105. T. Okumura and K.N. Tu, *J. Appl. Phys.*, **54**, 922 (1983).
106. T. Okumura and K.N. Tu, *J. Appl. Phys.*, **61**, 2955 (1987).
107. T. Okumura and T. Shibata, *Jpn. J. Appl. Phys.*, **27**, 2119 (1988).
108. T. Okumura and K. Shiojima, *Jpn. J. Appl. Phys.*, **28**, L1108 (1989).
109. T. Okumura, S. Yamamoto and M. Shimura, *Jpn. J. Appl. Phys.*, **32**, 2626 (1993).
110. T. Okumura, C. Kaneshiro, M. Shimura and S. Yamamoto, *Electrochemical Microfabrication, ECS Proceedings*, **94–32**, M. Datta, K. Sheppard and J.O. Dukovic (eds.), p.374 (*Electrochemical Soc.*, Pennington, 1995).
111. T. Okumura, *Inst. Phys. Conf. Ser.*, No. 149, p.287 (IOP, Bristol, 1996).
112. T. Okumura, Sci. Rep. of Research Institute of Tohoku University, **A44**, 187 (1997).
113. T. Okumura, SOTAPOCS XXXIII, *ECS Proceedings*, **98-2**, H.Q. Hou, R.E. Sah, S.J. Pearton, F. Ren and K. Wada (eds.), p.400 (*Electrochemical Soc.*, Pennington, 1998).
114. A. Olbrich, J. Vanceaq, F. Kreupl and H. Hoffmann, *Appl. Phys. Lett.*, **70**, 2559 (1997).
115. F.A. Padovani and R. Stratton, *Solid-State Electron.*, **9**, 695 (1966).
116. J.M. Palau, E. Testemale and L. Lassabatere, *J. Vac. Sci. Technol.*, **19**, 202 (1981).
117. J.M. Palau, E. Testemale, A. Ismail and L. Lassabatere, *J. Vac. Sci. Technol.*, **21**, 6 (1982).
118. H. Palm, M. Arbes and M. Schulz, *Phys. Rev. Lett.*, **71**, 2224 (1993).
119. S.J. Pearton, U.K. Chakrabarti, A.P. Perley and K.S. Jones, *J. Appl. Phys.*, **68**, 2760 (1990).
120. J. Pearton, W. Corbett and M. Stavola, *Hydrogen in Crystalline Semiconductors*, (Springer, Berlin, 1992).
121. D. Pons and J.C. Bourgoin, *J. Phys. C: Solid State Phys.*, **18**, 3839 (1985).
122. E.H. Rhoderick and R.H. Williams, *Metal-Semiconductor Contacts*, 2nd ed. (Clarendon Press, Oxford, 1988).
123. I.K. Robinson, R.T. Tung and R. Feidenhans'l, *Phys. Rev.*, **B38**, 3632 (1988).
124. W. Schottky, *Naturwiss.*, **26**, 843 (1938).
125. M. Schulülter, *Phys. Rev.*, **B17**, 5044 (1978).
126. G.P. Schwartz, G.L. Gualtieri, J.E. Griffiths, C.D. Thurmond and B. Schwartz, *J. Electrochem. Soc.*, **127**, 2488 (1980).
127. G.P. Schwartz and G.J. Gualtieri, *J. Electrochem. Soc.*, **133**, 1021 (1986).
128. A.S. Sedra and K.C. Smith, *Microelectronic Circuits*, 4th ed. p.604 (Oxford University Press, New York, 1998).
129. J.M. Shannon, *Appl. Phys. Lett.*, **24**, 369 (1974).
130. J.M. Shannon, *Appl. Phys. Lett.*, **25**, 75 (1974).
131. Y.C. Shih, M. Murakami, E.L. Wilkie and A. Callegari, *J. Appl. Phys.*, **62**, 582 (1987).
132. K. Shiojima and T. Okumura, *Jpn. J. Appl. Phys.*, **30**, 2127 (1991).
133. K. Shiojima and T. Okumura, *Proc. IEEE International Reliability Physics Symposium*, Las Vegas, p.234 (1991).
134. S.-Q. Shao, H.-T. Kaibe and T. Okumura, *Inst. Phys. Conf. Ser.*, No. 129, p.657 (IOP, Bristol, 1993).
135. A.K. Sinha and J.M. Poate, *Appl. Phys. Lett.*, **23**, 666 (1973).
136. A.K. Sinha, T.E. Smith, M.H. Read and J.M. Poate, *Solid-State Electron.*, **19**, 489 (1976).

137. A.K. Sinha and J.M. Poate, *This films–Interdiffusion and Reactions*, eds J.M. Poate, K.N. Tu and J.W. Mayer, Chap. 11 (Wiley, New York, 1978).

138. I.L. Singer, *Surf. Sci.*, **7**, 108 (1981).

139. H. Sirringhaus, E.Y. Lee, H. von Känel, *J. Vac. Sci. Technol.*, **B12**, 2629 (1994).

140. H. Sirringhaus, E.Y. Lee, U. Kafedar and H. von Känel, *J. Vac. Sci. Technol.*, **B13**, 1848 (1995).

141. W.E. Spicer, I. Lindau, P.E. Gregory, C.M. Garner, P. Pianetta and P.W. Chye, *J. Vac. Sci. Technol.*, **13**, 780 (1976).

142. W.E. Spicer, P.W. Chye, P.R. Skeath, C.Y. Su and I. Lindau, *J. Vac. Sci. Technol.*, **16**, 1426 (1979).

143. W.E. Spicer, P.W. Chye, P.R. Skeath, C.Y. Su and I. Lindau, *J. Vac. Sci. Technol.*, **17**, 1019 (1980).

144. W.E. Spicer, I. Lindau, P.R. Skeath, C.Y. Su and P.W. Chye, *Phys. Rev. Lett.*, **44**, 420 (1980).

145. W.E . Spicer, Z. Liliental-Weber, E. Weber, N. Newman, T. Kendelewicz, R. Cao, C. McCants, P. Mahowald, K. Miyano and I. Lindau, *J. Vac. Sci. Technol.*, **B6**, 1245 (1988).

146. K. Stiles, A. Kahn, D. Kilday and G. Margaritondo, *J. Vac. Sci. & Technol.*, **B5**, 987 (1987).

147. T. Sugino, Y. Sakamoto, T. Sumiguchi, K. Nomoto and J. Shirafuji, *Jpn. J. Appl. Phys.*, **32**, L1196 (1993).

148. S.M. Sze, *Physics of Semiconductor Devices*, 2nd ed. (Wiley, New York, 1981).

149. P.C. Tayler and D.V. Morgan, *Solid-State Electron.*, **19**, 473 (1976).

150. C. Tejedor, F. Flores and E. Louis, *J. Phys.*, **C10**, 2163 (1977).

151. J. Tersoff, *Phys. Rev. Lett.*, **52**, 465 (1984).

152. J. Tersoff, *Phys. Rev.*, **B32**, 6968 (1985).

153. C.D. Thurmond, G.P. Schwartz, G.W. Kammlott and B. Schwartz, *J. Electrochem. Soc.*, **127**, 1366 (1980).

154. E. Tokumitsu, *Jpn. J. Appl. Phys.*, **29**, L698 (1990).

155. R.T. Tung, J.M. Poate, J.C. Bean, J.M. Gibson and D.C. Jacobson, *Thin Solid Films*, **93**, 77 (1982).

156. R.T. Tung, *Phys. Rev. Lett.*, **52**, 461 (1984).

157. R.T. Tung, *Appl. Phys. Lett.*, **58**, 2821 (1991).

158. R.T. Tung, *Phys. Rev.*, **B45**, 13509 (1992).

159. M.S. Tyagi, *Proceedings of the 8th Conference on Solid State Device*, Tokyo, 1976 (1977), p.333.

160. R.G. Unsperger, O.J. Marsh and C.A. Mead, *Appl. Phys. Lett.*, **13**, 295 (1968).

161. L.M.O. Van Den Berghe, R.L. Van Merhaeghe, W.H. Laflere and F. Cardon, *Solid-State Electron.*, **29**, 1109 (1986).

162. E. Vlieg, A.E.M.J. Fischer, J.F. van der Veen, B.N. Dev and G. Materlik, *Surf. Sci.*, **178**, 36 (1986).

163. R.E. Viturro, S. Chang. J.L. Shaw. C. Mailhiot, L.J. Brillson, A. Terrasi, Y. Hwu, G. Margaritondo, P.D. Kirshner and J.M. Woodall, *J. Vac. Sci. & Technol.*, **B7**, 1007 (1989).

164. O. Wada, S. Yanagisawa and H. Takanashi, *Jpn. J. Appl. Phys.*, **12**, 1814 (1973).

165. O. Wada, S. Yanagisawa and H. Takanashi, *Appl. Phys. Lett.*, **29**, 263 (1976).

166. O. Wada and A. Majorfeld, *Electron. Lett.*, **14**, 125 (1978).

167. J.R. Waldrop, S.P. Kowalczyk and R.W. Grant, *J. Vac. Sci. Technol.*, **21**, 607 (1982).

168. J.R. Waldrop, *Appl. Phys. Lett.*, **41**, 352 (1982).

169. J.R. Waldrop, *J. Vac. Sci. Technol.*, **B2**, 445 (1984).

170. J.R. Waldrop, *Appl. Phys. Lett.*, **44**, 1002 (1984).

171. Y.-X. Wang and P.H. Holloway, *J. Vac. Sci. Technol.*, **B2**, 613 (1984).

172. Y.G. Wang and S. Ashok, *J. Vac. Sci. Technol.*, **A6**, 1548 (1988).

173. T. Wada, T. Kanayama, S. Ichimura, Y. Sugiyama and M. Komuro, *Jpn. J. Appl. Phys.*, **33**, 7228 (1994).

174. E.R. Weber, H. Ennen, V. Kaufmann, J. Windschief, J. Schneider and T. Wosinski, *J. Appl. Phys.*, **53**, 6140 (1982).

175. T. Wilson, W.R. Osicki, J.N. Gannaway and G.R. Booker, *J. Mater. Sci.*, **14**, 961 (1979).

176. C.S. Wu, D.M Scott, W.-X. Chen and S.S. Lau, *J. Electrochem. Soc.*, **132**, 918 (1985).

177. N.-J. Wu, H. Hasegawa and T. Hashizume, *Jpn. J. Appl. Phys.*, **33**, 936 (1994).

178. N.-J. Wu, T. Hashizume, H. Hasegawa and Y. Amemiya, *Jpn. J. Appl. Phys.*, **34**, 1162 (1995).
179. Y. Yamada, E. Tokumitsu, K. Saito, T. Akatsuka, M. Miya, M. Konagai and K. Takahashi, *J. Crystal. Growth*, **95**, 145 (1989).
180. L.L.-M. Yeh, Y.-J. Xie and P.H. Holloway, *J. Appl. Phys.*, **65**, 3568 (1989).
181. M.A. Zaikovskaya, A.E. Kiv and O.R. Niyazova, *Phys. Stat. Sol. A*, **3**, 99 (1970).
182. D. Coulman, N. Newman, G.A. Reid, Z. Liliental-Weber, E.R. Weber and W.E. Spicer, *J. Vac. Sci. & Technol.*, **A5**, 1521 (1987).

CHAPTER 7

Electrical Characterization of Defects Introduced in Epitaxially Grown GaAs by Electron-, Proton- and He-Ion Irradiation

F.D. AURET

Physics Department, University of Pretoria,
Pretoria 0002, South Africa

1. INTRODUCTION

Radiation of semiconductors has been studied for several decades and the two main thrusts of these investigations were, firstly, determining the effects of irradiation on semiconductor properties as a function of projectile type, energy and dose and, secondly, establishing the properties and structure of the defects introduced during irradiation. The results obtained from these studies served as a basis for several areas in semiconductor materials modification and device fabrication, such as doping, isolation and carrier lifetime control.

In many early studies, the focus was on the undesirable effects of radiation induced defects in electronic devices which have to function in radiation environments, for example, near radiation sources or in outer space where they are exposed to cosmic rays. Radiation induced defects were also shown to be particularly harmful during doping by ion implantation, where they compensate the doping effect. Some insight into the significance of this problem may be gained by considering that, when doping GaAs by implantation of light ions, for example Be^+ at 1 MeV, approximately 200 carriers are removed per implanted ion [1], i.e. significantly more than the

4 carrier per 150–200 keV protons [1]. Thus, even if all the implanted doping ions become electrically active, one residual defect per implanted ion is sufficient to totally compensate the doping effect. Consequently, much of the early research was directed towards determining the optimum annealing conditions required to remove the radiation damage, and thus allow the implanted atoms to dope the semiconductor. Whilst solving this problem, it was realized that the defects introduced during ion implantation could be valuable for device isolation applications, and subsequently there was a shift in emphasis towards optimizing ion implantation induced isolation procedures. In GaAs, for example, particle radiation was found to introduce defects which were effective compensating centers, stable at room temperature, and, if present in high enough concentrations, able to convert doped GaAs into its semi-insulating form [2, 3].

The realization that radiation induced defects can beneficiate electronic materials and devices by altering their properties, lead to new fields of research and novel applications. In the development of new processes for fabrication of Si integrated circuits, the use of ion implantation with energies in the MeV range has steadily increased. This includes, for example, formation of retrograde wells [4], luminescence centers for electro-optical devices [5], conductive silicides and insulating oxides and nitrides [6, 7]. Moreover, irradiation with high energy electrons, protons, alpha-particles and heavier ions has received growing interest as a technique to control the carrier lifetime in silicon power devices [8, 9]. Compared to gold diffusion traditionally used for lifetime control, irradiation can provide either a uniform lifetime tailoring (when using MeV electrons), or a limited-depth lifetime alteration (when employing keV – MeV ions). Localised layers with a high density of lifetime "killers" can be created at any depth within the device by varying the ion energy [10]. Most of these experiments have been conducted using Si, and in comparison, little attention has been paid to altering the properties, for whichever purpose, of GaAs. It has, however, recently been suggested that metastable defects created by particle irradiation of GaAs, may be significant in realizing optical non-linear devices [11].

For successfully realizing these applications, it is imperative to understand the effect of radiation on electronic materials and devices fabricated on them, firstly, in order to avoid the deleterious effects of some defects and, secondly, to utilize the beneficial effects of others, depending on the application. To fully understand the role of any defect in a particular electronic or optoelectronic device, several properties of the defect need to be known, namely its energy level(s) in the band gap, concentration, and thermal and optical capture and emission rates for electrons and holes [12]. This allows the effect of a defect on the properties of electronic materials and devices to be predicted by modelling. In addition, the structure, introduction rate, introduction mechanism and thermal stability of the defect should be determined so that it can be reproducibly introduced, avoided or eliminated, depending on the application.

Deep level transient spectroscopy (DLTS) [13, 14] has been instrumental in determining most of the above mentioned properties of defects introduced in several different semiconductors during crystal growth, radiation with different particle types and processing during device fabrication. DLTS has also provided important information pertaining to the behaviour of defects in electric and magnetic fields and under uniaxial stress. The latter type of study provides valuable information about the symmetry of defects, which in turn aids in determining their structure. In the case of GaAs, electron and proton irradiation have attracted the most interest and comprehensive reviews of electron irradiation induced defects have appeared [15, 16]. Finally, DLTS is particularly attractive because it can be used to characterize defects using various kinds of space charge based devices, ranging from simple Schottky barrier diodes (SBDs) and p-n junctions to device structures with higher degrees of complexities [17].

In Section 2, the terminology and key concepts underlying the field of particle-solid interaction are introduced, with special reference to the irradiation of GaAs with protons and He-ions. In Section 3, a description of some characteriztics of deep level defects and the DLTS technique is presented. Sections 4–6 deal with the defects introduced in GaAs by electron-, proton- and He-ion irradiation, respectively. Metastable irradiation induced defects in GaAs are discussed in Section 7, while the effect of an electric field on the emission of carriers from radiation induced defects in n- and p-GaAs is the topic of Section 8. In Section 9, the effects of radiation induced defects on GaAs free carrier density and device properties are elucidated. The main results discussed in these sections are summarized in Section 10.

2. PARTICLE IRRADIATION OF SEMICONDUCTORS

When energetic particles enter a material, they are decelerated and the energy thus transferred to the material can modify its structure and properties. In the case of crystalline semiconductors, particle induced materials modification, traditionally known as radiation damage, will occur as long as the projectile particles can transfer energy, E, larger than the displacement energy, E_d, of the lattice atoms. The extent to which a material is modified, depends on the energy deposited per unit volume. The main energy-loss mechanisms of energetic ions in a solid are discussed below.

2.1. Energy-loss Mechanisms

Energetic particles primarily transfer energy by two mechanisms: (1) screened Coulomb collisions with target atoms (nuclear stopping), and (2) interactions with bound or free electrons in the solid (electronic stopping). The relative effect of the two mechanisms depends on the energy and the mass of the incident particle, as well as the mass of the target atoms. Figure 1 depicts the velocity ($\propto, \varepsilon^{1/2}$) dependence of the two energy-loss processes, nuclear stopping $\nu = (d\varepsilon/d\rho)_n$ and electronic stopping $(d\varepsilon/d\rho)_e$, in terms of the reduced energy, ε. Both ε and the reduced path length, ρ [18], are dimensionless and can be expressed in terms of laboratory units E and x as:

$$\varepsilon = \frac{4\pi\varepsilon_0 a M_2}{Z_1 Z_2 e^2 (M_1 + M_2)} E, \tag{1}$$

and

$$\rho = N\pi a^2 \frac{4M_1 M_2}{(M_1 + M_2)^2} x, \tag{2}$$

where M_1 and M_2, and Z_1 and Z_2 are the mass numbers and atomic numbers of the incident particle and target atoms, respectively, e is the electronic charge, N is the number of atoms per unit volume, ε_0 is the permittivity of free space and a is the screening radius. Several expressions exist for the screening radius [19, 20], with the most frequently used being that given by [21]:

$$a = \frac{0.8853 a_0}{\left(Z_1^{2/3} + Z_2^{2/3} \right)^{1/2}}, \tag{3}$$

where $a_0 = 0.529$ Å is the Bohr radius. The values of E/ε (keV) required to quantitatively interpret Figure 1 are tabulated in Table I for some typical ions implanted into GaAs.

Using the reduced energy has two advantages, namely: (1) nuclear stopping and associated quantities, such as damage production and sputtering, can be approximated by a universal curve for all particle-target combinations (Figure 1), and (2) ion bombardment studies can be separated into two distinct regimes. In the first regime, called the ion implantation regime ($\varepsilon < 10$), both nuclear and electronic stopping significantly contribute to energy loss, and theoretical treatments in this regime are complex. In the second regime, referred to as the nuclear analysis regime ($\varepsilon \gg 10$), where ions are sufficiently fast so that they are virtually stripped of all their electrons, electronic stopping is the dominant energy-loss process. In this regime, the ion trajectory is almost linear, sputtering and collision cascade effects are practically negligible, and the Bethe-Bloch formalism can be used for determining energy loss and penetration [22].

Figure 1 shows that at low energies (relevant to sputtering and ion beam etching), nuclear stopping is the more important process and that it reaches a maximum value (ε_1) around $\varepsilon^{1/2} = 0.6$, and thereafter decreases. From Table I and Figure 1 it is seen that for protons and He-ions in GaAs this corresponds to energies of approximately 1.2 keV and 2.5 keV, respectively.

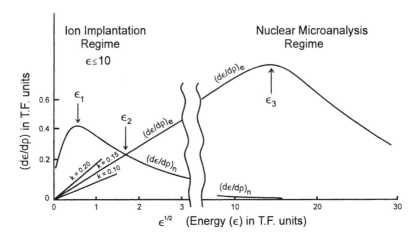

Fig. 1. Nuclear and electronic stopping versus the reduced energy of an implanted ion [1].

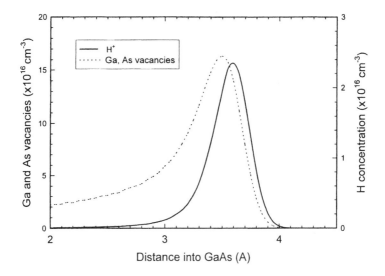

Fig. 2. Distribution of H (solid line), and Ga and As vacancies (dotted line) after implanting 10^{12} cm^{-2} H$^+$ into GaAs at an energy of 1 MeV, calculated using TRIM [20].

TABLE I
Ratio of laboratory energy, E (keV), to reduced energy, ε, for some ions in GaAs.

Ion	H	He	Be	B	C	N	Si
E/ε (keV)	3.3	7.1	15.7	20.4	25.1	30.4	76.1

It can further be seen from Figure 1 that electronic stopping becomes the dominant energy-loss mechanism for $\varepsilon^{1/2} > 2$ (ε_2) and that it increases linearly with velocity up to about $\varepsilon^{1/2} = 14$ (ε_3). It should be pointed out that whereas $(d\varepsilon/d\rho)_n$ is only a function of ε (i.e., independent of incident particle type and host atoms), the electronic stopping can be written as:

$$(d\varepsilon/d\rho)_e = k\varepsilon^{1/2}, \tag{4}$$

where k is a function of Z_1, Z_2, M_1 and M_2. Consequently, electronic stopping does not exhibit true ε-scaling and can consequently not be described by a universal curve.

2.2. Range Distribution of Implanted Ions

The deceleration of a projectile in an amorphous solid is a statistical process. The range, R, of the projectile is related to its mean track length before coming to rest, while the projected range R_p gives the mean penetration depth of the projectile relative to the surface and is the parameter of practical significance. The range, R, of a bombarding ion with initial energy, ε_i, is given by:

$$R = \int\limits_{\varepsilon_i}^{0} \frac{1}{(d\varepsilon/d\rho)_{total}} d\varepsilon, \tag{5}$$

where

$$(d\varepsilon/d\rho)_{total} = (d\varepsilon/d\rho)_n + (d\varepsilon/d\rho)_e. \tag{6}$$

For $M_1 \geq M_2$, the projected range, R_p, is obtained by multiplying the range, R, with the projection factor, approximately equal to $(1 + M_2/3M_1)^{-1}$ [21]. In the case of a monocrystalline target, the deviations in predicted and measured projected ranges can be significantly different (more than 20%) when channelling effects occur [23].

A more exact knowledge about the depth distribution calls for an evaluation of the range straggling. If the spatial distribution of implanted ions is approximately gaussian, then the straggle can be described by the standard deviation ΔR. According to the Lindhard–Scharff–Schiott (LSS) theory [24], the total straggling per ion path length, i.e. $\Delta R/R$, for $\varepsilon < 3$ approaches a

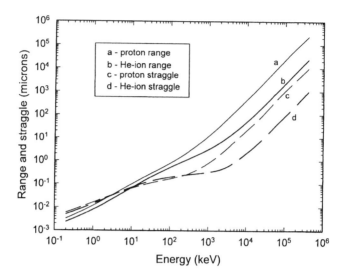

Fig. 3. Range and straggle of protons and He-ions in GaAs as function of energy, calculated using TRIM [20].

constant value of:

$$\Delta R/R = 0.35 \frac{2(M_1 M_2)^{1/2}}{M_1 + M_2} \tag{7}$$

As electronic stopping increases at higher energies, the relative straggling becomes smaller. The relative straggling in projected range, $\Delta R_p/R_p$, is almost identical to $\Delta R/R$ since the path length correction terms for the numerator and denominator are virtually identical.

Simulations of implantation profiles and ranges can be computed by using Monte Carlo based techniques, for example, the TRIM (TRansport of Ions in Matter) code of Biersack and Haggmark [20]. For example, Figure 2 depicts the distribution profiles of 1 MeV protons and the vacancies that they create as a function of distance into GaAs, calculated using TRIM. Figure 3 depicts the range and straggle of protons and He-ions in GaAs as a function of projectile energy, calculated using TRIM, and gives the ranges of 1 MeV protons and He-ions as 11.7 μm and 3.0 μm, respectively. An extensive review of computer simulations of ion-solid interactions can be found in [25]. Although being widely employed, these codes should be used with caution, especially in crystalline substrates, as they generally do not take channelling into account and do not make provision for possible diffusion effects during and after the implantation. One of the exceptions to this is the

Marlowe code [26, 27, 28], which takes account of the crystallinity of the material and includes channelling effects.

2.3.   Radiation Damage and its Distribution

Radiation damage will be created in crystalline materials as long as the energy that the projectile can transfer, E, exceeds the displacement energy, E_d, of the lattice atoms. Atomic replacements and interstitials can occur for $E_d < E < 2E_d$, while vacancies and interstitials are formed for $E > 2E_d$. In the case of electron irradiation, only E_d is required to form a vacancy since, after having knocked an atom off its site, the electron is mobile and rapidly diffuses away, leaving behind a vacancy. This is not necessarily the case for protons and heavier atoms.

The energy transferred in an elastic collision depends on the masses of the particles involved and the impact parameter. The maximum energy, $T_{\max}$, transferred in a head-on collision, is given (in the non-relativistic limit) by:

$$T_{\max} = \frac{4M_1 M_2}{(M_1 + M_2)^2} E, \tag{8a}$$

where E is the energy of the bombarding particle with mass M_1. In the case of electron irradiation, the relativistic expression:

$$T_{\max} = \left(\frac{2M_1}{M_2} \right) \left(\frac{E}{M_1 c^2} \right) (E + 2M_1 c^2) \tag{8b}$$

must be used, which can be written in the convenient form:

$$T_{\max} = \left(\frac{2147.8}{A} \right) (E + 1.022) E, \tag{8c}$$

where A is the atomic number of the target atom, E is in MeV and $T_{\max}$ in eV. If the energy of an atom which is knocked out of its lattice site is large enough, this atom may in turn displace other atoms, and if this process continues, a cascade of atomic collisions is created. This results in a distribution of vacancies, interstitial atoms and higher order defects in the region surrounding the ion track. The extent of collision cascades is governed by nuclear and electronic energy-loss processes. However, only the nuclear component, ν, contributes to the production of energetic recoil atoms. For $\varepsilon < 1$ (linear cascade regime for sputtering), a constant fraction of approximately 80% of the total energy ($\nu/\varepsilon \approx 0.8$) is consumed in recoil processes, whereas the ratio ν/ε decreases rapidly when $\varepsilon > 1$.

The number of displaced atoms, N_d, in a collision cascade can be estimated from $\nu(E)$ through the Kinchin–Pease relationship [29]:

$$N_d = \xi \frac{\nu(E)}{2E_d}, \tag{9}$$

where E_d is the displacement energy threshold of the lattice, which for GaAs is approximately 9–10 eV [30] and $\nu(E) = E\nu/\varepsilon$ is the total energy deposited in atomic motions. For the original derivation of Eq. (9), hard sphere collisions were assumed in which case $\xi = 1$. The value of $\xi \approx 0.8$ was introduced by Sigmund [31] to account for the effect of more realistic (not hard sphere) interatomic potentials. It is further important to note that Eq. (9) often overestimates the amount of disorder, since it overlooks the effects of ion channelling and of vacancy-interstitial recombination within the cascades. The amount of energy deposited into energetic recoils and the total number of vacancies produced for a given ion-target interaction can also be simulated using TRIM. Since an ion requires an amount of energy greater than $2E_d$ for the production of vacancies and interstitials via elastic collisions, the maximum of the damage distribution profile is always closer to the surface than the maximum of the ion depth distribution, as indicated in Figure 2.

An additional mechanism of creating defects, which has not received much attention, occurs when positively charged ions entering a solid have enough energy to surmount the coulombic energy barrier of the host nuclei. The threshold energy for a positively charged ion with atomic and mass numbers, Z_1 and A_1, respectively, to enter the nucleus with atomic and mass numbers Z_2 and A_2, respectively, is the energy, E_p, required to surmount the coulombic barrier and can be written as:

$$E_p = \frac{e^2}{4\pi\varepsilon_0} \frac{Z_1 Z_2}{R} \tag{10}$$

where $R = R_0(A_1^{1/3} + A_2^{1/3})$ with $R_0 = 1.2 \times 10^{-15}$ m [32]. Calculations based on this, show that protons and He-ions require about 6 MeV and 11 MeV, respectively, to enter the nuclei of Ga and As nuclei. Once in the nucleus, a multitude of nuclear reactions are possible, including, for example, neutron emission by (p, n) reactions. The fact that neutrons can be emitted is important, because the mechanism for damage formation by neutron irradiation is different to that by charged particles, and consequently some different defects can expected to be formed.

3. DEFECT CHARACTERIZATION BY DEEP LEVEL TRANSIENT SPECTROSCOPY

Defects in the crystal lattice are the consequence of the intentional or unintentional introduction of impurities or damage into the crystal structure. These imperfections introduce electronic states in the band gap which lie further away from the band edges than dopant levels, and these states are commonly referred to as "deep levels". Deep level defects are present in all semiconductors in concentrations that depend on the semiconductor and the method by which it was grown and processed, and these deep levels affect the electronic properties of semiconductors. Sometimes deep level defects are intentionally introduced in order to modify a specific materials property. For example, it has been shown that radiation induced defects can be used to reduce the lifetime of carriers in Si [8, 9, 33], thereby enhancing the switching time of high frequency oscillators fabricated on it. The challenge in the area of radiation induced materials modification lies in innovative defect engineering through which the deleterious effects of defects can be avoided, and their beneficial effects harnessed to tailor the properties of materials and devices for specific needs and niche applications. Clearly, this requires an intimate knowledge of the properties and behaviour of defects.

In the next few sections, the most important aspects of deep level defects and of deep level transient spectroscopy (DLTS), used for the determination of the electronic properties of these defects, are reviewed.

3.1. Characteristics of Deep Level Defects

Defects with deep states in the band gap are often referred to as traps, recombination centers, generation centers or deep level defects. In a neutral semiconductor an electron trap can be defined as a defect for which the electron capture rate, c_n, is much larger than the hole capture rate, c_p. In contrast, a recombination center is one for which both c_n and c_p are about the same. The capture cross section, σ_n, for electron capture is related to the electron capture rate, c_n, by [12]:

$$c_n = \sigma_n \langle v_n \rangle n \qquad (11)$$

where n is the electron concentration and $\langle v_n \rangle$ its average thermal velocity. An analogous expression holds for c_p in terms of σ_p. Note that the capture rates depend on the doping concentration of the material. Traps are further categorised as majority or minority carrier traps. An electron trap is a majority carrier trap in n-type material and a minority carrier trap in p-type material.

The thermal emission rate, e_n, of carriers from traps is proportional to a Boltzmann factor, $\exp(-\Delta E / kT)$, and, for traps which emit electrons to the conduction band, e_n can be written as [34]:

$$e_n = \frac{\sigma_n \langle v_n \rangle N_C}{g} \exp\left[-\frac{\Delta E}{kT} \right] \qquad (12)$$

where $\Delta E = E_C - E_T$, with E_C and E_T the energies of the conduction band and the trap respectively, N_C is the effective density of states in the conduction band, g is the degeneracy of the defect level and T is the absolute

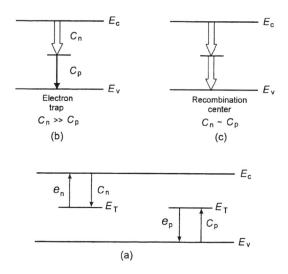

Fig. 4. (a): Thermal emission and capture processes at deep levels, and (b) and (c): difference between electron trap and recombination center.

temperature. An analogous expression holds for hole emission to the valence band. The quantity $\langle v_n \rangle N_C$ varies as T^2, and therefore, if e_n can be measured as function of temperature, an Arrhenius plot of e_n/T^2 vs $1/T$ will yield ΔE and σ_n. The thermal emission and capture processes at deep levels are illustrated in Figure 4(a).

The electron and hole capture cross sections, σ_n and σ_p, are generally temperature dependent [12, 34]. In the case of deep level defects, this is often the result of carrier capture by multiphonon emission via lattice relaxation [35, 36], in which case the capture cross section, σ, has the form:

$$\sigma = \sigma_\infty \exp\left[-\frac{\Delta E_\sigma}{kT}\right] \tag{13}$$

where ΔE_σ is the thermal activation energy of the capture cross section. Thus, a more general expression for the thermal emission rate of electrons to the conduction band is:

$$e_n = \frac{\sigma_n \langle v_n \rangle N_C}{g} \exp\left[-\frac{(\Delta E + \Delta E)}{kT}\right] \tag{14}$$

The thermal activation energy for emission of an electron to the conduction band, $\Delta E_a = \Delta E + \Delta E_\sigma$, determined from an Arrhenius plot, therefore, has two components: the energy difference between the trap level and the bottom of the conduction band, ΔE, and the thermal activation energy of the capture cross section, ΔE_σ (Figure 5).

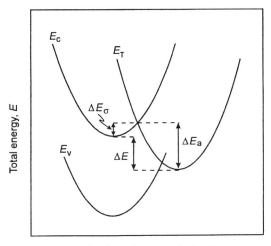

Fig. 5. Configuration co-ordinate (CC) diagram depicting the energy level of the defect below the conduction band, ΔE, the thermal activation energy of the capture cross section, ΔE_σ, and the total energy an electron requires to escape from the trap level to the conduction band, ΔE_a.

The activation energy for thermal emission is the most commonly used parameter to characterize a deep level. However, the physical parameter ΔE in Eqs. (12) and (14) is the Gibbs free energy:

$$\Delta E = E_C - E_T \equiv \Delta H - T \Delta S \qquad (15)$$

where ΔH and ΔS are the changes in enthalpy and entropy due to the change in charge state of the level. When combining Eqs. (12) and (15), it follows that:

$$e_n = \frac{\sigma_n \langle v_n \rangle N_C}{g} \exp\left[\frac{\Delta S}{k}\right] \exp\left[-\frac{\Delta H}{kT}\right] \qquad (16)$$

Therefore, the slope of an Arrhenius plot yields the enthalpy of the deep level, not the free energy, which can only be determined from optical measurements (34, 37, 38). However, from Eq. (16) it can be seen that the activation energy obtained from the slope of an Arrhenius plot is equal to the free energy obtained at $T = 0$ K. This is generally not the same as the free energy at the measurement temperature, due to the temperature dependence of the band gap energy. For temperatures greater than absolute zero, therefore, the energy position of the trap can only be accurately determined when the temperature dependence of the band gap energy is known [12].

Finally, in addition to being emitted by thermal excitation, carriers may also be emitted from defect states by optical excitation. Optical emission rates can be determined from photocurrent or photocapacitance measurements. The two optical cross sections involved, $\sigma_n(h\nu)$ and $\sigma_p^o(h\nu)$, which are functions of the photon energy, and define the optical properties of a defect state. In n-type material $\sigma_n(h\nu)$ is the optical absorption spectrum of the defect and $\sigma_p^o(h\nu)$ is equivalent to the photoluminescence spectrum, while in p-type materials the roles of $\sigma_n^o(h\nu)$ and $\sigma_p^o(h\nu)$ are reversed [34].

3.2. Deep Level Transient Spectroscopy

In this section the concepts required for DLTS measurements and the interpretation of the results will be briefly reviewed.

With DLTS, defects in semiconductors are analyzed by probing the space-charge layer, also referred to as the depletion region, which exists in a p-n junction or Schottky barrier diode [34, 39], as illustrated in Figure 6. The width, w, of the depletion region of a SBD or p^+-n junction varies with the applied voltage according to:

$$ w = \sqrt{\frac{2\varepsilon(V_{bi} + V)}{qN}} \qquad (17) $$

where ε is the dielectric constant of the depleted semiconductor, V_{bi} is the built-in voltage of the junction, V is the externally applied voltage, q is the charge on the electron, and N is the density of the ionized impurities due to dopants and other imperfections. The junction capacitance due to the depletion layer is:

$$ C = \frac{\varepsilon A}{w} = A\sqrt{\frac{q\varepsilon_N}{2(V_{bi} + V)}} \qquad (18) $$

where A is the area of the junction. It is clear from Eqs. (17) and (18) that if N changes in the depletion region, w and C will also change, and therefore the junction capacitance is a direct measure of the total charge. If the concentration of electrons or holes trapped at deep levels is changed by the thermal or optical emission of carriers to the conduction or valance bands, this change can be monitored by measuring the variation in the junction capacitance at constant applied voltage [40, 41]. The variation in junction capacitance as a result of the temperature dependent variation of N forms the basis of capacitance-based DLTS.

Carrier capture and emission is shown schematically in Figure 7 for a SBD having an electron trap in the upper half of the band gap. In equilibrium, when a defect level is below the Fermi level, it is filled with electrons, while it is empty when above the Fermi level. Under a steady-state

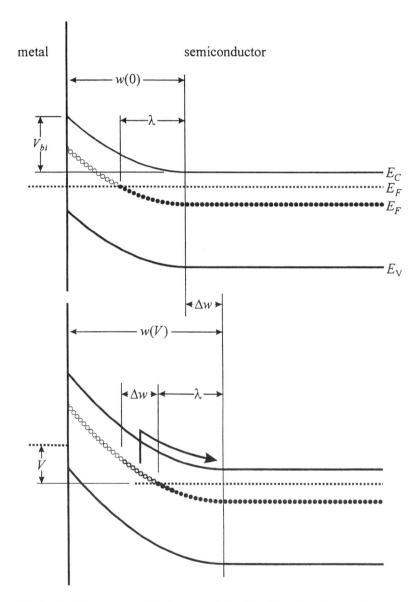

Fig. 6. Depletion region of a Schottky barrier diode with built-in voltage V_{bi}, on n-GaAs under zero bias and reverse bias, V. The widths of the depletion layer (w) and transition region (λ) are defined in Eqs. (17) and (23), respectively.

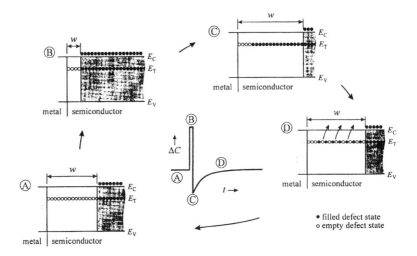

Fig. 7. Variation of depletion region width and trap population of an electron trap in n-type semiconductor for a DLTS bias and filling pulse cycle. (A) and (B) are during reverse bias and the filling pulse, respectively, whereas (C) and (D) are directly, and at a time t, after removing the filling pulse. The resultant capacitance transient is qualitatively shown in the center of the figure.

reverse bias voltage (A), the traps in the depletion region, above the Fermi level, are therefore empty. Reducing the applied voltage to $V - V_p$, (B), reduces the width of the depletion region and allows electrons to be trapped at the deep levels. Electron capture into an initially empty trap is given by:

$$N(t) = N_T[1 - \exp(-c_n t)] \qquad (19)$$

where N_T is the trap density and c_n is the capture rate as defined in Eq. (11). When the voltage is restored to its original steady-state value, V, the filled traps lie within the depletion region, as shown in (C). The emission of trapped electrons to the conduction band, where they are instantaneously swept away by the junction electric field, is observed as a majority carrier capacitance transient (D).

Experimentally, the electron emission rate can be determined from the time dependence of the capacitance transient. The density of occupied traps at time t after removing the filling pulse is:

$$N(t) = N_T \exp(-e_n t) \qquad (20)$$

where e_n is the thermal emission rate and N_T is the trap concentration, assuming all the traps to be initially occupied. From Eqs. (12) and (20) it can be shown that the change in trap population gives rise to a corresponding change in diode capacitance, which is time dependent, and if $N_T \ll N_D$, can be expressed as:

$$C(t) = C_o - \Delta C_o \exp(-e_n t) \tag{21}$$

Here C_o is the equilibrium reverse bias (V) capacitance and ΔC_o the change in capacitance directly after removal of the filling pulse.

The procedure for determining the energy level, E_T, and capture cross section, σ_n, of a defect is to extract e_n from the transient, $C(t)$, in Eq. (21) at several temperatures and then use Eq. (12) to obtain E_T and σ_n, commonly referred to as the defect's "signature". The key step in this process is to obtain e_n from the transient for which several analogue and digital methods exist. Most of the early DLTS experiments were performed using boxcar averagers [13, 14] or lock-in amplifiers [34] to analyze the transient resulting from the application of a repetitive bias/pulse sequence. For both these methods, DLTS proceeds by scanning the sample temperature and feeding the repetitive thermal emission transient, which varies with temperature, into the analyzer (boxcar or lock-in amplifier). A DLTS spectrum is obtained by plotting the output signal of the analyzer vs temperature. The resulting signal is a maximum when the thermal emission rate matches the "rate window" of the analyzer. Each peak on the DLTS spectrum is indicative of an energy level with a specific E_T and σ_n. The emission rate of the transient at the peak temperature is a function of the analyzer used. For example, for a lock-in amplifier with a sine wave mixing function, the emission rate is $e_n = f/0.424$ (provided that the filling pulse width is much shorter than the pulsing period), where f is the pulse frequency [42]. By varying the frequency of the lock-in amplifier (or "rate window" in the case of the boxcar analyzer), the peak temperature value shifts, and thus the emission rate can be measured as a function of temperature. From Eq. (12) it follows that an Arrhenius plot of $\log(e_n/T^2)$ vs $1/T$ yields E_T and σ_n. Note that the σ_n thus obtained, is an apparent capture cross section. The real capture cross section may be temperature dependent and has to be determined by measuring the transient at a fixed temperature when the pulse width is varied [13].

An attractive alternative to these techniques is to digitize the full transient, which is readily achievable using modern electronics. This approach forms the basis of isothermal DLTS measurements, and various methods to analyze the transients have been reported [43, 44]. A digital method that is rapidly gaining interest, is the inverse LaPlace transform analysis, which yields a substantial improvement in energy resolution, thereby facilitating the separation of closely spaced defect levels [45].

The sensitivity of the junction capacitance to trapped charge depends on the location of the charge in the depletion region. The relative capacitance change, $(\Delta C/C)_x$, due to $n(x)$ trapped electrons in the interval Δx at a depth $x < w$ below the interface, is given by [39]:

$$\left(\frac{\Delta C}{C}\right)_x = -\frac{n(x)x\Delta x}{Nw^2} \tag{22}$$

Note that the sensitivity of the junction to trapped charge increases linearly from zero at the junction ($x = 0$) to a maximum at the edge of the depletion region ($x = w$).

In order to obtain the defect concentration, N_T, the total signal due to a majority carrier pulse of amplitude V_p, superimposed onto the reverse bias, V, and duration long enough to fill all traps below the Fermi level, has to be determined by integrating Eq. (22). The limits of this integration are determined by the location in the space-charge layer where emission takes place (Figure 6), i.e. from $w(V_p) - \lambda$ to $w(V) - \lambda$, where ε is the width of the transition region:

$$\lambda = \sqrt{\frac{2\varepsilon(E_F - E_T)}{q^2 N}} \tag{23}$$

For a constant free carrier density, N_T can be related to the magnitude of the total signal, $\Delta C/C$, thus found by [46]:

$$N_T(x_m - \lambda) = 2N(x)\left(\frac{\Delta C}{C}\right)_{t=0}\left[\left(\frac{x - \lambda}{x}\right)^2 - \left(\frac{x_p - \lambda}{x}\right)^2\right]^{-1} \tag{24}$$

where x and x_p are the depletion widths during application of the reverse and filling pulses, respectively, and $x_m - \lambda$ is the mean distance from the interface to the region from which traps emit carriers. If the edge region is neglected and $C(V) \ll C(V_p)$, Eq. (24) reduces to the frequently used approximation [14, 34]:

$$N_T \approx 2\frac{\Delta C}{C}N \tag{25}$$

However, neglecting the edge region may cause significant underestimation ($\approx 50\%$) of the trap concentration, especially at low reverse bias voltages. In a noise-free environment, $\Delta C/C$ values of $10^{-5} - 10^{-6}$ can readily be achieved, implying that defect concentrations of as low as 10^{10} cm^{-3} can be detected.

When determining spatially varying trap concentrations, it is essential to take into account the effect of the transition region. Using Eq. (24), the deep level concentration profile can be measured through the variation of the depletion width with applied bias, if the shallow level concentration profile is known. Trap concentration profiles can be measured either by using a filling pulse that extends to a constant voltage level and varying the reverse bias voltage, or by using a constant reverse bias and varying the filling pulse amplitude [39].

The average electric field of typically 10^5 V m^{-1} $-$ 10^7 V m^{-1} in the depletion layer can influence the shape of defect potentials as well as enhance charge emission from potential wells, by, for example, the Poole-Frenkel effect [47]. Detecting the Poole-Frenkel effect establishes an important characteriztic of a defect: it evidences the presence of donor-like electron traps in an n-type semiconductor or acceptor-like hole traps in a p-type semiconductor, i.e. states which are neutral when occupied and charged when empty (Section 8). It has, for example, been shown that oxygen-related thermal donors in Si exhibit Poole-Frenkel enhanced emission [17]. A phonon assisted tunnelling model has been proposed to explain electric field enhanced electron emission from deep states in GaAs which do not exhibit Poole-Frenkel enhanced emission [48, 49]. Both models have been used to investigate electron irradiation induced traps in GaAs [50, 51, 52].

When the thermal emission rate is strongly field dependent, non-exponential transients occur due to the variation of the electric field in the space-charge region. This must be carefully considered when interpreting emission rate data and comparing the electronic properties of defects (Section 8). Ideally, any comparison of defect properties should be done under equal electric field conditions in a narrow spatial region to limit the electric field variation. This is the essence of the double DLTS (DDLTS) measurement technique [53].

For a complete electrical characterization of a defect, its majority as well as minority carrier capture cross sections must be determined. Minority carrier capture in SBDs is usually achieved either optically, by illuminating semi-transparent Schottky contacts with above band gap light, or electrically, by applying a large forward bias to the SBD. For p-n diodes, minority carrier injection occurs when the filling voltage pulse causes the diode to be forward biased. After minority carrier injection, the occupation of the traps in the depletion region is the net result of the capture of both types of carriers. Note that the sign of the resulting capacitance transient is negative for the emission of majority carriers, as in Figure 7(e), and positive for minority carriers, since their charge has the opposite sign [34], making it easy to distinguish between the type of carriers being emitted.

From a knowledge of the energy level, E_T, concentration, N_T, and capture cross section, σ_n and σ_p, of a defect for minority and majority carriers, it is possible to establish whether the defect will act as a trap (and thus cause carrier reduction according to Eqs. (12) and (20)), or whether it will act as a recombination center. Markvart *et al.* [54] have demonstrated that a modified version of DLTS, referred to as recombination DLTS, can be used to identify defects which act as recombination centers. These centers act as "stepping stones" for carriers and contribute to the current-voltage characteristics of rectifying junctions (Section 9) at a recombination rate, U, [55]:

$$
U = \frac{\sigma_p \sigma_n v_{\text{th}} (pn - n_i^2) N_T}{\sigma_n [n - n_i \exp\{(E_T - E_i)/kT\}] + \sigma_p [n + n_i \exp\{-(E_T - E_i)/kT\}]}
\tag{26}
$$

where E_i and n_i are the intrinsic Fermi level and intrinsic carrier density, respectively, and the other symbols have previously been defined. Clearly, the most efficient recombination centers are those with levels close to the middle of the band gap, and with approximately equal capture cross section for electrons and holes.

3.3. Summary

The spectroscopic nature of DLTS allows independent studies of different defect species in the same semiconductor. Furthermore, because DLTS is a space-charge related technique, it facilitates control of the charge state of defects in the space charge region which enables charge state dependent processes to be studied and metastable defect configurations to be revealed. DLTS is further attractive because it can be used to characterize defects using various kinds of space charge based devices. This includes simple Schottky barrier diodes (SBDs), as well as device structures with higher degrees of complexity. Finally, sensitivity of DLTS for detecting defects in concentrations of 10^{10} cm^{-3} is superior to any other electrical characterization technique.

Since its introduction, DLTS has become a widely accepted technique to determine the electronic properties of defects in semiconductors. It has been applied to study growth induced defects, impurities, radiation induced defects and metastable defects in various semiconductors. Particle irradiation induced defects of III–V compounds have been studied intensively using DLTS and several extensive reviews have appeared [15, 56, 57]. In these studies the electronic properties of numerous radiation induced defects were determined.

Finally, it should be pointed out that DLTS in itself *cannot directly* provide the structure of a defect. This has to be obtained by structure-sensitive techniques, such as electron paramagnetic resonance (EPR) or local vibrational mode (LVM) spectroscopy. However, structural defect identification via DLTS is possible by comparing defect properties, e.g., annealing kinetics, which can be measured by DLTS as well as by structure-sensitive techniques.

4. ELECTRON IRRADIATION INDUCED DEFECTS IN GaAs

This section begins with a brief overview of some aspects related to defect introduction in GaAs by MeV electron irradiation, and the characterization of these defects by DLTS. Whereas most investigations of defects introduced by high energy electron irradiation to date have been performed by electron irradiation in Van de Graaff accelerators, it will be shown that the same defects can be produced by beta-particle (electron) irradiation using radio-nuclides and linear accelerators. Finally, defect introduction by low energy (0.3–20 keV) electrons is demonstrated and discussed.

4.1.   Overview

Using DLTS, five electron traps (E1–E5) and one hole trap (H1) were first reported by Lang [58] after 1 MeV electron irradiation at 300 K of LPE grown GaAs. After analyzing the dependence of the introduction rates of E1–E3 and H1 on the crystallographic direction of the incident electron beam, Pons and Bourgoin [59] concluded that these defects arise from displacements in the As sublattice. In a review on electron irradiation induced defects in GaAs, Pons and Bourgoin [15] reported the introduction of several more hole traps, H0 and H2–H5. From the results of several studies, it was concluded [16] that E1–E5 in n-GaAs and H0 and H1 in p-GaAs are intrinsic primary defects because: (1) the total measured introduction rate (4 cm^{-1}) of all electron radiation induced defects in n-GaAs is equal to the calculated introduction rate, assuming that the defects all belong to the same sublattice [15], (2) all the main defects can be introduced at 4 K [15], (3) they are all removed by annealing at about 220°C with the same first order kinetics [60]; and (4) the energy liberated during annealing, 8 eV per defect [61], is equal to the theoretical estimation of the energy stored in a vacancy-interstitial pair [62]. More specifically, it has been proposed that defect levels E1 and E2 are the $-/0$ and $0/+$ charge states of the isolated arsenic vacancy, V_{As}, [15, 63] and that E3 is related to close arsenic vacancy — interstitial pairs, V_{As}–As_i, [15, 16]. More recently, von Bardeleben [64] proposed, based on EPR measurements, ab initio self-consistent pseudopotential calculations and self-consistent tight binding Green's function calculations, that the E1 and E2 levels belong to the distorted arsenic antisite defect, As_{Ga} (or As^-As_3).

It was reported that the introduction rates of E1–E5 in n-GaAs are independent of the dopant used [15], implying that all defects thus observed are "pure" lattice defects and do not involve foreign atoms. In p-GaAs, however, only H0 and H1 were found to be impurity independent while the presence of H2–H5 was found to depend on the nature of the material [65, 66]. It was argued that the H2–H5 complexes may involve native impurities and, possibly, the arsenic interstitial [66]. In particular, H2 has thus far only been observed in p-GaAs containing copper in detectable concentrations before irradiation [66, 67]. The proposed defect assignments are summarized in Tables III and IV for n- and p-GaAs, respectively.

In the light of the above mentioned defect assignments, it was not clear why the anion antisite, As_{Ga}, or defects in the Ga sublattice, which must be created at about the same rate as defects on the As sublattice, could not be observed by DLTS. In a search for antisite related defects, Lim et $al.$ [68] irradiated n-GaAs to a relatively high electron dose (10^{17} $e^- cm^{-2}$). They used highly doped (3×10^{17} cm^{-3} – 1×10^{18} cm^{-3}) GaAs to ensure that sufficient carriers remain after irradiation to allow for accurate DLTS measurements. From their DLTS results, combined with theoretical considerations, they proposed that E4 (at 0.75 eV below the conduction band [68]) is the energy level of the As_{Ga} - V_{As} complex.

The effect of annealing on radiation induced carrier removal in n-GaAs, irradiated at low-temperatures (5 K–77 K) with 1 MeV electrons, was studied by Thommen [69] using electrical resistivity measurements. This study revealed that annealing recovered the carrier density in three temperature stages, centered at 235 K ("stage I"), 280 K ("stage II") and 520 K ("stage III"). Using DLTS, it was shown that the E1–E5 can be removed by annealing at about 520 K [60], corresponding to the "stage III" of Thommen. In this temperature range, the annealing of E1–E5 obeys first order kinetics, and their removal are accompanied by the introduction of the P1–P3 defects [60]. Lang [70] showed that E1 and E2 can be removed at lower temperatures under recombination conditions. In order to study the defects responsible for the annealing stages I and II of Thommen, Rezazadeh and Palmer [71] irradiated Al SBDs on Sn-doped (to 2×10^{15} cm^{-3}), VPE grown n-GaAs, with 1 MeV electrons at 80 K. They found that irradiation at 80 K and subsequent annealing to 140 K, produced an electron trap, E(0.23), resulting in a broad energy level centered at 0.23 eV below the conduction band, which can be removed by heat treatment in the 235 K range (stage II). They proposed that this defect is a plural defect, such as the (V_{Ga}–V_{As}) divacancy, or the combination of such a divacancy with a trapped interstitial, and that annealing at 235 K involves internal recombination and/or dissociation of E(0.23) into simpler defects. A further effect of the 80 K–300 K heating was that the concentrations of E1 and E2 increased, probably as a result of internal recombination or dissociation of plural defects.

Irradiation at 300°C, where primary defects are unstable, introduced several defects of which the I1, with an energy level and capture cross section very close to that of EL2, is the most prominent [72]. The introduction rate of I1 is about 1 cm^{-1}, i.e., about half that of E1 and E2 for room temperature irradiation, and about two orders of magnitude higher than those of the P1–P3 traps present after annealing between 200°C–300°C, i.e., after removing E1–E5. From a technological point of view, this is an important observation, because defects with levels near midgap are generally more efficient recombination centers than shallow levels. I1 can therefore be expected to be important when irradiating GaAs with electrons with the purpose of reducing its free carrier concentration. From the results obtained after irradiating at elevated temperatures, Lim et al. [73] concluded that the introduction of the P2 and P3 defects, which accompanies the removal of the E1–E5 defects, result from the creation of multi-displacement defects and the recombination of V_{As}–As_i Frenkel pairs.

4.2.   Electron Irradiation Using a ^{90}Sr Radio-nuclide

In this section we describe defect introduction in GaAs by β-particle (electron) irradiation using a ^{90}Sr radio-nuclide, and demonstrate that this introduces the same defects as 1–2 MeV electron irradiation in Van de Graaff accelerators. In Table II, some electron irradiation sources and typical conditions employed to introduce defects, as well as some prominent irradiation induced defects using these sources and the nomenclatures used to identify them, are summarized.

In the first reported DLTS investigation of defects produced by electron irradiation from radio-nuclides, epitaxially grown layers with free carrier densities of 4×10^{14} cm^{-3} (undoped) and 1×10^{16} cm^{-3} (Si-doped), grown by OMVPE on n^{++}-type bulk-grown 10^{18} cm^{-3} doped GaAs substrates, were used [74]. Pd SBDs, fabricated on these layers, were irradiated with β-particles (electrons) emitted from a 20 mCi ^{90}Sr radio-nuclide, at a dose rate of 1.9×10^9 e$^-$cm^{-2}s^{-1} and a dose of 1.1×10^{14} e$^-$cm^{-2}. For a dose rate as low as this, the temperature and Fermi level remain unperturbed during irradiation, preventing any dose rate effects on the introduction rate during irradiation. The DLTS spectrum in curve (a) of Figure 8 shows the presence of three well defined radiation induced defect peaks, labelled Eβ1, Eβ2 and Eβ4, in undoped GaAs. The electronic properties of these defects, measured in the undoped material under low electric field conditions to avoid electric field-enhanced emission rate, are summarized in Table III, where they are compared to the properties of the same defects determined in doped n-GaAs under higher electric fields. The properties of the E1–E5 irradiation induced defects reported after pioneering investigations using DLTS, as well as defects observed in recent studies involving electron-,

TABLE II

Defects, particles, radiation sources and radiation conditions for typical studies of radiation induced defects in GaAs.

Defects	References	Particles, radiation sources and radiation conditions				
		Particles	Energy	Dose (cm^{-2})	Dose rate $(cm^{-2} s^{-1})$	Source
E1 – E5	[15, 16, 56,8, 58, 120]	above-threshold	0.5–2 MeV	10^{13}–10^{16} $0.1 \mu a$	$0.1 \mu A cm^{-2}$ $2nA cm^{-2}$	Van de Graaff accel.
H0 – H5	[15, 16, 58, 66]	electrons	0.5–2 MeV	10^3–10^{16}	$0.1 \mu A cm^{-2}$	
Eβ1 – Eβ5	[74]		< 2 MeV	10^{12}–10^{14}	1.9×10^9	^{90}Sr radio-nuclide
Hβ1 – Hβ5	[76]		< 2 MeV	10^{12}–10^4	1.9×10^9	^{90}Sr radio-nuclide
EE1 – EE5	[77, 86]		3–10 MeV	3–5×10^{11}	2–3×10^{10}	Linear accelerator
Eel – Ee5	[86]	sub-threshold	0.5–3 keV	10^{17}		Electron gun,
		electrons	2–20 keV	10^{15}–10^{16}		EB deposition
E1 – E5	[56, 58, 100, 104]	protons	400 keV – 2 MeV	10^{10}–10^{13}	$1.5 nA cm^{-2}$	Van de Graaff accel.
E1 – E5	[94]		60 keV	10^{11}–10^{14}		Ion implanter
HO – H3, HX1 – HX3	[93]		95–130 keV	10^{10}–10^{12}		Ion implanter
Ep1 – Ep6	[98, 99]		40 keV - 1 MeV	10^{10}	5×10^9	Pelletron implanter
Ep7	[105]		200 MeV	10^{12} – 10^{13}	Cyclotron	
Ep8	[106]		5 keV	10^{12}	10^{11}	ion gun
E1 – E5	[56, 58, 71]	α-particles	O.5–4 MeV	10^9–10^{12}	5×10^9 –??	Van de Graaff accel.
Eα1 – Eα10	[74]		5.4 MeV	10^8–10^{10}	2×10^4	^{241}Amradio-nuclide
Hα1 – Hα5	[67]		5.4 MeV	10^{10}–10^{11}	7×10^6	^{241}Amradio-nuclide
EHe1 – EHe5	[116]	He-ions	1–5 keV	10^{11}	5×10^{10}	ion gun

proton- and α-particle irradiation, are also included in Table III. From a detailed comparison of their properties [74] it was found that Eβ1, Eβ2 and Eβ4 are the same as E1, E2 and E3, respectively, introduced by electron irradiation in Van de Graaff accelerators [15, 58]. Using the effective dose of electrons with energies above 250 keV [74], the introduction rates, η = (defect density)/dose, of Eβ1 and Eβ2 were found to be 2 cm^{-1}, while that of Eβ4 was 0.7 cm^{-1}. These introduction rates agree well with those reported by Pons et al. [15] for E1–E3.

The DLTS results obtained by irradiating the 1×10^{16} cm^{-3} doped GaAs with electrons from the ^{90}Sr source also show the presence of Eβ1, Eβ2 and Eβ4, but with broadened DLTS peaks (curves (b) and (c) in Figure 8), displaced to lower temperatures due to electric field enhanced emission [50].

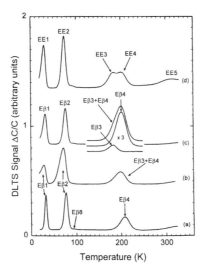

Fig. 8. DLTS spectra of electron irradiated OMVPE grown n-GaAs. All spectra were recorded at a lock-in frequency of $f = 46$ Hz, (a decay time constant of 9.23 ms). Spectra (a)–(c): β-particles from a ^{90}Sr source [74]. (a): Undoped GaAs with a free carrier density of $N_D = 4 \times 10^{14}$ cm^{-3}. The reverse bias and pulse were $V_r = 8$ V and $V_p = 4$ V. (b): Si-doped (to 1×10^{16} cm^{-3}) GaAs with $V_r = 4$ V and $V_p = 2$ V. (c): De-convolution of the spectra revealing the presence of the metastable Eβ3 [75]. (d): Spectrum of Si-doped (to 1×10^{16} cm^{-3}) GaAs irradiated with 10 MeV electrons in a linear accelerator [77].

Even though small bias and pulse conditions were used to determine the signatures of the defects in doped GaAs, Table III nevertheless shows that significant differences in E_T and σ_a occur due to the larger electric field in the doped material. This effect is particularly severe for defects with levels close to the conduction band, e.g., Eβ1. Electric field enhanced emission of these defects will be discussed in detail in Section 8. The near-midgap levels introduced by electron irradiation, Eβ6 and Eβ7, and their relation to E4 and E5, will be discussed in Section 4.4.

The introduction rates of Eβ1, Eβ2 and Eβ4 in doped GaAs, calculated from depth profiles, were the same as those in undoped GaAs [74], confirming that these electron traps are point defects and do not involve dopant atoms. Recently, it was shown [75] that, in addition to Eβ1, Eβ2 and Eβ4, electron irradiation of GaAs, doped with Si to 10^{16} cm^{-3}, introduces a defect Eβ3 (curves (c) in Figure 8), which has a DLTS peak in the same region as Eβ4 (E3) and an introduction rate of about 0.14 cm^{-1}, i.e. about one fifth that of Eβ4. The most striking property of Eβ3 is its metastability (Section 7), which together with its DLTS signature [75] and behaviour in an electric

TABLE III

Electronic properties of prominent particle induced defects in undoped and doped epitaxially grown n-GaAs. Error margins can be found in the referenced texts.

Defect label	$E_C - E_T$ (eV)	σ_a (cm^2)	T_{peak}[a] (K)	η (cm^{-1})	Ref. (eV)
			$N_D \leq 1 \times 10^{15}$ cm^{-3}		
E1	0.045	2.2×10^{-15}	34[d]	1.5	[161]
Eβ1	0.041	6.8×10^{-16}	34	2	[74]
EE1	0.045	1.1×10^{-15}	35	2	[86]
Ep1	0.042	8.0×10^{-16}	34	–	[98]
Eα1	0.141	5.4×10^{-16}	34	–	[74]
EHe1	–	–	–	–	–
E2	0.140	1.2×10^{-13}	77[d]	1.5	[161]
Eβ2	0.139	9.3×10^{-14}	77	2	[74]
EE2	0.142	1.3×10^{-13}	78	2	[86]
Ep2	0.140	1.0×10^{-13}	78	–	[98]
Eα2	0.141	1.2×10^{-13}	77	–	[74]
EHe2	–	–	–	–	–
Eβ3					
EE3			Not detected		
Ep3			in		
Eα3			undoped GaAs		
EHe3					
E3	–	–	–	–	–
Eβ4	0.378	1.6×10^{-14}	208	0.7	[74]
EE4	0.375	1.3×10^{-14}	209	0.7	[86]
Ep4	0.376	1.4×10^{-14}	208	–	[98]
Eα4	0.378	1.6×10^{-14}	208	–	[74]
E4	–	–	–	–	–
Eβ5	–	–	–	–	–
EE5	–	–	–	–	–
Ep5	–	–	–	–	–
Eα5	0.591	1.9×10^{-14}	312	–	–
EHe5	–	–	–	–	–
E5	–	–	–	–	–
Eβ6, E5a	–	–	–	–	–
EE6	–	–	–	–	–
Ep6	–	–	–	–	–
E5b	–	–	–	–	–
Eβ7	–	–	–	–	–
Ep7	–	–	–	–	–
Ep8	–	–	–	–	–
$I1$[c]	–	–	–	–	–

[a] At a lock-in amplifier frequency of 46 Hz, i.e. an emission rate of 107 s^{-1} (time constant of 9.23 ms.)

[b] Introduction rate decreases rapidly below surface

[c] Introduced during electron irradiation at 300 °C.

[d] Calculated from the a and E, values in references in table.

TABLE III (Continued)
Electronic properties of prominent particle induced defects in undoped and doped epitaxially grown n-GaAs. Error margins can be found in the referenced texts.

$N_D \le 1 \times 10^{16}$ cm^{-3}					
$E_C - E_T$ (eV)	σ_a (cm^2)	$T_{peak}^{(a)}$ (K)	η (cm^{-1})	Ref.	Notes and suggested assignment
0.031	4.8×10^{-17}	32	2	|74|	V$_{As}$ |15|,
0.032	6.8×10^{-17}	31	–	|77|	As$_{Ga}$(or As$^-$As$_3$) |64|,
0.029	2.7×10^{-17}	31	1056	|99, 105|	C3v symmetry |104|
0.027	1.6×10^{-17}	30	2963	|99, 116|	
0.027	1.2×10^{-17}	30	(b)	|116|	
0.14	1.1×10^{-13}	77$^{(d)}$	1.5	–	
0.127	2.9×10^{-14}	75	2	|74|	
0.129	4.3×10^{-14}	75	–	|77|	V$_{As}$ |15|,
0.123	1.8×10^{-4}	74	1706	|99, 105|	As$_{Ga}$ (or As$^-$As$_3$)' |64|,
0.122	1.8×10^{-4}	74	9638	|99, 116|	C3v symmetry |104|
0.128	2.7×10^{-4}	74	(b)	|99, 116|	
0.36	1.0×10^{-13}	184	0.14	|75|	
0.357	1.0×10^{-13}	183	1.2	|77|	Meta stable |117|
0.355	8.8×10^{-14}	183	–	–	In Si:n-GaAs |74|, but
0.36	1.2×10^{-14}		3060	|99|	not in S:n-GaAs |137|
0.341	2.8×10^{-14}	184	(b)	|116|	
0.35	6.2×10^{15}	202$^{(d)}$	0.4	|161|	
0.361	1.0×10^{-14}	204	0.7	|74|	
0.350	4.8×10^{-14}	205	0.9	|77|	V$_{As}$–As$_i$ |15|,
0.362	1.9×10^{-14}	199	470	|99|	C3v symmetry |104|
0.353	1.2×10^{-14}	198	3060	|99|	
0.617	3.1×10^{-14}	318$^{(d)}$	0.08	|161|	
	Seems to be same as EE6 and E4				
0.614	2.5×10^{-14}	311	–	|77|	As$_{Ga}$ – V$_{As}$ |68|
0.630	1.2×10^{-13}	309	127	|99|	
0.635	1.6×10^{-13}	305	1737	|99|	
0.525	3.9×10^{-15}	299	(b)	|105|	Defect band |116|
0.83	1.9×10^{-12}	358$^{(d)}$	0.1	|161|	
	T_{peak} = 249K (at 1 mHz), i.e. the same as for EE6			|77|	
0.783	1.1×10^{-12}	344$^{(d)}$	–	|77|	
0.77	9.4×10^{-13}	341	–	|105|	
0.840	2.3×10^{-12}	357$^{(d)}$	–	|77|	
	T_{peak} = 260K (at 1 mHz), i.e. the same as for E5b			|77|	
0.68	4.0×10^{-12}	290	–	|105|	"U-band" |107|
0.650	4.5×10^{-13}	301	–	|105|	
0.72	1.2×10^{-14}	380	≈ 1	|72|	Similar to EL2 |72|

$^{(a)}$ At a lock-in amplifier frequency of 46 Hz, i.e. an emission rate of 107 s^{-1} (time constant of 9.23 ms.)

$^{(b)}$ Introduction rate decreases rapidly below surface

$^{(c)}$ Introduced during electron irradiation at 300 °C.

$^{(d)}$ Calculated from the a and E, values in references in table.

field [51], serves as an unique fingerprint, clearly distinguishing it from defects with similar signatures, such as Eβ4 (E3). The electronic and metastable transformation properties of Eβ3 are identical to those of the alpha-particle irradiation induced defect, Eα3 (Sections 6 and 7), implying that the two defects are identical. Finally, Eβ3 was observed only in Si-doped epitaxially grown GaAs, but not in undoped and S-doped GaAs, suggesting that it is related to Si or impurities accompanying its introduction during crystal growth.

Beta-particle irradiation of p-GaAs was studied by irradiating Al SBDs on a MBE grown Be-doped (to 1.8×10^{16} cm^{-3}) GaAs layer with β-particles from a ^{90}Sr radio-nuclide to a maximum dose of 10^{15} cm^{-2} at a dose rate of 1.9×10^9 cm^{-2}s^{-1} [76]. Using DLTS, the radiation induced hole traps Hβ1–Hβ7 were studied (curves (a) and (b) in Figure 9). The three most prominent defects, Hβ1, Hβ4 and Hβ5, have energy levels at 0.08, 0.20 and 0.30 eV, respectively, above the valence band. For comparison, the spectrum of the same material irradiated with α-particles is included (curve (c)). The electronic properties and introduction rates of these defects are summarized in Table IV where they are compared to the properties of defects reported after early investigations using DLTS, as well as to defects observed more recently created during electron-, proton- and α-particle irradiation. From Table IV it can be seen that the Hβ1 and Hβ5 have the same "signatures" as H0 and H1 [15], respectively, but Hβ4 has thus far only been observed in Be-doped p-GaAs. Table IV also shows that the introduction rate of Hβ1 (H0) is almost the same as that of Eβ1 (E1) and Eβ2 (E2), and it may therefore be a third energy level of the same defect.

4.3.   Electron Irradiation in Linear Accelerators

Linear accelerators are standard equipment for radiation therapy and usually provide electrons with energies of 2–20 MeV at dose rates of typically $2 - 3 \times 10^{10}$ e$^-$cm^{-2}s^{-1}. Although these accelerators are not as versatile as Van de Graaff accelerators in terms of the selectivity of electron energy and dose rate, they can, nevertheless, be useful in studying electron irradiation induced damage in semiconductors. The results presented below were obtained using a Philips SL75 linear accelerator at the Pretoria Academic Hospital in Pretoria. Palladium SBDs fabricated on a Si-doped (to 1×10^{16} cm^{-3}) epitaxially grown GaAs layer, were irradiated with 10 MeV electrons at a dose rate of 2.4×10^{10} e$^-$cm^{-2}s^{-1} to a dose of 3.3×10^{13} e$^-$cm^{-2}. Curve (d) in Figure 8 shows that the main defects in the GaAs are EE1–EE6 [77]. Table III shows that the EE1, EE2 and EE4 are identical to Eβ1 (E1), Eβ2 (E2) and Eβ4 (E3), respectively [74, 77]. The near-midgap level, EE6, will be discussed in more detail in the next section.

There are two important differences between the spectra of defects produced by irradiation in linear accelerators and those obtained after

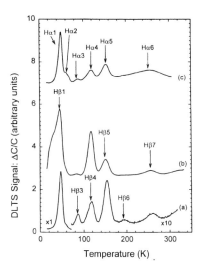

Fig. 9. DLTS spectra of Be-doped (to $1 - 2 \times 10^{16}$ cm^{-3}) MBE grown p-GaAs irradiated with β-particles (curves (a) and (b) [76], and α-particles (curve (c)). All spectra were recorded at $f = 46$ Hz, i.e., a decay time constant of 9.23 ms and $V_r = 2$ V. The filling pulses for curves (a), (b) and (c) were $V_p = 0.4$ V, 2.4 V and 0.4 V, respectively.

irradiating SBDs with electrons from the ^{90}Sr source. Firstly, the EE3 level (same as Eβ3) is introduced at a rate of 5–6 times that of Eβ3, i.e., at roughly the same rate as Eβ4. Secondly, the EE6 introduced by 12 MeV electrons is not the same as the E5 introduced by 1 MeV electrons.

4.4. Near Mid-gap Levels Introduced by High Energy Electron Irradiation

In this section the focus is on the E4 and E5 electron irradiation induced defects in Si-doped (to 1×10^{16} cm^{-3}) n-GaAs. First, consider these defects in GaAs irradiated with 1 MeV electrons in a Van de Graaff accelerator (curve (a) in Figure 10). The E4 and a broad E5 can clearly be seen together with a shoulder on the high temperature side of E5 belonging to the EL2, which is always present in OMVPE grown n-GaAs. The signatures of E4 and E5 (Table III) are in good agreement with those reported before [161]. The slight asymmetry of E5 suggests that that it may be the superposition of two closely spaced defect peaks, say E5a and E5b. Curve (b) of this figure depicts the spectrum of the same GaAs irradiated with electrons from a ^{90}Sr source which emits electrons with energies of up to 2 MeV. Eβ5 is the same as E4 but in this case the E5 peak is broader than in curve (a). A spectrum of β-particle irradiated GaAs recorded at 2 mHz showed a clear split of the

TABLE IV
Electronic properties of prominent paticle induced defects in epitaxially grown p-GaAs. Error margins can be found in the referenced texts.

Defect label	η (cm^{-1})	$E_T - E_V$ (eV)	$\sigma_a \times 10^{-16}$ (cm2)	T_{peak}[a] (K)	Dopant	Refs.	Proposed Assignment
H0	0.8	0.06	1.6×10^{-16}	43[b]	Zn	[15, 66, 123]	
Hβ1	1.5	0.08	2.8×10^{-15}	49	Be	[76]	V_{As} [66]
Hα1	4705	0.08	2.6×10^{-15}	49	Be	[67]	
Hβ2	0.1			60	Be	[76]	??Also in Zn-doped GaAs
Hα2	900			60	Be	[67]	
Hβ3	0.03	0.16	1.7×10^{-14}	87	Be	[76]	??Also in Zn-doped GaAs
Hα3	490	0.15	5.9×10^{-15}	87	Be	[67]	
Hβ4	0.05 – 0.20	0.20	1.2×10^{-15}	118	Be	[76]	Only in Be-doped GaAs
Hα4	1050 – 4335	0.20	1.2×10^{-15}	118	Be	[67]	
H1	0.1 – 0.7	0.29	5×10^{-15}	155[b]	Zn	[15, 123]	
Hβ5	0.2	0.30	1.2×10^{-4}	154	Be	[76]	V_{As}–As_i [66]
Hα5	2100	0.30	1.6×10^{-14}	154	Be	[67]	
H2	[c]	0.41	2×10^{-16}	250[b]	Zn	[15]	Only in Cu contaminated GaAs [16, 66]
Hβ7	0.04	055	7.4×10^{-14}	256	Be	[76]	Broad peak: closely spaced defects or extended defects?
H3	0.2	0.71	1.2×10^{-14}	346[b]	Zn	[15]	

[a] At a lock-in amplifier frequency of 46 Hz, i.e. a decay time constant of 9.23 ms.
[b] Calculated from the σ_a and E_t values in [15].
[c] Depends on Cu concentration.

E5 peak [77], indicating that it consists of two discrete levels, labelled Eβ6 (E5a) and Eβ7 (E5b). The spectrum of GaAs irradiated with 12 MeV electrons (curve (c) of Figure 16) shows that EE6 (E5a) is the dominant E5 level.

If we assume that the defects E5a, Eβ6 and EE6 are the same and that E5b and Eβ7 are the same, then it follows from curves (a), (b) and (c) that the introduction rate of E5a increases relative to that of E5b with increasing electron energy. A possible explanation for this may be found by considering the collision cascades caused by energetic Ga or As atoms which are displaced if the electron energy is large enough to transfer more than E_d to them. Simple calculations using Eqs. (8a) or (8b) show that the maximum kinetic energy of displaced Ga and As atoms will be about 63 eV for 1 MeV electrons and 4.8 keV for 12 MeV electrons. Although only very few displaced atoms will have this maximum kinetic energy, it is nevertheless clear that the atoms displaced by 12 MeV electrons will have significantly more kinetic energy

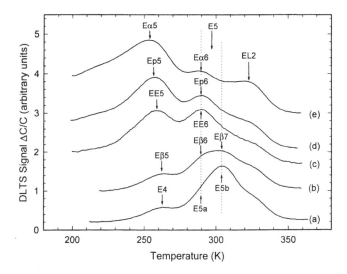

Fig. 10. DLTS spectra [77] of Si-doped (to 1×10^{16} cm^{-3}) OMVPE grown n-GaAs irradiated with: (a): 1 MeV electrons at a dose of 1×10^{15} cm^{-2} in a Van de Graaff accelerator, (b) 1–2 MeV electrons at a dose of 1×10^{15} cm^{-2} from a ^{90}Sr radio-nuclide, (c) 12 MeV electrons at a dose of 1×10^{15} cm^{-2} in a linear accelerator, (d) 1 MeV protons at a dose of 1×10^{11} cm^{-2} in a Van de Graaff accelerator, and (e) 5.4 MeV α-particles at a dose of 1×10^{10} cm^{-2} from an ^{241}Am radio-nuclide. All curves were recorded at a frequency of $f = 0.2$ Hz, i.e., a decay time constant of 8.49 s, using a filling pulse width and amplitude of $t_p = 5$ ms and $V_p = 0.8$ V, respectively, and a quiescent bias of $V_r = 1$ V.

than atoms displaced by 1 MeV electrons. Therefore, the secondary defects produced by 12 MeV electrons will be more extensive than, and therefore be different from, defects produced by 1 MeV electrons.

4.5.   Sub-threshold Energy Electron Irradiation

The threshold electron energies for defect production by elastic collisions in GaAs is about 200 keV [15, 56]. It has, however, been known for some time that irradiation with electrons having energies well below the threshold energy affects semiconductor properties [78, 79, 80]. The defects responsible for these effects were not observed until 1984 when the first reports emerged regarding electrically active defects introduced by exposure of semiconductors to sub-threshold electrons. Kleinhenz *et al.* [81] reported the properties of electron traps introduced in GaAs during exposure to 2–20 keV electrons, and Auret and Mooney reported the properties of electron traps introduced in p-Si [82] and hole traps created in n-Si [83] during electron beam evaporation of Schottky contacts. This was followed by more reports of "sub-threshold" defects introduced during electron beam deposition of SBDs on GaAs [84] and exposure of GaAs to 0.2–3 keV electrons [85]. In the latter report it was concluded that some defects introduced in GaAs by sub- and above threshold energy electrons have "very similar" properties. However, the introduction rates of sub-threshold irradiation induced defects were found to be much lower than those of defects produced by above threshold energy electron irradiation, suggesting that the introduction mechanisms of these defects are different.

In a follow-up experiment, SBDs were fabricated before low energy electron irradiation to ensure that only electrons, and no heavier particles, reach the GaAs [86]. To minimise electric field enhanced emission (Section 8) when determining defect properties, and thus to increase the accuracy of the measured defect signatures, undoped n-type epitaxial layers (with a free carrier density of 2.5×10^{14} cm^{-3}), grown by OMVPE, were used. The Ti SBD contacts, 0.73 mm in diameter and 40 nm thick, which were fabricated on the epitaxial layers, were thin enough to allow electrons with energies as low as approximately 1.7 keV to pass through them [87], and were irradiated in three different ways. The first set of diodes were irradiated with 3 keV (sub-threshold) electrons to a dose of about 10^{17} e$^-$cm^{-2} in an AES system. The second set of diodes were irradiated in a linear accelerator, using an irradiation geometry identical to that in the AES system, with 3 MeV (above-threshold) electrons at a dose of 3.3×10^{11} cm^{-2}. The third set of diodes were first irradiated with 3 keV electrons in the AES system and thereafter with 3 MeV electrons in the linear accelerator at the same doses as above.

In Figure 11, the spectra of high and low energy electron irradiation induced defects observed in this experiment are compared. Curve (c) in Figure 11 shows that the main defects introduced after exposure to 3 keV electrons are Ee1, Ee3, Ee4 and Ee5 with activation energies of 0.049, 0.130, 0.195 and 0.305 eV below the conduction band, respectively. The signatures of these defects are the same as those of defects introduced during electron beam deposition of Pt and Ti on the same GaAs epitaxial layer [88]. Curve (b) in Figure 10 shows that the defect spectrum of the 3 MeV electron irradiated GaAs contains three defects, EE1, EE2 and EE4, described in Section 4.3, with energy levels at 0.045, 0.142, and 0.375 eV below the conduction band, respectively. The energy values of EE1 and EE2 are, within an experimental error of 1 K in temperature measurement, the same as those reported by Pons *et al.* [15] for the E1 and E2 introduced during high energy electron irradiation. Finally, curve (a) clearly shows that there are two sets of defects present in the material irradiated by both sub- and above threshold energy electrons. The peaks of the first set occur at the same temperatures as those of the sub-threshold defects created by 3 keV irradiation which were shown in curve (c), while the peaks of the second set coincide

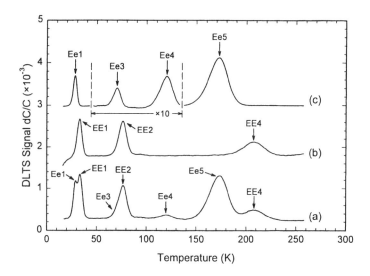

Fig. 11. DLTS spectra [86] of undoped OMVPE grown n-GaAs irradiated with a combination of 3 keV and 3 MeV electrons (curve (a)), 3 MeV electrons (curve (b)) and 3 keV electrons (curve (c)). All three spectra were recorded at $f = 46$ Hz, i.e., a decay time constant of 9.23 ms, $V_r = 2$ V and $V_p = 1.2$ V.

with those of defects introduced during high energy electron irradiation as shown in curve (b). The conclusions drawn from this experiment are, firstly, that sub-threshold electrons introduce defects in GaAs and, secondly, that the properties of these defects in undoped GaAs are similar to, yet clearly discernible from, those of defects introduced by above-threshold electron irradiation of the same GaAs.

The existence of defects produced by sub-threshold electrons have been clearly demonstrated, and several mechanisms for their introduction have been reviewed and proposed [89, 90]. However, a satisfactory model that accounts for all the DLTS observations is still lacking. It has been proposed by Christensen *et al.* [91] that the defects introduced in Si during electron-beam evaporation of Pt are not created by the low electrons, but by negatively charged Pt ions which are accelerated from the filament region towards the sample. This is an attractive suggestion because Pt ions with an energy of a few keV are certainly capable of producing defects. However, it is difficult to reconcile these results with those obtained by Auret *et al.* [86], discussed above. Clearly, this is an area which needs to be explored further, especially in view of the important role that electron beam processing plays in the microelectronics industry.

4.6. Summary

The electronic properties of the main defects introduced in GaAs by MeV electron irradiation are accurately known. Initially, it has been argued that E1 and E2 are two charge states of the isolated arsenic vacancy, V_{As}, and that E3 is related to V_{As}–As_i pairs. This assignment of E1 and E2 has been strongly questioned by Von Bardeleben [64], who proposed that E1 and E2 may belong to the distorted arsenic antisite, As^-As_3. The introduction rates of the main defects in room temperature irradiated GaAs (the shallow levels E1 and E2) are about 2 cm^{-1}. Irradiation at elevated temperatures (300°C) introduced an electron trap, $I1$, at about one half the introduction rate of E1 and E2, with approximately the same signature as the EL2 (which has been shown to be an effective compensation center in n-GaAs).

It was demonstrated that radio-nuclides (^{90}Sr) and linear accelerators are useful sources for providing electrons to study irradiation induced defects, since they produce the same set of defects as electron irradiation in Van de Graaff accelerators. Using electrons from these sources, it was found that electron irradiation of epitaxially grown n-GaAs introduced a metastable defect which seems to be related to the presence of Si dopant atoms, or accompanying impurities, in GaAs. The 1–2 MeV electrons from a ^{90}Sr source introduce this metastable defect at about one fifth the introduction rate of E3, while 12 MeV electron irradiation of the same material in a linear accelerator introduces this defect at the same rate as E3. Whereas the E4 signature was the same for 1–2 MeV electrons from the ^{90}Sr source and 12 MeV electrons from a linear accelerator, the E5 defect peak was shown to consist of two components, E5a and E5b, of which the introduction rate of E5a increases with increasing electron energy. This is speculated to be the result of collision cascades, produced by displaced Ga and As atoms, of which the extent increases with increasing electron energy.

Electron irradiation introduces several impurity related defect complexes in p-GaAs: a Be-related hole trap was observed in Be-doped MBE grown GaAs and Cu-related radiation induced hole traps have been observed in Zn-doped OMVPE grown GaAs containing Cu. In n-GaAs, the Eβ3 and EE3 defects could only be observed in Si-doped GaAs, but not in undoped or S-doped GaAs.

It has been demonstrated that sub-threshold (3–20 keV) electron irradiation introduced several defects in epitaxially grown GaAs. These defects have similar, but not identical, signatures as the defects introduced by high energy (above-threshold) electron irradiation. The important role that low energy electrons play in the microelectronics manufacturing industry necessitates further investigations to determine the structure and introduction mechanism of the defects introduced by sub-threshold electrons.

5. PROTON IRRADIATION

5.1. Overview

High energy ($\approx$ MeV) proton irradiation of GaAs removes of the order of 10^3 times more carriers per incident particle than electrons with the same energy [71]. This has obvious advantages in applications such as device isolation where it is important to create defects at a high rate to ensure a high processing throughput. On the other hand, due to its high carrier removal rate, proton irradiation has a much more severe influence than electron irradiation on device operation where devices have to operate in a radiation environment, for example, in outer space. Most initial studies of proton irradiation of semiconductors focused on carrier removal and the effect of annealing on carrier recovery. These studies were typically conducted using protons with energies from 60 keV to a few MeV. TRIM calculations show that the proton range and straggle increase from 0.48 μm and 0.10 μm, respectively, at 60 keV to about 33.1 μm and 1.3 μm, respectively, at 2 MeV (Figure 3). These numbers illustrate that, in contrast to high energy electrons, high energy protons can be used to modify GaAs up to a depth which can be selected by using protons with a suitable energy.

One of the first DLTS investigations of defects responsible for carrier removal in GaAs by proton irradiation was reported by Lang [58], who reported that the same set of defects (E1–E5) were introduced in LPE grown GaAs by 1 MeV electrons, 400 keV protons and 1.8 MeV He ions, but in different relative concentrations. Thereafter, several authors reported on DLTS characterization of defects introduced in GaAs by room temperature proton irradiation [92–94]. Some of the defects thus observed at proton irradiation energies of 60 keV and above, were found to be similar to those introduced by electron irradiation [71], but appeared to have different introduction rates and somewhat different properties [94]. In particular, Rezazadeh and Palmer [71] found that the introduction rate of E4 for 1 MeV protons, compared with those of E1–E3, is much higher for proton than for electron irradiation, leading them to conclude that E4 is formed during high energy ion recoils or by a combination of simple defects.

Apart from the E1–E5 defects, other defect levels have also been observed after proton irradiation. One reason for this was suggested to be that their introduction depends on the crystal quality and impurities therein [95]. Guillot *et al.* [93] reported that 95 keV protons introduce a level EX1 at 300 K in Cl-doped VPE grown n-GaAs which was not observed by electron irradiation. Li *et al.* [96] have reported on the increase of defect density with increasing proton dose, and found that an additional hole trap was introduced upon increasing the proton dose from 10^{11} cm^{-2} to 10^{12} cm^{-2}. Auret *et al.* [97] reported that 200 keV proton irradiation of VPE grown

GaAs introduced a shallow level with an energy of 0.035 eV below the conduction band and a DLTS signal partly superimposed onto that of the E1. It was also found [98, 99] that proton irradiation of Si-doped OMVPE grown n-GaAs introduced a defect, Ep3, which was not observed in undoped n-GaAs. Ep3 exhibits the same electronic and metastable properties as EE3 and Eβ3 (Section 6), which has subsequently been linked to the presence of Si, or impurities accompanying its introduction during crystal growth, in GaAs (Section 7). After irradiating p-GaAs at 300 K with 130 keV protons, Guillot *et al.* [93] reported that the DLTS spectrum was much less definitive than after electron irradiation of the same material. In addition to the main electron irradiation induced defects, H0–H3, they found that proton irradiation introduces additional defects, HX2 and HX3, thought to be related to impurities in the GaAs.

A recent interesting observation was that, under certain irradiation conditions, proton irradiation introduced the EL2 defect in LPE grown n-type GaAs. Using DLTS, Brunkov *et al.* [167] reported that 6.7 MeV proton irradiation at a dose of 1×10^{10} cm^{-2} introduced the EL2, exhibiting the persistent photoquenching effect. When increasing the dose to 1×10^{11} cm^{-2}, they observed a monotonic increase in the EL2 concentration. However, at a dose of 1×10^{12} cm^{-2} and above, the defect interactions intensified and the electronic properties of the EL2 seem to have changed.

Several investigations of proton irradiation at low temperatures were also conducted. Guillot *et al.* [93] reported that the EX2 level, introduced by 95 keV proton irradiation at 77 K, could be removed after a 15 minute anneal at 300 K. Siyanbola and Palmer [100] irradiated n-GaAs at 85 K–120 K with 1 MeV protons and, using capacitance-voltage measurements, have identified three stages of thermal annealing of the irradiation induced electron traps, at 180 K, 235 K and 280 K, with the latter two stages corresponding to the Thommen stages I and II. From these experiments they concluded that defects that anneal in the 280 K stage, have electron trapping levels lying near or below the middle of the GaAs band gap. The same authors [101] have found that the concentration of a proton irradiation induced hole trap, HNI(0.42), with a level at $E_V + 0.42$ eV in n-GaAs, was reduced by successive temperature scans to about 300 K. Siyanbola and Palmer [102] subsequently demonstrated that, after proton irradiation of n-GaAs at 120 K, HNI(0.42) and another radiation induced hole trap, HNI(0.25$'$), are removed in the 280 K anneal stage. They concluded that HNI(0.42) is the same as the H2 defect in irradiated p-GaAs which was found to be stable at 300 K in that material [66]. Irvine and Palmer [103] used capacitance-voltage measurements to study defect annealing in the 85 K–500 K range after irradiating n-type GaAs and n-type Al$_x$Ga$_{1-x}$As ($x = 0.22$) at 85 K with 1 MeV protons. Their results suggest the involvement of Ga atom defects in the Thommen annealing stage II, centered near 280 K, leading them to conclude that such defects are not involved in the stage I annealing near 235 K. Their results also reveal a correlation between Thommen stage III annealing at 450–500 K and the removal of As-related defects.

The most interesting recent results pertaining to proton irradiation were presented by Hartnett and Palmer [104] who investigated the effect of uniaxial stress up to 0.4 GPa on the majority carrier DLTS spectra of 1 MeV proton irradiated n-GaAs. They found that for E1, E2 and E3, uniaxial stress applied along the $\langle 100 \rangle$ direction caused an increase, but no broadening, of the DLTS peak and an increase of the mean ionization energy of the defect, but that $\langle 110 \rangle$ applied stress produces both broadening of the peak and an increase in the mean ionization energy of the defect. For the E1 defect level, the effect of 0.4 GPa $\langle 110 \rangle$ stress is to split its peak into two clear peaks. By a detailed analysis of the E1, E2 and E3 peak shapes they deduced that $\langle 110 \rangle$ stress causes splitting of each of the respective electronic levels into two levels of equal population. This $\langle 100 \rangle$ and $\langle 110 \rangle$ uniaxial stress data strongly suggest that the E1, E2 and E3 defects have C3v crystallographic symmetry, i.e., that each has a structure that contains a single $\langle 111 \rangle$ symmetry axis. For the E3 defect the data are consistent with the proposal of previous work [15], namely that this defect is an arsenic Frenkel pair, provided that the interstitial-vacancy direction is along the $\langle 111 \rangle$ direction. However, for E1 and E2 the data of Hartnett and Palmer do not support the identification of these defects as simple arsenic vacancies of Td symmetry in different charge states [15]. Note, however, that these results for E1 and E2, are not in conflict with the model proposed by Von Bardeleben [64], namely that E1 and E2 are levels of the distorted arsenic antisite As_{Ga} (i.e., $As^- As_3$).

5.2. Dependence of Defect Introduction on Proton Energy

In this section, we discuss DLTS results obtained after irradiating OMVPE grown n-type GaAs with 5 keV to 200 MeV protons [105]. In Table II, some proton irradiation sources and typical conditions employed to introduce defects, as well as some prominent proton irradiation induced defects and the nomenclatures used to identify them, are summarized.

First, consider 40 keV–1 MeV proton irradiation performed in a 1.7 MeV Pelletron implanter (ANU, Canberra, Australia). Pd SBDs (130 nm thick) on a Si-doped (to 10^{16} cm^{-3}) epitaxial layer, grown by OMVPE, were irradiated with 1×10^{10} H$^+$cm^{-2} at a dose rate of 5×10^9 H$^+$cm^{-2}s^{-1} at energies of 1 MeV, 400 keV, 95 keV and 40 keV. Typical DLTS spectra recorded using these diodes are depicted in curves (b)–(e) of Figure 12, and show that proton irradiation in this energy range introduces the defects Ep1–Ep6, with relative concentrations depending on the irradiation energy. The Ep6 cannot clearly be observed in the 1 MeV and 400 keV samples, due to the fact that its introduction rate decreases with increasing proton energy (more below), and consequently its peak is masked by the much larger EL2 peak to its high temperature side. Defects Ep1, Ep2, Ep4 and Ep5 are identical to the E1, E2, E3 and E4 defect levels (Table III), respectively, introduced by 1 MeV electron irradiation [15]. Ep3 is metastable and electronically identical to the Eα3/Eβ3 defect introduced by α-particle/β-particle irradiation of the same material (Sections 4, 6 and 7). The activation energy, E_T, and capture cross section, σ_a, of Ep6 (0.78 eV and 1×10^{-12} cm^2, respectively) are similar, but not identical, to that of the L2 defect detected by Yuba et al. [94] in electron and proton irradiated LPE grown n-GaAs. Ep6 has the same signature as EE6 (E5a) introduced by 12 MeV electron irradiation (Section 4.4 and Figure 16, curves (c) and (d)).

Next, consider the defects introduced in the same GaAs by 200 MeV proton irradiation at a dose of 10^{13} H$^+$cm^{-2} in a cyclotron (National Accelerator Center, Faure, South Africa) [105]. The DLTS spectrum (curve (a) in Figure 12) shows that the most significant difference between 200 MeV and 40 keV–1 MeV proton irradiation is that the 200 MeV spectrum contains a broad asymmetric peak, Ep7, just below 300 K. In contrast to 40 keV–1 MeV proton irradiation, the Ep7 peak height is larger than that of Ep3 and Ep4, and is about the same as the Ep1 and Ep2 peak heights. The activation energy, E_T, and capture cross section, σ_a, of Ep7 were determined as 0.68 eV and 4×10^{-12} cm^2, respectively. This is very similar to that of the En5 [106] and the "U-band" [107], observed in neutron-irradiated n-GaAs. The similarity between Ep7 and the neutron introduced defects, En5 and the "U-band", suggests that Ep7 may be created by neutrons or other nuclear fragments produced in GaAs when high energy protons penetrate, or pass through, the Ga and As nuclei. As has been shown in Section 2, the threshold energy for a proton to enter Ga and As nuclei is about 6 MeV. This means that 200 MeV protons can easily penetrate Ga and As nuclei and thus cause nuclear reactions, including neutron emission by, for example, (p, n) reactions. The neutrons thus created can, in turn, introduce Ep7.

Low energy (5 keV) proton bombardment of n-GaAs was investigated using a section of the same wafer used above for the irradiations [105]. It was first bombarded with 5 keV protons at a dose of 1×10^{12} cm^{-2}, and thereafter Pd SBDs were deposited onto it by resistive evaporation. The range and straggle of 5 keV protons calculated using TRIM are 586 Å and 297 Å, respectively (Figure 3), which means that all the implanted protons reside within the depletion region probed by DLTS, close to the surface. Spectra 1, 2 and 3 in curve (f) of Figure 12 were recorded at specific bias and pulse conditions, chosen to probe different regions below the GaAs surface. These curves show the presence of the defect peaks Ep2, Ep3, Ep5 and Ep8. Note that Ep8, with $E_T = 0.65$ eV $\sigma_a = 5 \times 10^{-13}$ cm^2, has a peak in the same temperature region as Ep5 and Ep6, but its signature is discernibly different. Ep3 and Ep8 are the main defects, and no As Frenkel pairs, E3 (Ep3) could

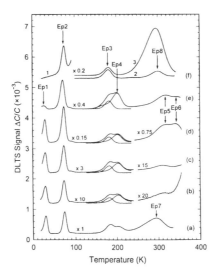

Fig. 12.   DLTS spectra [105] of Si-doped (to 1×10^{16} cm^{-3}) OMVPE grown n-GaAs irradiated with protons of different energies. For curve (a) the proton dose and energy was 10^{13} cm^{-2} and 200 MeV, respectively. Curves (b)–(e) were recorded after irradiating with 10^{10} H$^+$cm^{-2} at energies of 1 MeV, 400 keV, 95 keV and 40 keV, respectively. For curve (f) the dose and energy was 10^{11} H$^+$cm^{-2} and 5 keV. All spectra were recorded at $f = 46$ Hz, i.e., a decay time constant of 9.23 ms. For curves (a)–(e), $V_r = 2$ V and $V_p = 1$ V were used, while several different values of V_r and V_p had to be used to reveal all the peaks in curve (f).

be detected in measurable concentrations. Although a conserted effort was made to detect Ep1 (E1), by recording data at various depths below the interface *under as low as possible electric field conditions* to prevent peak shifts, it could not be detected. Because E1 was detected after 5 keV He-ion bombardment of the same GaAs (Section 6), it would seem as if the E1 can be passivated by H in its vicinity, which will be the case in low energy proton bombarded GaAs.

Several qualitative trends can be identified from Figure 12. Firstly, the Ep1/Ep2 ratio decreases from almost unity for 200 MeV proton irradiation to zero for 5 keV proton irradiation. We suggest that the charge state occupation giving rise to Ep1 is suppressed for some reason, possibly similar to the suppression of the population of the $V_2^{0/-}$ state of the divacancy in Si by stress fields resulting from extended defects [108]. In the case of 5 keV H$^+$-ions, the E1 suppression may result from H-passivation. Secondly, the ratios of the Ep5 and Ep6 concentrations to those of Ep1 and Ep2 increase with decreasing irradiation energy from 1 MeV to 40 keV. Because Ep5 is the same as E4, this increase in Ep5 concentration with decreasing energy is understood in terms of the suggestion by Rezazadeh and Palmer [71] that E4 may be formed by high energy ion recoils, of which the concentrations will increase as the proton energy decreases, because the nuclear stopping increases with decreasing energy. Thirdly, it is instructive to note that Ep5–Ep8 have capture cross sections larger than 10^{-13} cm^2. This suggests that, either these defects are physically large, or that they contain multiple positive charges, which in turn seems to suggest that Ep5–Ep8 may be related to defect clusters of different sizes. Finally, no Frenkel pairs (E3, i.e. Ep4) could be observed in 5 keV proton bombarded GaAs. The absence of Frenkel pairs in the end-of-range region of 5 keV protons indicates, either, that they are not formed at all, possibly due to the high energy transfer rate in this region resulting in spike formation, or, alternatively, that after their formation, Frenkel pairs are immediately transformed into, or trapped by, larger defect clusters.

The spatial variation of the irradiation induced defect concentrations was qualitatively assessed using the method of Zohta and Watanabe (Section 3, [46]). The profiles thus obtained revealed that, whereas the introduction rates of defects created by 200 MeV, 1 MeV and 400 keV protons are constant in the region analyzed by DLTS (first 0.7 μm below the surface), the concentration of defects introduced by 5 keV, 40 keV and 95 keV decreases strongly away from the surface into the GaAs. This is illustrated in Figure 13 for Ep2, Ep3 and Ep4 introduced by 40 keV and 95 keV protons. A comparison of the profiles of Ep4 introduced by 40 keV and 95 keV protons (Figure 13), shows that the maximum Ep4 concentration for 95 keV protons is about 0.5 μm below the surface. This compares well with the proton range (0.6 μm into GaAs) and the maximum of the vacancy profile (0.5 μm into the GaAs), calculated from TRIM, for 95 keV protons incident on the Pd/n-GaAs SBD. For 40 keV protons, Figure 13 shows that the maximum Ep4 concentration is somewhere within the first 0.2 μm. This, too, is in agreement with the proton range (0.1 μm into the GaAs) and the maximum of the vacancy profile (0.04–0.08) μm, calculated using TRIM. The defects introduced by 5 keV protons are too close to the surface to be profiled accurately up to their expected maximum positions. The profiles obtained (not shown here), nevertheless indicate that the average Ep2 concentration in the first 0.1 μm below the surface is about two orders of magnitude lower than that of Ep3 in the same region.

The introduction rates of proton irradiation induced defects, as a function of distance into the irradiated GaAs, were calculated from their concentration profiles, as discussed above. These profiles showed that closer to the surface (e.g. between 0.2 and 0.3 μm), the introduction rates increase with decreasing proton energy for the 40 keV–200 MeV range. This is illustrated in Figure 14 where the average introduction rate at 0.2–0.3 μm below the surface is plotted as function of proton energy. The trends in Figure 14 are in qualitative agreement with Figure 1, which shows that in this energy range, a decrease

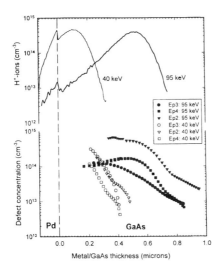

Fig. 13. Depth profiles [105] of the Ep2 (E2), Ep3 and Ep4 (E3) defects introduced by 40 keV and 95 keV proton irradiation, calculated using the method of Zohta and Watanabe [46].

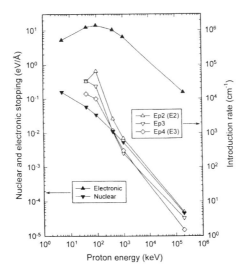

Fig. 14. Nuclear and electronic stopping [105] (solid symbols, left hand axis), and average introduction rate at 0.2 μm–0.3 μm below the surface of the Ep2, Ep3 and Ep4 proton irradiation induced defects (open symbols, right hand axis), as function of proton energy.

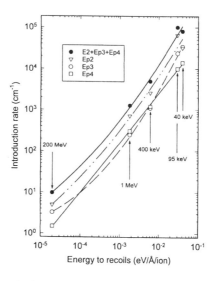

Fig. 15. Average introduction rate [105] at 0.2 μm–0.3 μm below the surface of the Ep2, Ep3 and Ep4 proton irradiation induced defects as function of energy transferred to Ga and As atoms during nuclear stopping.

in proton energy is accompanied by an increase in nuclear stopping, which, in turn, leads to an increase in defect concentration (Figure 14). Note that it is not possible to extend these trends to lower energies, since it is not possible to accurately obtain defect concentration profiles closer than 0.1 μm below the surface, i.e., in the region where the 5 keV proton defect concentrations reach a maximum. To further illustrate the dependence of defect introduction on nuclear stopping, the average defect introduction rate measured at 0.2–0.3 μm below the surface was plotted as a function of the average energy transferred from protons to recoiling Ga and As atoms, calculated from TRIM. Figure 15 shows that, for the energy range of 40 keV to 200 MeV, the measured defect introduction rate increases monotonically with the energy transferred during nuclear stopping, which in turn increases with decreasing proton energy. As mentioned above, it was not possible to extract any meaningful values for introduction rates of defects produced by 5 keV proton irradiation.

5.3. Summary

The defects Ep1, Ep2, Ep3, Ep4 and Ep5, introduced by proton irradiation of n-GaAs, are the same as the 1–2 MeV electron irradiation induced defects E1, E2, Eβ3 (EE3), E3 and E4, respectively, whereas Ep6 seems to be the same as the EE6 (E5a) introduced during 12 MeV electron irradiation. However, the relative concentrations of the proton induced defects differ from those introduced by electron irradiation. In addition, proton irradiation introduces the Ep7 and Ep8 defects, of which the introduction rates strongly depend on the proton energy. It was speculated that Ep7 is caused by nuclear reactions triggered by protons with energies above 6 MeV — the threshold energy for protons to penetrate Ga and As nuclei. Ep3 is metastable and is the same as Eβ3 and Eα3 in electron and α-particle irradiated Si-doped n-GaAs.

Concerning the structure of the radiation induced defects, it is important to note that whereas the E1 and E2 have for a long time been assumed to be two charge states of the isolated V_{As}, this assignment has recently been shown to be incorrect by a combination of uniaxial stress measurements and DLTS, which revealed that E1 and E2 cannot be simple arsenic vacancies of Td symmetry. However, for the original assignment of E3, the uniaxial stress data are consistent with the model that this defect is an As Frenkel pair, e.g. $V_{As}–As_i$, provided that the interstitial-vacancy direction is along the $\langle 111 \rangle$ direction.

Proton irradiation in the 5 keV to 200 MeV energy range showed that the relative concentrations and introduction rates of proton irradiation induced defects depend on the proton energy and on the depth at which they are measured in the GaAs. In particular, whereas the E1/E2 peak height ratio decreases from almost one for proton energies above 1 MeV to zero for 5 keV protons, the introduction rates of Ep3, Ep5 and Ep6 increase with decreasing proton energy down to 40 keV, below which no accurate introduction rates can be calculated. In 5 keV proton bombarded GaAs, the metastable Ep3 and the Ep8 are the most prominent defects. Further, the formation of Ep7 and Ep8 depends on the proton energy: Ep7 is only detected for proton energies of 200 MeV, and Ep8 was only observed in 5 keV proton bombarded GaAs. Finally, in the end-of-range region of low energy (5 keV) proton bombarded GaAs, no Frenkel pair defects (E3) could be observed.

6. He ION IRRADIATION

6.1. Overview

The effect of high energy He-ion (α-particle) irradiation on material and device properties have been reported [71, 109, 110], but not as extensively as for electron and proton irradiation. One of the most important findings of these investigations was that the carrier removal rate of high energy He-ions is about four orders of magnitude higher than for electrons and about 3 times higher than for protons [71]. He-ion implantation, therefore, would seem to be more effective for device isolation than proton irradiation. Another application of high energy He-ion irradiation was suggested by Zakharenkov et al. [112] who found that high energy (15–20 MeV) α-particle irradiation can be used for transmutation doping of GaAs, yielding a total transmutation factor, K_{tr}, of about 2×10^{-4} cm^{-1}. This is considerably higher than for proton irradiation of the same material under the same conditions for which $K_{tr} \approx 10^{-6}$ cm^{-1} was measured. For Si, it has recently been demonstrated that lifetime control can be achieved by the voids induced by He-ion implantation [113]. Naturally, for successfully realizing these applications, it is essential to know the properties of defects thus introduced so that their influence on the electronic properties of GaAs and on the characteriztics of devices fabricated on it, can be predicted, for which the defect properties are required.

The first DLTS results of defects produced by He-ion implantation of GaAs were reported by Lang [56] who observed that 1.8 MeV He ions introduced the same set of defects as 1 MeV electrons (E1–E5), but that the defect peaks were broader than for the electron irradiation induced defects. Rezazadeh and Palmer [71] observed (from DLTS spectra recorded between 77 K and 380 K) that 4.1 MeV α-particle irradiation of Sn-doped VPE grown n-GaAs introduced the E2–E4 levels, but they could not observe the E5 level which Lang observed in LPE grown GaAs. The reason for this is most probably that the E5 concentration is too low compared with that of the EL2, which has a peak near that of E5, and therefore masks the E5 peak. The E4 concentration, compared to that of E2 and E3, was found to be much higher than for electron irradiation of the same material. This is important because E4 lies much deeper in the band gap than E1–E3. It is therefore expected to be an important compensating center at room temperature, and as such play an important role in the performance of depletion region based devices.

6.2. Alpha-particle irradiation using an ^{241}Am radio-nuclide

For most of the studies of He-ion irradiation of semiconductors, Van de Graaff accelerators or ion implanters were used to produce the He-ions. In this section it is demonstrated how α-particles (He-ions), liberated by an ^{241}Am radio-nuclide, can be used to produce and study radiation induced defects. These radio-nuclides are inexpensive and small, and can easily be incorporated in a cryostat for low temperature irradiation. Another major advantage of these sources is that the dose rate at which they liberate particles is at least 10^3 times lower than that in accelerators and implanters (Table II). The consequences of this are, firstly, that the temperature and Fermi level position remain constant during irradiation, and secondly, that low doses, required for irradiating low carrier density semiconductors for DLTS studies, can easily be achieved.

For the first reported DLTS investigation of defects produced by α-particles from a radio-nuclide, Rezazadeh and Palmer [71] used a 150 μCi ^{241}Am source to irradiate Al/n-GaAs Schottky devices fabricated on a VPE grown n-GaAs layer (doped to 2×10^{15} cm^{-3} with Sn). They investigated the free carrier concentration profiles at 85 K and 293 K after α-particle irradiation at the same temperatures and employed DLTS to detect the defects introduced by the α-particles. The range of these α-particles in GaAs is about 13 μm, which is considerably larger than the region (1–4 μm) probed by C-V and DLTS measurements. The carrier trapping (free carrier reduction) rate was found to be $(1.2 \pm 0.4) \times 10^4$ cm^{-1} and $(2.0 \pm 0.6) \times 10^4$ cm^{-1} at 293 K and 85 K, respectively. DLTS measurements at temperatures between 85 K and 380 K revealed that α-particle irradiation introduces the E2, E3 and E4 electron traps, also reported by Lang [56] for He-ion irradiation, but little or no indication of the E5 defect was observed. The concentration ratio of E4/E3 was about 0.7–0.8 compared to 0.1 for electron irradiation of similar material.

In a more recent experiment, Auret *et al.* [74] compared the defects introduced by α-particle irradiation in undoped and Si-doped GaAs. For this study, Pd SBDs, fabricated on OMVPE grown n-GaAs layers with free carrier densities of 4×10^{14} cm^{-3} (undoped) and 1×10^{16} cm^{-3} (doped with Si), were irradiated with 5.4 MeV α-particles from an ^{241}Am radio-nuclide at a dose rate of 2×10^4 α cm^{-2}s^{-1}. The DLTS spectrum in curve (a) of Figure 16 shows the presence of three well defined defect peaks labelled Eα1, Eα2 and Eα4, as well as a defect with a less well defined and asymmetric peak, Eα5, which seem to be superimposed on a skewed baseline. A comparison of the spectra of α-particle irradiated GaAs (curve (a) in Figure 16) and electron irradiated GaAs (curve (b) in Figure 16), and of the DLTS signatures of these defects (Table III), showed that Eα1, Eα2 and Eα4 are identical to Eβ1, Eβ2 and Eβ4, respectively [74], which in turn have been shown to be the same as E1, E2 and E3 [15], respectively. Eα5 has approximately the same signature as Ep5 and EE5, which in turn appear to be the same as E4.

The DLTS spectrum for Si-doped GaAs irradiated with α-particles, curve (c) in Figure 16, shows peak shifts to lower temperatures due to the higher electric field in doped than in undoped GaAs. This is accompanied by a lowering of the barrier to emission, and consequently in E_T (Table III), which is particularly noticeable for Eα1 and Eα2. This spectrum further shows that α-particle irradiation of Si-doped GaAs introduces a metastable defect, Eα3, which can be reversibly removed and re-introduced by hole injection at low temperatures and annealing above 200 K, respectively, (Section 7), as shown in curve (d) of Figure 16. Eα3 could only be observed in Si-doped, but not in S-doped or undoped GaAs, suggesting that Eα3 is related to Si or impurities that accompany its introduction during crystal growth. It will be shown in section 6.4 that Eα3 is *not* introduced *during* α-particle irradiation, *but thereafter* at temperatures above 285 K. The energy level of Eα5 in the doped GaAs is 0.64 eV below the conduction band (Table III), and its signature seems to be the same as that of the E4 [161], Ep5 [105] and EE5 [77] in GaAs with approximately the same doping levels.

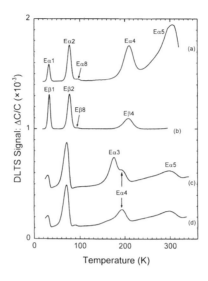

Fig. 16. DLTS spectra [74] of undoped and Si-doped (to 1×10^{16} cm^{-3}) OMVPE grown n-GaAs irradiated with α-particles (curves (a), (c) and (d)) and β-particles (curve (b)), recorded at $f = 46$ Hz, i.e., a decay time constant of 9.23 ms. For curves (a) and (b) $V_r = 8$ V and $V_p = 4$ V were used, while curves (c) and (d) were recorded using $V_r = 2$ V and $V_p = 1.9$ V. Curve (d) were recorded after applying a forward current density of 5.6 A cm^{-2} at 105 K for 10 s.

In addition to the much higher concentration of Eα5 in α-particle irradiated GaAs than Eβ5 (E4) in electron irradiated GaAs, two other differences between the spectra of electron and α-particle irradiated undoped GaAs deserve attention. Firstly, whereas the Eβ1 and Eβ2 concentrations are the same in electron irradiated GaAs, the Eα1 peak in α-particle irradiated GaAs is significantly smaller than that of the Eα2. This was observed for all irradiation doses of Si-doped and undoped GaAs, which raises the following apparent anomaly: Eα1 and Eα2 have identical electronic properties as E1 and E2, respectively, which in turn have been proposed to be two different charge states of the same defect. If this is true, and if the defect concentration is much lower than the free carrier density (as is the case here), to prevent non-exponential transients and incomplete filling due to Fermi level lowering, then the two different charge states should have the same concentrations and should be equally populated. However, this is clearly not the case for Eα1 and Eα2. As for proton irradiation, we suggest that the charge state occupation giving rise to Eα1 is suppressed for some reason, possibly similar to the suppression of the population of the $V_2^{0/-}$ state of the divacancy in Si by stress fields resulting from extended defects [108]. Secondly, an additional feature of the spectrum of α-particle irradiated GaAs is the skewing of the baseline which is seen for all α-particle irradiated curves in Figure 16. It was verified that the baseline skewing was not caused by interface damage during irradiation, but that it is characteristic of α-particle irradiation [74]. This effect was also observed in the DLTS spectra reported by Lang [56] and Rezazadeh and Palmer [71] for He-ion irradiated n-GaAs.

Finally, we compare the near-midgap levels introduced by electron-, proton- and α-particle irradiation of Si-doped, epitaxially grown, n-GaAs. It is clear from in Figure 10 that electrons, protons and α-particles introduce a defect with properties very similar to E4. However, the EE6, Ep6 and Eα6 (i.e., the E5a) introduced by 12 MeV electrons, protons and α-particles, respectively, are clearly not the same as the (Eβ6 + Eβ7) peak combination introduced by 1–2 MeV electrons or the E5b introduced by 1 MeV electrons. Since the probability of forming collision cascades is higher during 12 MeV than during 1 MeV electron irradiation, and because EE6 is observed in the 12 MeV spectrum, but not in the 1 MeV spectrum, it seems possible that E5a (=Eα6=Ep6=EE6) is related to collision cascades. These cascades occur easily for protons and α-particles because, owing to their larger mass, they transfer much more energy during collisions than electrons with the same energy.

Alpha-particle irradiation of p-GaAs was studied by irradiating Al SBDs on a MBE grown Be-doped (to 1.8×10^{16} cm^{-3}) layer using α-particles from an ^{241}Am radio-nuclide with an activity of 192 μCi cm^{-2} at a dose rate of 7.1×10^6 cm^{-2}s^{-1} at room temperature [67]. As curve (a) of Figure 17 shows, this material did not contain any defects in significant concentrations before irradiation. Curve (c) in Figure 17 shows that room temperature α-particle irradiation introduced the defects Hα1–Hα7. Hα1 and Hα5 have the same "signatures" (Table IV) as H0 and H1 [15], respectively, but Hα4 has thus far only been observed in irradiated Be-doped p-GaAs and is the same as the Hβ4 introduced by electron irradiation of the same material (Section 4 and Table IV). It is further interesting to note that the unirradiated GaAs showed no sign of the Cu-related peaks often observed in Zn-doped p-GaAs, and that spectra from the irradiated sample do not show the presence of H2, proposed to be a radiation induced defect involving Cu [66].

6.3. Low Energy He Ion Bombardment

Low energy ions play an important role in several areas of semiconductor device fabrication. During any form of ion beam processing, the ions impinging on the semiconductor produce crystal damage which affects the electro-optical properties of semiconductors [114] and the characteristics of devices fabricated on them, such as the barrier height of SBDs [115]. In the case of low energy ion beam induced defects in GaAs, most DLTS

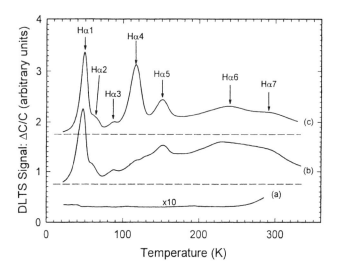

Fig. 17. DLTS spectra [67] of Be-doped (to $1 - 2 \times 10^{16}$ cm^{-3}) MBE grown p-GaAs: (a): unirradiated, (b) and (c): irradiated with α-particles. All three curves were recorded at $f = 46$ Hz, i.e., a decay time constant of 9.23 ms using $V_r = 1$ V and $V_p = 0.25$ V. Curve (c) was recorded after room temperature irradiation and curve (b) was recorded directly after irradiation at 20 K.

studies focused on the defects created in n-type GaAs by heavy noble gas ions. One of the reasons for choosing n-GaAs is that high quality (high barrier height and low reverse current) SBDs, which are required to detect low energy ion induced defects by DLTS as deep as possible in the band gap, are more easily fabricated on n-GaAs than on p-GaAs. The frequent use of heavy ions, e.g., Ar, in these studies is based on their higher etch rates and established application in industry. Not much is yet known regarding the defects introduced during low mass/low energy ion bombardment of GaAs and, in particular, very little information is available regarding low energy ion induced defects in p-GaAs. In this section some recent results for low energy He-ion bombardment of n-GaAs are discussed.

For investigating low energy He-ion bombardment of n-GaAs, a 1×10^{16} cm^{-3} doped (with Si) layer, grown by OMVPE on an n$^+$-type bulk-grown 10^{18} cm^{-3} doped GaAs substrate, was used [116]. After chemical cleaning, the sample was bombarded with 5 keV He ions at a dose of 1×10^{11} cm^{-2} at an angle of $11°$ with respect to the normal on the sample. Thereafter, Pd SBDs were resistively deposited through a metal contact mask. In curve (a) of Figure 18, the DLTS spectrum of n-GaAs bombarded with 5 keV He-ions is compared with the deconvoluted spectra recorded from GaAs irradiated with 5.4 MeV α-particles (He-ions) [curves (b)] [117]. The

DLTS spectra of both samples were recorded under identical bias and pulse conditions ($V_p = 2.6$ V and $V_r = 2$ V), selected for detecting defects from very near to the surface up to a depth of about 0.5 μm below the surface. Curve (a) shows the presence of three low energy He-ion bombardment induced electron traps, EHe2, EHe3 and EHe5, with the EHe5 peak being the most prominent. Since the spectra in Figure 18 were recorded under identical measuring conditions, it is tempting to conclude that EHe3 in the 5 keV bombarded sample may be the same defect as Eα3 in the 5.4 MeV irradiated sample, and that no other similarities exist between low and high energy He-ion bombardment induced defects.

The obvious criticism to this seemingly straightforward comparison is that, although the spectra were recorded under identical bias and pulse condition, defects in different spatial regions are being analyzed. Whereas defects introduced by high energy α-particles are distributed almost uniformly throughout the first few microns, defects created by the low energy He-ions are located close to the metal-semiconductor interface. Since the electric field in a SBD increases from zero at the edge of the depletion region to maximum at the interface, electric field enhanced emission [47] will be much more pronounced for defects close to the interface than for defects located deeper in the depletion layer. Because near surface defects experience a higher field than defects which are uniformly distributed throughout the depletion layer, the conclusions drawn from curves (a) and (b) of Figure 18 should thus be re-examined.

The inaccuracy in comparing properties of defects within different spatial regions can be reduced by recording their DLTS signals in a *narrow region*, at *equal distances* below the surface, under *as low as possible electric field*. To achieve this, the smallest possible DLTS pulse, which still results in an accurately detectable signal, has to be used in conjunction with a reverse bias which sets the depletion region edge within, or as close as possible to, the region damaged by low energy ions. Following this strategy, the spectra in curves (c) and (d) of Figure 18 were recorded. The maximum electric field here was 2.6×10^6 Vm^{-1}, whereas it was 8.4×10^6 Vm^{-1} under the conditions used for curves (a) and (b). It is clear that the EHe2 peak appears at a higher temperature than in curve (b) and, in addition, another peak which was not present in curve (b), EHe1, is now observed. Both EHe1 and EHe2 were found to be strongly field dependent, which explains why EHe2 was observed 15 K lower in curve (b) than in curve (d) and why EHe1 could not be detected at all under the conditions used to record curve (b). Most important, for the bias and pulse conditions used here, the EHe1 and EHe2 defect peaks now appear within 1 K of the peak temperatures of Eα1 and Eα2 in 1×10^{16} cm^{-3} doped GaAs, respectively (Table III).

The same pulse conditions used to record the spectra in curves (c) and (d), were used to determine the DLTS signatures of EHe1–EHe3 and Eα1–Eα3.

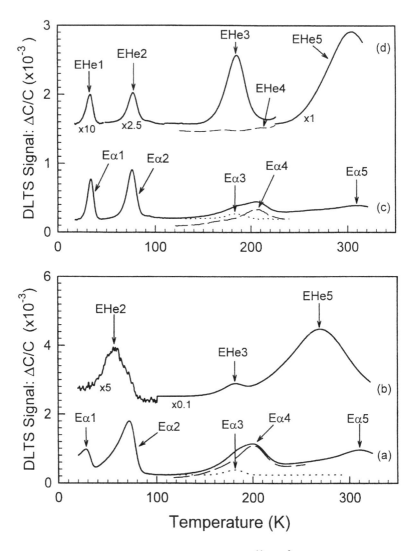

Fig. 18. DLTS spectra [116] of Si-doped (to 1×10^{16} cm^{-3}) OMVPE grown n-GaAs bombarded with 5 keV He-ions (curves (a) and (c)) and 5.4 MeV α-particles (curves (b) and (d)), recorded at $f = 46$ Hz, i.e., a decay time constant of 9.23 ms. The bias and pulse conditions for curves (a) and (b) were $V_r = 2$ V and $V_p = 2.6$ V, while for curves (c) and (d) these parameters were $V_r = 0.3$ V and $V_p = 0.3$ V. The deconvoluted broken and dotted line spectra were obtained after using the metastable transformation cycle of Eα3 [117].

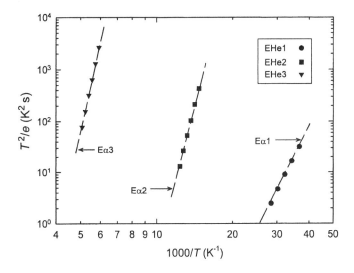

Fig. 19. DLTS Arrhenius plots to obtain the signatures of EHe1–EHe3 and $E\alpha1$–$E\alpha3$ in Si-doped (to 1×10^{16} cm^{-3}) OMVPE grown n-GaAs [116]. The bias and pulse conditions used for obtaining the data were $V_r = 0.3$ V and $V_p = 0.3$ V.

Figure 19 illustrates that, for the small bias and pulse conditions used, the signatures of EHe1, EHe2 and EHe3 are the same as those of $E\alpha1$, $E\alpha2$ and $E\alpha3$, respectively (Table III) [117]. This, in turn, implies that EHe1 and EHe2 are the same as E1 ($E\alpha1$) and E2 ($E\alpha2$), originally associated with the isolated V_{As} [15], but recently shown to be two charge states of a defect with C3v symmetry (i.e., *not* V_{As}) [104]. Figure 19 further shows that EHe3 is the same as $E\alpha3$, which involves, amongst others, Si atoms or impurities accompanying Si during doping of OMVPE GaAs growth [117, 118]. It is instructive to note that, as for 5 keV proton irradiation, the EHe4 (E3, i.e., Frenkel pairs) concentration is almost below the DLTS detection limit.

The results presented in this section showed that at least three of the main defects introduced by low energy (5 keV) and high energy (5.4 MeV) bombardment of n-GaAs are identical, but that these defects are present in different relative concentrations.

6.4. Low Temperature Experiments

The first detailed results of low temperature α-particle induced defects in n-GaAs, after the initial investigations by Rezazadeh and Palmer [71], were reported by Goodman *et al.* [119] using an undoped n-type GaAs layer with a free carrier concentration of 1.1×10^{15} cm^{-3}, grown by OMVPE on an n^+-GaAs substrate. An ^{241}Am radio-nuclide foil with an activity of 192 μCi cm^{-2}, mounted at a variable position inside a cryostat, was used to irradiate the samples at 15 K with 5.4 MeV α-particles. The DLTS peaks of two defects, Eα7 and Eα9, which are not detected after room temperature irradiation, could be observed at 55 K and 129 K, respectively, (at a lock-in frequency of 46 Hz). The energy level and apparent capture cross section of Eα7, calculated from conventional DLTS Arrhenius, plots are (0.07 ± 0.01) eV and 9.7×10^{-16} cm^2, while for Eα9 these parameters are (0.19 ± 0.01) eV and 6.7×10^{-16} cm^2.

Isochronal and isothermal annealing showed that Eα7 and Eα9 annealed out in the 225 K–245 K temperature range (Thommen stage I), following first order kinetics with activation energies for removal of (0.86 ± 0.05) eV and (0.88 ± 0.05) eV, respectively, and pre-exponential factors of 1×10^{15} s^{-1} and 1.7×10^{17} s^{-1}, respectively. Such large prefactors imply that the annealing processes in Eα7 and Eα9 involve only a small number of lattice jumps by the defects. From the annealing results presented here, it seems that defects Eα7 and Eα9 are the same as defects E7 and E9 reported by Pons and Bourgoin [15] after 1.0 MeV electron irradiation at 4 K which annealed at 250 K. The Eα9 also has similar electronic properties as a trap detected by Rezazadeh and Palmer [120] at (0.20 ± 0.03) eV below the conduction band after electron irradiation at 80 K, which completely annealed in the 220 K–300 K temperature region.

Low temperature irradiation of p-GaAs was investigated by irradiating Al SBDs on a MBE grown Be-doped (to 1.8×10^{16} cm^{-3}) layer, using α-particles from an ^{241}Am radio-nuclide with an activity of 192 μCi cm^{-2} at 20 K [121]. The first spectrum recorded up in temperature after α-particle irradiation at 20 K (curve (b) in Figure 17) shows the presence of Hα1–Hα3, and Hα5–Hα7. The Hα4 peak is totally absent and the Hα5 (H1) peak is significantly smaller than on the spectrum recorded after room temperature irradiation (curve (c) in Figure 17). Further, the "baseline skewing" in curve (b) is much more pronounced than in curve (c). The spectrum recorded after heating the sample to 340 K was found to be the same as curve (c) in Figure 17. This indicates, firstly, that Hα4 is introduced *after heating* of the irradiated sample to between 120 K and room temperature, and secondly, that heating in this temperature range also increased the concentration of Hα5 (H1). It was previously proposed that H1 is a primary defect related to defects in the As-sublattice [16, 66]. However, the results presented here shows that at least part of its concentration is the result of annealing at or above 150 K (the H1 peak position).

6.5.   Impurity Related α-particle Irradiation Induced Defects

In n-GaAs, the main irradiation induced defects, E1–E5, have been found to be independent of impurities. However, α-particle irradiation of undoped

n-GaAs and n-GaAs doped with Si to 2×10^{15} cm^{-3}, 1×10^{16} cm^{-3} and 8×10^{16} cm^{-3} showed that the concentration of the metastable Eα3 increases with Si concentration while that of Eα4 remains constant [117]. Eα3 could also be detected in Si-doped MBE grown GaAs. Whereas Si-doping in OMVPE growth is realized by using silane (which contains hydrogen), Si-doping during MBE is accomplished using a solid source. These results strongly suggest that Si is involved in the formation of Eα3. In S-doped n-GaAs, grown by OMVPE, α-particle irradiation introduced a metastable defect, Eα10 (Section 7), which could not be observed in irradiated undoped and Si-doped GaAs. A defect, EX2, with a similar "signature" as Eα10, has been observed by Guillot et al. [93] after irradiating Cl-doped n-GaAs with protons. If Eα10 and EX2 are the same defect, then these results indicate that they are related to impurities which are not present in Si-doped GaAs.

For p-GaAs, it has been shown that the H2 is related to Cu and that H3–H5 are related to other impurities in the GaAs. In Be-doped p-GaAs, which contained no measurable concentrations of Fe and Cu, irradiation did not introduce the H2–H5 defects. Instead, electron- and α-particle irradiation of Be-doped p-GaAs introduced the Hα4 defect (Figure 17), which was not detected in irradiated Zn-doped GaAs.

6.6. Summary

High energy α-particle irradiation introduces the same set of defects in GaAs as high energy (12 MeV) electrons and protons, but in different relative concentrations. In particular, the E1/E2 peak height ratio is smaller than for electron and proton irradiation and α-particle irradiation also results in a pronounced "skewing" of the baseline, thought to be due to defect clusters. Alpha-particle irradiation using an ^{241}Am radio-nuclide was shown to be a convenient and powerful, yet inexpensive, method to introduce and study radiation induced defects in several experiments, especially during low temperature irradiation. The first prominent metastable irradiation induced defect reported for epitaxially grown GaAs, Eα3, was observed after α-particle irradiation of Si-doped material. Another, less prominent metastable defect, Eα10, was observed after α-particle irradiation of S-doped GaAs.

In low energy (keV) He-ion irradiated n-GaAs, the metastable defect, EHe3 (same as Eα3), is the most prominent defect, and the E1 concentration is much lower than that of E2. Further, as in the case of low energy proton irradiation, the As Frenkel defect, E3 (Eα4), could not clearly be observed in low energy He-ion irradiated GaAs.

Low temperature α-particle irradiation of n-GaAs introduces two electron traps, Eα7 and Eα9, which anneal below room temperature. The metastable Eα3 can be introduced by irradiating Si-doped n-GaAs at low temperatures and annealing above 280 K (Section 7). In low temperature (20 K) irradiated p-GaAs, one of the main hole traps, Hα4, was shown to be introduced during annealing in the 120 K–290 K range, and the concentration of H1 (Hα5) was significantly increased by annealing between 150 K and 290 K.

Finally, He-ion irradiation introduced several impurity related defects in n- and p-GaAs. In n-GaAs, the Eα3 and Eα10 were detected after irradiating Si and S-doped GaAs, respectively, with α-particles. In p-GaAs, H2 was shown to be related to Cu, H3–H5 are related to other impurities in p-GaAs and Hα4 could only be observed in Be-doped GaAs.

7. METASTABLE RADIATION INDUCED DEFECTS IN GAAS

7.1. Background

Metastable defects have been observed, using DLTS, in as-grown GaAs [124], as well as in hydrogen- [125] and deuterium-passivated [126] epitaxially grown GaAs. Under certain experimental conditions, usually bias-on/bias-off annealing cycles, these defects are configurationally transformed to different energy states, which can be detected by DLTS. Electron irradiation introduced several metastable defects in Si [127] and InP [128]. It has recently been reported that electron irradiation also introduces metastable defects in GaAs [129, 130]. Using DLTS, Kol'chenko and Lomakoi [129] observed that electron irradiation introduces a metastable defect in bulk-grown Te-doped GaAs and that the introduction rate of this defect increased by at least an order of magnitude when increasing the electron irradiation energy from 0.9 MeV to 3.5 MeV. Since they could not observe this defect in irradiated epitaxially grown GaAs, these authors concluded that this metastable defect is a radiation induced complex involving either Te or some other impurity present in bulk-grown GaAs, but not in epitaxially grown GaAs.

In the next section, we present some results obtained after electron, proton and α-particle irradiation of epitaxially grown n-GaAs, doped with different dopants.

7.2. Silicon-doped Epitaxially Grown n-GaAs

In the first report regarding a metastable radiation induced defect in epitaxially grown GaAs, Auret *et al.* [131] reported the electronic and metastable transformation properties of Eα8 (Figure 16), introduced by 5.4 MeV α-particle from an [241]Am radio-nuclide. Eα8, with an energy level of 0.18 eV below the conduction band, has been observed after electron-, proton- and α-particle irradiation of doped and undoped n-GaAs, indicating that it is not related to dopant impurities. Eα8 can be reversibly removed and re-introduced during first order processes by annealing under zero bias to above 100 K, and by annealing under reverse bias to above 120 K, respectively [131] Eα8 is not a major defect — it is present in about 0.04 times the E2 concentration, and as such it is only of academic interest.

For a detailed investigation of other metastable defects [117], Pd SBDs on a Si-doped (to 1.1×10^{16} cm^{-3}) n-GaAs layer, grown by OMVPE on an n$^+$-GaAs substrate, were irradiated with 5.4 MeV alpha-particles from an ^{241}Am source at a fluence of 10^{11} cm^{-2}. Curve (a) in Figure 20 is a typical DLTS spectrum recorded directly after irradiating n-GaAs with α-particles at 280 K and, together with the inset in Figure 20, depicts the presence of the prominent radiation-induced defects Eα1 (E1), Eα2 (E2), Eα4 (E3) and Eα5 (Section 6). Upon annealing at room temperature for several hours, the Eα3 peak appears as a "shoulder" on the low temperature side of Eα4 and further annealing at 360 K maximised the Eα3 peak (curve (b) in Figure 20).

The metastable behaviour of Eα3 was demonstrated and studied by employing bias-on/bias-off cooling cycles [132] in conjunction with minority carrier (hole) injection cycles. Hole injection was achieved by applying a forward current, as well as by optical stimulation [117]. After cooling from 300 K to 20 K at a reverse bias of 2 V, a DLTS spectrum was recorded from 20 K to 340 K [Figure 20, curve (b)]. The spectrum recorded from 20 K to 340 K after cooling at zero bias were identical to that of curve (b). Subsequently, a spectrum was recorded from 20 K to 300 K after first applying a strong forward current density of 5.6 A cm^{-2} at 105 K followed by cooling to 20 K [identical to Figure 20, curve (a)]. Curves (a) and (b) reveal that Eα3 is present after reverse bias as well as zero bias cool down cycles, but that it is absent after applying the forward current at 105 K. Subtraction of curve (a) from curve (b) yields a well defined Eα3 peak as depicted in curve (c). The Eα3 peak could also be removed by optical injection through semi-transparent SBDs at low temperatures using a 1.45 eV laser diode. The peak heights of the other defects were found to be independent of these hole injecting conditions. Scans recorded after first heating the sample to above 200 K and then cooling it, re-introduced Eα3 and yielded spectra identical to curve (b).

The reversible disappearance and re-appearance of the Eα3 level in the band gap, together with the manner in which it is invoked, suggests that Eα3 exhibits charge-state controlled metastability [132]. The Eα3 removal can be explained qualitatively in terms of hole injection: minority carrier injection ratio into the depletion region of a SBD is small but finite, and it increases with increasing forward current density [133]. Thus, when inducing a large enough forward current, holes are introduced into the depletion layer of the SBD. It was proposed that holes injected into the depletion region at 105 K during forward pulsing, are trapped by Eα3 or another level belonging to the same defect. After this, Eα3 is transformed to a different configuration, say Eα3*, with an energy level different to that of Eα3. Since the Eα3 $\rightarrow$ Eα3*

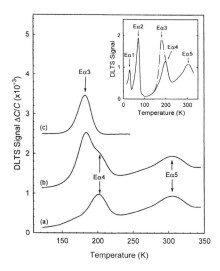

Fig. 20. DLTS spectra of α-particle irradiated Si-doped n-GaAs grown by OMVPE ($f =$ 46 Hz, $V_r = 2$ V and $V_p = 1.9$ V): (a) first scan after irradiation at 280 K; (b) after annealing at 360 K; (c): curve (b) minus curve (a), revealing in the Eα3 peak [117].

process results from hole capture, it is reasonable to expect that the opposite process Eα3* $\rightarrow$ Eα3 should stem from hole emission. The involvement of holes in this process was verified by thermally stimulated capacitance (TSCAP) and free carrier density measurements [117]. The latter showed that the increase in free carrier density (i.e., hole capture) after application of the injection pulse, was of a similar magnitude as the Eα3 concentration, namely $(2.0 - 2.4) \times 10^{14}$ cm^{-3}.

The isochronal and isothermal annealing characteristics of the Eα3 removal, i.e., of the Eα3 $\rightarrow$ Eα3* transformation, indicated that this process obeys second order kinetics and that the Eα3 removal rate can be expressed as:

$$\nu(T) = (2.1 \pm 0.5) \times 10^4 \exp[-(0.04 \pm 0.01)/kT] \qquad (27)$$

The rate constant $\nu_o = 2.1 \times 10^4$ s^{-1} is much smaller than that expected for carrier capture, namely $\nu_o = 10^7$ s^{-1} [135]. Since the term $\sigma_p p \, \nu$th dominates the rate constant, where σ_p is the capture cross section, p the minority carrier concentration and ν_{th} the hole thermal velocity, this discrepancy can be explained by noting that the minority carrier concentration p is small compared to the majority carrier concentration in n-type GaAs.

Isochronal and isothermal annealing indicated that $E\alpha 3$ is reintroduced at above 160 K under zero bias, and above 190 K under reverse bias. This transformation follows first-order kinetics which can be summarized as:

$$E\alpha 3^* \rightarrow E\alpha 3 : \nu(T) = (6 \pm 2) \times 10^8 \exp[-(0.40 \pm 0.01)/kT] \quad \text{(zero bias)} \tag{28}$$

$$\nu(T) = (6 \pm 2) \times 10^9 \exp[-(0.53 \pm 0.01)/kT] \quad \text{(2 V reverse bias)} \tag{29}$$

The rate constants above are smaller than expected for carrier emission, but slightly larger than expected for carrier capture [135]. The activation barrier for the transformation under zero-bias, $\Delta E = (0.40 \pm 0.01)$ eV, is lower than the $\Delta E = (0.53 \pm 0.01)$ eV of the process under reverse bias. The fact that there is an abundance of electrons at zero bias supports the proposed model, namely that $E\alpha 3^* \rightarrow E\beta 3$ occurs via hole release, i.e., electron capture.

The metastable transformation cycle of $E\alpha 3$ was used to isolate its DLTS peak by subtracting spectra with and without $E\alpha 3$ (e.g., Figure 20, curves (b) and (a)). These spectra, in conjunction with DLTS Arrhenius plots, were used to calculate the $E\alpha 3$ energy level and capture cross section as $E_T = E_C - 0.35$ eV and $\sigma_\infty = 9 \times 10^{-14}$ cm^2, respectively. Variable pulse-width measurements [13] revealed that the electron capture cross-section of $E\alpha 3$ is temperature dependent and that it changes from 3.0×10^{-19} cm^2 at 150 K to 1.6×10^{-18} cm^2 at 193 K. Figure 21 shows that electron capture by $E\alpha 3$ is thermally activated according to:

$$\sigma(T) = \sigma_\infty \exp(-\Delta E_\sigma/kT) \tag{30}$$

where k is Boltzmann's constant and ΔE_σ is the electron capture barrier, as discussed in Section 3 and illustrated in Figure 5. This equation indicates that electron capture by $E\alpha 3$ occurs by multiphonon emission [36]. From Figure 21 it was calculated that $\Delta E_\sigma = (0.049 \pm 0.003)$ eV and $\sigma_\infty = (1.3 \pm 0.4) \times 10^{-15}$ cm^2, and therefore that the real position of the $E\alpha 3$ level is close to 0.30 eV below the conduction band. A configuration-co-ordinate (C-C) diagram of the $E\alpha 3$ defect, encompassing its electronic and transformation properties, is suggested in Figure 22 where the positions of the $E\alpha 3$ and $E\alpha 3^*$ levels relative to the conduction band and valence band, are shown. The energy barrier for the transition from $E\alpha 3$ to $E\alpha 3^*$ is 0.04 eV and for the reverse transition, 0.40 eV. A large degree of lattice relaxation on the part of $E\alpha 3^*$ is implied by these results.

To investigate the $E\alpha 3$ introduction kinetics, a 10^{16} cm^{-3} Si-doped epitaxially grown n-GaAs layer was irradiated at 280 K and immediately thereafter, without heating, transferred to the DLTS cryostat where it was isochronally annealed for one hour periods. The DLTS scans recorded after each anneal cycle (Figure 23) showed that $E\alpha 3$ is introduced above 285 K, and further that its concentration approaches a maximum after annealing

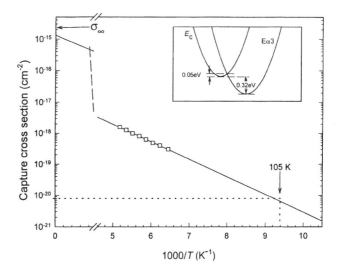

Fig. 21. Variation of Eα3 electron capture cross-section as function of temperature. Measurement conditions: $V_r = 4$ V and $V_p = 4.2$ V [117].

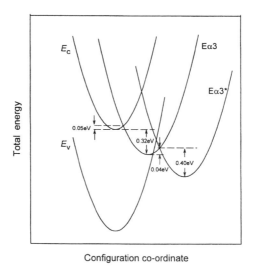

Configuration co-ordinate

Fig. 22. Configuration co-ordinate (C-C) diagram for the Eα3 defect [117].

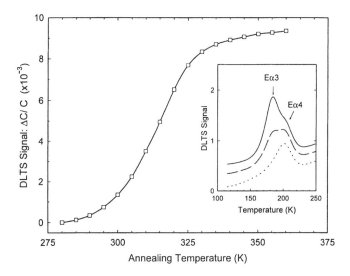

Fig. 23. Isochronal (1 hour periods) annealing behaviour of Eα3. The inset shows DLTS spectra recorded after annealing at 280 K (dots), 315 K (dashes) and 350 K (solid lines) [117].

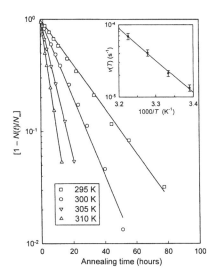

Fig. 24. Isothermal annealing plots for the introduction of Eα3 in α-particle irradiated Si-doped n-GaAs from which $\nu(T)$ is calculated. The inset shows the Arrhenius plot from which the jump frequency, ν_0, and activation energy, E_a, were calculated [117].

above 360 K. Isothermal anneal cycles in the temperature range 295 K–310 K (Figure 24) showed that the Eα3 introduction follows first order kinetics with an introduction rate, $\nu(T)$, given by:

$$\nu(T) = 10^{10\pm2} \exp(-(0.88 \pm 0.01)/kT) \qquad (31)$$

The activation energy of 0.88 eV is similar to that of defect migration by diffusion [56]. The pre-exponential factor (attempt frequency), however, is about one thousand times smaller than the lattice frequency of GaAs, indicating that a large number of interatomic jumps are required to complete the process.

Usually, the introduction of a defect as a result of annealing is the consequence of other defects being annihilated or transformed. In the case of the Eα3 introduction, no reduction of any of the major discrete level defect peaks could be detected. However, the slope of the baseline decreased with increasing anneal temperature. If this baseline skewing is the result of extended defects or defect clusters, as previously speculated, then it appears as if the increase in Eα3 concentration occurs at the expense of a reduction in the concentration of these clusters. This would imply that Eα3 is formed by products of the annihilation or transformation of the larger defect clusters. The small value of the pre-exponential constant and the fact that the Eα3 concentration increases with increasing Si-doping concentration [117], suggest that fragments dislodged from larger defect clusters diffuse to sinks, such as interstitial or substitutional Si centers, where they recombine to form Eα3. It is also instructive to note that the introduction of Eα3 coincides with the Thommen annealing stage III, which was shown by Irvine and Palmer [103] to involve Ga defects. It is therefore tempting to think of Eα3 as comprising Ga vacancies or interstitials combined with Si. Finally, Eα3 can be permanently removed by means of thermal annealing at above 500 K, with the annealing following predominantly first order kinetics [136].

Recently, it was shown that electron irradiation from a ^{90}Sr radio-nuclide of Si-doped GaAs induces a metastable defect Eα3 (in addition to main defects Eβ1, Eβ2 and Eβ4) with a DLTS peak in the same region as Eβ4 (E3) [75]. The electronic and metastable transformation properties of Eβ3 were found to be the same as those of Eα3, implying that Eβ3 and Eα3 are identical. However, the ratio of the Eβ3 to Eβ4 introduction rates is much lower than that of Eα3 to Eα4 (Section 4). Since the introduction rate of extended defects is lower in electron- than in α-particle irradiated GaAs, it is conceivable that the Eβ3 (Eα3) production may be related to extended radiation induced defects. Subsequently, it was shown that electron irradiation in a linear accelerator at energies of 3–12 MeV introduced the metastable defect EE3 (the same as Eβ3, Section 4) in approximately the same concentration as that of Eβ4 (E3).

It will be shown in Section 8 that electric field assisted emission from $E\alpha3$ ($E\beta3$) is not nearly as severe as for $E\alpha4$ (E3, $E\beta4$). This, together with the metastability of $E\alpha3$ ($E\beta3$), serve as a unique fingerprint to distinguish $E\alpha3$ ($E\beta3$) from defects with similar signatures and DLTS peaks in the same temperature region, such as E3.

7.3.   Sulphur-doped Epitaxially Grown GaAs

Pd SBDs on a S-doped (to 3×10^{16} cm^{-3}) n-GaAs layer, grown by OMVPE on an n$^+$-GaAs substrate, were irradiated with 5.4 MeV alpha-particles at a fluence of 1×10^{11} cm^{-2} [137]. Curves (a) and (b) in Figure 25 show the DLTS spectrum recorded after reverse and zero bias cool down cycles. Subtraction of these spectra yielded curve (c) which clearly shows the presence of the metastable defects $E\alpha8$ and $E\alpha10$. The main difference between the DLTS spectra for S-doped and Si-doped (Figure 20) GaAs are: 1) $E\alpha3$ is present in Si-doped GaAs but not in S-doped GaAs, and 2) $E\alpha10$ is present in S-doped GaAs, but not in Si-doped GaAs. The energy and apparent capture cross section of $E\alpha10$ were determined as 0.26 eV and $(2-5) \times 10^{-13}$ cm^2, respectively. From these observations it is tempting to conclude that $E\alpha3$ and $E\alpha10$ are Si- and S-related, respectively. Conventional zero-bias/reverse-bias anneal cycles revealed that $E\alpha10$ is metastable and can be reproducibly removed and re-introduced. Isochronal annealing shows that $E\alpha10$ is removed during zero bias annealing at temperatures of 200–280 K and is re-introduced during reverse bias (4 V) annealing in the same temperature range (inset in Figure 25). Finally, the $E\alpha10$ concentration is gradually reduced by DLTS temperature cycling up to 300 K and it can be permanently removed by annealing above 300 K.

A defect, EX2, with a similar "signature" as $E\alpha10$, was detected by Guillot et $al.$ [93] after irradiating Cl-doped VPE grown GaAs at 77 K with protons. These authors did, however, not mention metastability, but observed that EX2 could be removed after annealing at 320 K for 15 min. Since the annealing characteristics of $E\alpha10$ and EX2 are not identical, it is uncertain whether these defects are identical. However, if $E\alpha10$ is the same as EX2, then the results discussed here provide no definite proof that either of these defects are S- or Cl-related — they may also be related to impurities which are incorporated during crystal growth. However, from the data discussed above, it is evident that $E\alpha10$ and EX2 cannot involve Si.

7.4.   Summary

The results presented in this section revealed that irradiation of Si- and S-doped n-GaAs introduces the metastable defects $E\alpha3$ and $E\alpha10$, respectively. The more prominent of the two defects, $E\alpha3$, exhibits charge-state controlled

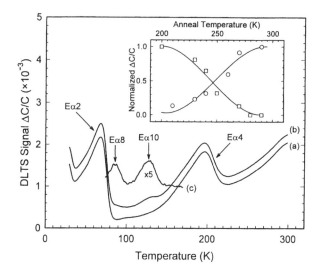

Fig. 25. DLTS spectra of α-particle irradiated S-doped n-GaAs recorded after cooling at (a) zero bias, and (b) reverse bias (4 V). Spectra were recorded at $f = 46$ Hz, $V_r = 3$ V and $V_p = 2.8$ V). The inset shows the reversible introduction and removal of Eα10 in [137] during isochronal annealing under reverse and zero bias, respectively.

metastability and can be reversibly transformed using bias-on/bias-off temperature cycles in conjunction with hole injection. This implies that Eα3 provides the means of controllably altering the carrier density of n-GaAs between 100 K and 250 K via a combination of optical and electrical excitation.

8. EFFECT OF AN ELECTRIC FIELD ON EMISSION FROM RADIATION INDUCED DEFECTS

8.1. Background

The effect of an electric field on the emission properties of carriers from deep level defects in semiconductors has been extensively studied and it has been suggested that the sensitivity of the emission rate to an applied electric field can be used to probe the nature and range of a defect potential [138, 139]. On the application side, it has been reported that electric field enhanced emission assists to reduce the degradation in energy resolution of GaAs γ-ray detectors, by allowing traps in their active region to emit carriers during operation (i.e., under a reverse bias) at lower temperatures as would

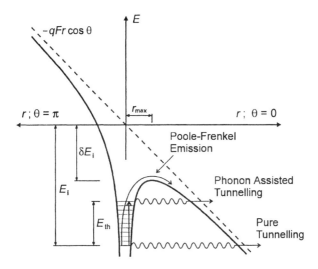

Fig. 26. Electric field assisted emission enhancement from a coulombic well for the Poole-Frenkel effect [47], phonon assisted tunnelling [142] and pure tunnelling [142]. E_i is the depth of the potential well in the absence of an electric field, δE_i is the Poole-Frenkel lowering, E_{th} is the effective depth of the well during phonon assisted tunnelling, and pure tunnelling can take place from the bottom of the well.

be the case in the absence of the field [140]. During the investigation of field assisted emission, several mechanisms, including barrier lowering induced by the Poole-Frenkel effect of coulombic wells [47, 141] and square wells [141], phonon assisted tunnelling [36, 48, 142] and pure tunnelling [142] of carriers from potential wells, were considered. For illustrative purposes only, these three processes are schematically indicated for emission from a coulombic well in Figure 26. The most frequently investigated of these models is the Poole-Frenkel effect for emission from a one dimensional coulombic well, with the characteristic (derived from Eq. (36)) that the log of the emission rate (e) is directly proportional to the square root of the electric field (F). As such, it has been used by to distinguish between donor and acceptor type defects. Examples of deep level defects that display the Poole-Frenkel effect for electron emission are oxygen-related thermal donors in silicon [17], chalcogen double donors in silicon [143] and hydrogen-related defects in OMVPE grown in GaAs [124, 144]. In the next sections, the effect of an electric field on emission from some prominent radiation induced electron and hole traps in n- and p-GaAs is demonstrated and discussed.

8.2. Defects in n-GaAs

The first detailed interpretation of experimental results regarding electric
field dependent emission from defects in GaAs was reported by Pons *et al.*
[48] for the E3 defect introduced by electron irradiation. They showed that
the field assisted emission enhancement of electrons from E3 is the result of
phonon-assisted tunnelling whereby carriers are first excited to phonon states
in the potential well and then they tunnel through the barrier (Figure 26).
According to this model [48], the ionisation rate, e_F, for a carrier trapped in
a Dirac well is given by:

$$e_F = \sum_p \prod_p \Gamma(\Delta_p)(1 - f_{1,p}) \tag{32}$$

where Δ_p is the depth of the phonon level in the well ($= E_C - E_T + p\hbar\omega$)
and $\prod_p$ is the probability of state p being occupied:

$$\prod_p = \left(1 - e^{-\hbar\omega/kT}\right) \sum_{n=0}^{+\infty} e^{-n\hbar\omega/kT} J_p^2\left(2\sqrt{S(n + 1/2)}\right) \tag{33}$$

where J_p is a Bessel function of the first kind. The factor $(1 - f_{1,p})$ is the
Fermi-Dirac probability of finding an empty conduction band state to which
the electron can tunnel and $\Gamma(\Delta_p)$ is the probability of an electron from state
p tunnelling out of the well, given by:

$$\Gamma(\Delta_p) = \gamma \frac{\Delta_p}{q K_p} \exp\left(-\frac{4}{3} \frac{(2m^*)^{1/2}}{\hbar F} \Delta_p^{3/2}\right) \tag{34}$$

where the factor Δ_p is related to the shape of the defect potential well and
K_p is

$$K_p = \frac{4}{3} \frac{(2m^*)^{1/2}}{\hbar F} \Delta_p^{3/2} \tag{35}$$

for a triangular barrier in a uniform field F. Fitting the experimental data
to Eq. (32), with S as variable, yielded a Frank-Condon shift of $S\hbar\omega = 95 \pm 10$ mV for E3 (S is the Huang-Rhys factor). This $S\hbar\omega$ value for E3 is
in good agreement with other results obtained using a quantum treatment of
phonon assisted tunnel ionisation [49].

In order to investigate electric field assisted emission from the irradiation
induced defects E1–E3 for a wide electric field range, Goodman *et al.*
[50] irradiated epitaxially grown n-GaAs with free carrier densities ranging
from 1×10^{14} cm^{-3} to $1 - 2 \times 10^{16}$ cm^{-3} with electrons from a ^{90}Sr
radio-nuclide. To accurately obtain the dependence of the emission rate on
electric field, DLTS spectra which were recorded at filling pulse amplitudes
V_p and $V_p + \delta V_p$ (using a fixed quiescent reverse bias V_r) were subtracted
to yield difference DLTS (DDLTS) spectra [77]. Figure 27 shows the effect

of an electric field on the activation energies of E1–E3 calculated from DLTS Arrhenius plots. It is clear that the measured activation energies strongly depend on the electric field. These results demonstrate the strong dependence of DLTS "signatures" on the electric field, and consequently, on how electric field assisted emission can cause incorrect defect identification, or non-identification of defects.

The emission enhancement from E1–E3 was further investigated by modelling the experimental data according to the Poole-Frenkel and phonon-assisted tunnelling models [119]. According to the one-dimensional Poole-Frenkel formalism, the field enhanced emission rate, $e(F)$, from a one-dimensional coulombic well is [47]:

$$e(F) = e(0) \exp[\beta(F)^{1/2}/kT] \qquad (36)$$

Here $e(0)$ is the emission rate at zero electric field, k is Boltzmann's constant, T is the absolute temperature and:

$$\beta = (q^3/\pi\varepsilon)^{1/2} \qquad (37)$$

From Eq. (36) it follows that a plot of $\log[e(F)]$ vs $F^{1/2}$ will yield a straight line with slope β/kT. Figure 27 illustrates the experimental and modelled results for the electric field dependence of the emission from E2 [119]. The plotted data points represent the field dependent measured emission rate minus the low field or thermal emission rate, resulting in the field-assisted emission rate. Figure 28 clearly shows that the enhancement of the emission rate from the E2 defect cannot be explained by the one dimensional Poole-Frenkel effect for a coulombic well, but that it can be explained by phonon-assisted tunnelling with a Huang-Rhys factor of $S = 4.5 \pm 0.5$.

Goodman *et al.* further demonstrated that neither the Poole-Frenkel effect nor the phonon-assisted tunnelling model yield a satisfactory fit to the experimental data for defect E1. They suggested that electric field assisted carrier emission from this shallow level may occur through a hopping mechanism, similar to that suggested by Pons and Bourgoin [145] for carrier capture from shallow levels.

As discussed in Section 6, α-particle irradiation introduces the same defects as electron irradiation, but introduces the metastable defect Eα3 (Eβ3) in higher concentrations with respect to the other defects. To compare the field dependence of the Eα3 and Eα4 (E3) emission rates, their DLTS peaks were first separated using the reversible transformation characteristics of Eα3 [117], illustrated in Figure 20. From Figure 29(a), where the DLTS peak temperatures of Eα3 and Eα4 are plotted as a function of electric field, it is clear that Eα4 exhibits a much larger peak shift than Eα3 for the same variation in electric field strength. It is worthwhile to note that whereas the Eα4 peak occurs at a higher temperature than that of Eα3 at low fields, it appears at the same temperature as Eα3 at a field of

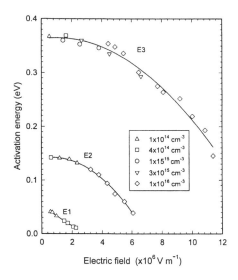

Fig. 27. Activation energies of defects E1–E3 in *n*-GaAs with different free carrier densities as a function of the electric field strength [50].

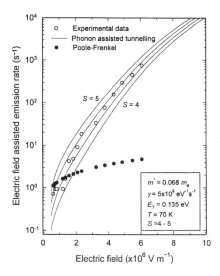

Fig. 28. Electric field dependence of the emission rate (e_n) of defect E2. The solid lines represent the phonon assisted tunnelling modelled results for different Huang-Rhys factors, *S*. The modelled Poole-Frenkel emission dependence is represented by solid symbols and the experimental emission data are represented by open symbols [50].

about 1.2×10^7 V m^{-1} (at a lock-in amplifier frequency of 46 Hz) and at higher fields it is observed at a lower temperature than Eα3. It should be emphasized that this only becomes evident after separating the peaks by using the metastability transformation cycle of Eα3. To investigate the origin of the field induced emission enhancement, the Poole-Frenkel [47] and phonon-assisted tunnelling models [36, 48, 142] were considered.

It is evident from Figure 29(b) that the log(e) data for Eα3 and Eα4 do not show an $F^{1/2}$, but rather an F^n ($n > 1$) dependence. This rules out the possibility that enhanced emission from Eα3 and Eα4 is caused by emission from a one-dimensional coulombic well. Since the shape of the log(e) vs. F relation for a three-dimensional treatment of the coulombic well is qualitatively similar to that of the one-dimensional well, it implies that the three-dimensional coulombic well also cannot account for the experimentally observed e vs. F data. By fitting the phonon-assisted tunnelling model to the experimentally measured emission rate data, using $E_T = (0.31 \pm 0.05)$ eV for Eα3 and $E_T = (0.34 \pm 0.05)$ eV for Eα4, the values $S = (11.4 \pm 1.0)$ and $\gamma = (1.0 \pm 0.5) \times 10^9$ eV^{-1}s^1 for Eα3 and $S = (8.3 \pm 1.0)$ and $\gamma = (1.4 \pm 0.7) \times 10^{11}$ eV^{-1}s^{-1} were obtained for Eα4. Figure 29(b) shows the excellent agreement between the experimentally obtained emission rate enhancement and the fitted curves. The larger Huang-Rhys factor obtained for Eα3 indicates a larger coupling between Eα3 and the lattice phonon vibronic modes than in the case of Eα4, while the large difference between the γ values of the two defects furnishes further evidence as to the different nature of the potentials associated with them. Finally, it should be pointed out that the Frank-Condon shift (≈ 110 meV) arising from $S = 11$ for Eα3, is more than double that found from measurements on the thermal activation of the Eα3 capture cross section [117].

8.3.   Defects in p-GaAs

For investigating electric field assisted emission from hole traps in p-GaAs, Al Schottky contacts on a Be doped (to $1 - 2 \times 10^{16}$ cm^{-3}) GaAs layer, grown by MBE on a p^+-GaAs substrate, were irradiated with electrons from a ^{90}Sr radio-nuclide [51]. The inset in Figure 30 shows DLTS spectra of two electron-irradiation induced defects, Hβ4 and Hβ5, in p-GaAs, recorded at different electric field conditions. Curve (a) of this inset depicts the DLTS spectrum at low-field conditions, whereas curve (b) was obtained under high-field conditions. The low-field defect DLTS signatures, (E_T, σ_{pa}), of the defects were calculated from the conventional DLTS Arrhenius plots for the lowest possible electric fields. The signatures of Hβ4 and Hβ5 are (0.20 eV, 1.2×10^{-15} cm^2) and (0.30 eV, 1.2×10^{-14} cm^2), respectively. The signature of Hβ5 is the same as that of H1, which is thought to be related to correlated V$_{As}$–As$_i$ pairs [16]. Hβ4 has

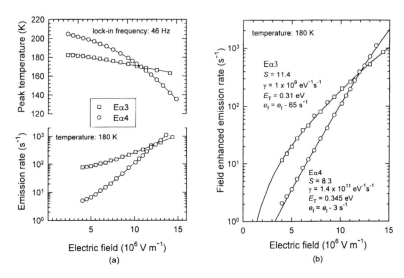

Fig. 29. (a) Peak temperature (at $f = 46$ Hz) and emission rate (at $T = 180$ K) of Eα3 and Eα4 as a function of electric field strength in 1.2×10^{16} cm^{-3} Si-doped OMVPE grown GaAs [52]. (b) Electric field enhanced emission rates as a function of the electric field for the Eα3 and Eα4 defects in the same GaAs. The fitted curves (solid lines) were calculated using the phonon-assisted tunnelling model assuming a Dirac well with parameters as shown in the graph.

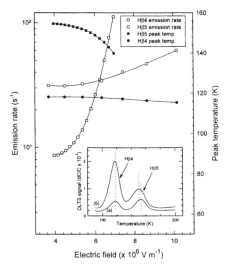

Fig. 30. Variation in the DLTS peak temperature and defect emission rate of Hβ4 and Hβ5 as function of electric field strength (at a lock-in amplifier frequency of 46 Hz). Curves (a) and (b) in the inset show the DLTS spectra of electron irradiated p-GaAs recorded under low- and high-field conditions, respectively.

up to now only been detected in Be-doped GaAs, suggesting that it involves Be or other impurities in Be-doped MBE grown GaAs [67].

From Figure 30, where the peak temperatures of Hβ4 and Hβ5 are plotted as function of electric field, it is clear that emission from Hβ5 is more strongly enhanced by an electric field than from Hβ4. The Poole-Frenkel effect and phonon-assisted tunnelling models were explored to explain the field assisted emission from Hα4 and Hα5. For both defects the emission rates calculated using coulombic potentials were orders of magnitude higher than the experimentally measured values. Furthermore, the generally expected linearity between log(e) and $F^{1/2}$ for a coulombic well could not be observed in the experimental data, indicating that field enhanced emission from Hβ4 and Hβ5 is not due to the Poole Frenkel effect.

In contrast to the slowly varying potential of a coulombic center discussed above, a spherically symmetric square well potential of radius b gives rise to a more rapid variation of emission rate with electric field [141]:

$$e(F) = e(0)[(kT/2qFb)\{\exp(qFb/kT) - 1\} + 1/2] \qquad (38)$$

Figures 30(a) and (b) depict the experimentally determined emission rates, as well as the emission rates calculated by assuming a square well. For comparative purposes, the modelled curves for the 3-dimensional Poole-Frenkel enhancement assuming a coulombic potential for Hβ4 and Hβ5 are also depicted in Figures 31(a) and (b), respectively. Figure 31(a) shows that a value of $b = 10$ Å results in a reasonable fit to the low-field experimental data for Hβ4, while the high-field data points are adequately described by $b = 25$ Å. Numerical modelling assuming a gaussian-shaped potential [146], yielded well widths of about the same range. Since electron irradiation is known to create mainly point defects, this value of b is physically acceptable, and indicates the localised nature of Hβ4. For Hβ5 (Figure 31(b)), a value of $b = 35$ Å had to be used to fit the low-field experimental data, while a reasonable fit for the high-field data was only possible when increasing b to 162 Å — a physically unacceptably large well radius for point defects or defect pairs. In addition, the depth of the well had to be increased from 0.30 eV to 0.35 eV for the high-field data. Modelling the data using a gaussian well yielded similar difficulties. Therefore, it is evident that the experimentally observed field enhanced emission for Hβ5 cannot be explained by emission from a square well.

Phonon assisted tunnelling, which has been shown to be responsible for enhanced emission from several defects in n-GaAs [36, 48, 142], is characterized by the onset of a field enhanced emission above a critical field F_c:

$$F_c = \frac{2SkT\hbar\omega}{\hbar q} \left(\frac{2m^*}{E_C - E_T} \right)^{1/2} \qquad (39)$$

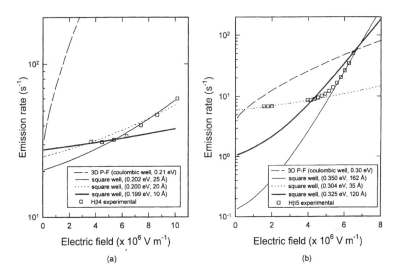

Fig. 31. Modelled curves for a square well with three different dimensions (as indicated in the inset) for (a) Hβ4, and (b) Hβ5 [51]. For comparison, one curve representing the three-dimensional Poole-Frenkel effect for a coulombic well is included. The symbols represent the experimental emission rate data for Hβ4 and Hβ5 at temperature of 112 K and 140 K, respectively.

Here S is the Huang-Rhys factor and $\hbar\omega$ is the phonon frequency ($=$ 10 meV for GaAs). From the Hβ5 data it appears that at electric fields of $5 - 6 \times 10^6$ Vm^{-1} the emission rate starts to increase rapidly. Assuming that this is the critical field strength (F_c), Eq. (33) yields an S factor between 3 and 4. To ascertain if this increase in emission rate is indeed due to phonon assisted tunnelling, values of $S = 3 - 4$ were used in conjunction with the expressions for transmission through a triangular barrier [48] to obtain the ionization rate e_F for a carrier trapped in a delta function potential. However, unrealistically high values of γ in Eq. (34) had to be used to obtain tunnel rates of the same order of magnitude as the measured emission rate. This indicates that the emission enhancement of Hβ5 cannot satisfactorily be explained by phonon-assisted tunnelling.

No single potential model could satisfactorily describe the enhanced emission of holes from Hβ5. A rapid increase of emission rate with electric field, such as observed for Hβ5, may among others be the result of the superposition of two different potentials [147], or may be indicative of a potential, which changes with electric field strength. A possible combination of the field enhanced emission of Hβ5 is the Poole-Frenkel model assuming a square well combined with a structural term [147], which is due to the position dependence of this defect. The structural term would imply that the structure of the defect changes shape with increasing electric field. When considering the existing model for Hβ5 (H1), namely that it is a combination of correlated and uncorrelated V$_{As}$–As$_i$ pairs [16], it is feasible to expect that the shape and extent of its potential may be influenced by an external electric field, leading to a field dependent structural contribution to the Hβ5 potential.

8.4. Summary

In this section it was demonstrated that the main radiation induced defects in n- and p-GaAs exhibit electric field enhanced emission, with the degree of emission enhancement depending on the individual defect. This emphasises the importance of using the same electric field when attempting to establish the similarity of defects via a comparison of their DLTS signatures. None of the defects in n-GaAs exhibited the Poole-Frenkel emission enhancement and it was, therefore, not possible to use the field enhanced emission to establish their donor or acceptor character. Field assisted emission from several radiation induced defects in n-GaAs can be explained by phonon assisted tunnelling, which allowed for a determination of their Frank-Condon shifts. For p-GaAs it was shown that Hβ4, only observed in Be-doped GaAs, behaves like a point defect with a short range potential, exhibiting the Poole-Frenkel effect. On the other hand, Hβ5 (H1), thought to be related to correlated V$_{As}$–As$_i$ pairs, has a more complex potential associated with it, probably containing a field dependent structural term.

9. EFFECT OF IRRADIATION ON GaAs FREE CARRIER DENSITY AND SBD CHARACTERISTICS

Defects affect the properties of semiconductor materials and devices in several ways. Firstly, they act as carrier traps and thereby reduce the free carrier density. This has ramifications for devices of which the operation depends on the free carrier density in the depletion layer, such as field effect transistors (FETs). For example, upon removal of carriers from the region below the gate of a MESFET, the active layer thickness will be modified resulting in a reduction in the drain-source current and a change in the pinch-off characteristics. The efficiency of carrier trapping depends on the defect's electronic properties and temperature (Section 3), and the most efficient traps are those with levels deep in the band gap so that thermal emission is minimised (Eq. (14)). Secondly, defects act as recombination centers and, as such, alter the I–V characteristics of SBD and p-n diodes. Defects with energy levels near the middle of the band gap and approximately equal electron and hole capture cross sections (Eq. (15)) are theoretically the most efficient recombination centers. Thirdly, defects act as scattering centers and thus influence the mobility of carriers. This is an important consideration for devices of which the operation relies on a high mobility, such as high electron mobility transistors (HEMTs). Fourthly, defects reduce the lifetime of carriers in semiconductors. This is desirable in devices which rely on a high switching speed, such as ultra-high frequency oscillators, but is disastrous in devices which require a long lifetime, such as solar cells. A reduction in the non-equilibrium carrier lifetime in semiconductors also results in deterioration of the energy resolution of particle detectors [160].

In the following sections, the effect of electron-, proton- and He-ion irradiation induced defects on two of these factors, namely the free carrier concentration of epitaxially grown n-GaAs and the characteristics of SBDs fabricated on it, is illustrated.

9.1. Low Energy Electrons

One of the main applications of low energy electrons in the semiconductor industry is in electron beam (EB) evaporation, which is popular for the deposition of metals on semiconductors, especially because of its ability to evaporate high melting point metals. Further, it yields a controllable deposition rate which can be varied from fractions of an angstrom per second to tens of angstroms per second. It has also been found that EB deposition enhances the adhesion of the deposited layer onto the substrate [148]. Unfortunately, EB evaporation also has some disadvantages. Most commercially EB evaporators utilize electrons with energies of between 2 keV and 20 keV to melt the material. When these electrons strike the molten metal in the crucible, x-rays of various energies and intensities are produced which reach the substrate on which the metal is deposited [149]. In addition, stray electrons from the filament (the primary source of high energy electrons), as well as electrons backscattered from the molten metal target and elsewhere in the vacuum chamber, also reach the substrate [149]. Although several reports regarding the electronic properties of defects introduced during EB processing have been forthcoming [81–86], only a few accounts have been given regarding the influence of these defects on the properties of SBDs (150, 151).

Myburg et al. [150] reported on the effect of stray electrons from the EB filament on the properties of the SBDs formed by EB deposition. In this investigation, undoped and doped n-type GaAs epilayers with different free carrier concentrations (1×10^{14}, 4×10^{14}, 1×10^{15} and 1×10^{16} cm^{-3}), and thickness of about 6 μm, grown on n^+-GaAs by OMVPE, were used for SBD fabrication. Ni-AuGe-Au ohmic contacts were formed on the n+ backsides of the samples. After chemical etching, circular Pt contacts, 0.73 mm diameter and 50 nm thick, were deposited in a vacuum system at a pressure of 10^{-8} mbar onto the GaAs at rates of 0.1, 0.3, 1.0, 3, and 10 Å s^{-1} through a metal contact mask by EB evaporation. During these depositions, the GaAs was not screened from stray electrons originating at the filament. Control diodes were fabricated at a rate of 1 Å s^{-1}, but with a metal shield placed in such a position as to prevent stray electrons originating at the filament from reaching the GaAs. The electron dose on the sample was estimated by measuring the current through the sample holder during evaporation. The electron dose reaching the sample decreased from 3×10^{16} e$^-$cm^{-2} for a deposition rate of 0.1 Å s^{-1} to 9×10^{14} e$^-$cm^{-2} when depositing at 10 Å s^{-1}. Note that these doses impinge on the metal-on-GaAs combination, and are not the effective electron doses reaching the GaAs, which will be reduced as the thickness of deposited metal, which partially shields the GaAs from the electrons, increases during metal deposition.

Log (I)-V curves, measured at room temperature (298 K), revealed that the control diodes (shielded from stray electrons) have lower reverse leakage currents, smaller ideality factors, and larger barrier heights than the unshielded diodes, in other words, diodes fabricated during exposure to stray electrons during metallization. The values of the ideality factor, n, and the effective barrier height, ϕ_e were obtained by analyzing the I-V data under the assumption that the dominant current transport mechanism is thermionic emission, according to which the current, I, is related to the voltage, V, by:

$$I = I_S \exp(qV/nkT - 1), \qquad (40)$$

Here I_S is the saturation current which is given by:

$$I_S = AA^{**}T^2 \exp[-q(\varphi_b - \Delta\varphi_{bi})/kT], \qquad (41)$$

where $\varphi_e = \varphi_b - \Delta\varphi_{bi}$ is the effective barrier height, $\Delta\varphi_{bi}$ the image-force lowering of the barrier, A^{**} the Richardson constant, A the diode area, and T the measurement temperature.

The results of the diode characteristics are summarized in Figure 32, which shows the ideality factor, n, and effective barrier height, φ_e, of the SBDs as a function of deposition rate for the different carrier density GaAs epilayers. A number of important deductions can be made from this figure. Firstly, the quality (in terms of n and φ_e) of all diodes fabricated without the electron shield is poorer than that of the diodes shielded from electrons during metallization. Secondly, the diodes deposited at lower evaporation rates (exposed to higher doses of electrons during metallization) exhibit poorer characteristics than those deposited at higher rates (exposed to lower electron doses). For example, for SBDs deposited on GaAs with a free carrier density of 1×10^{16} cm^{-3}, n decreased from 1.33 at a deposition rate of

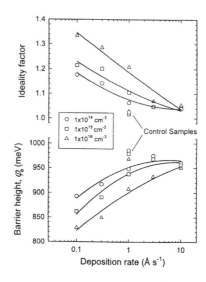

Fig. 32. Ideality factor, n, and effective barrier height, φ_e, of Pt SBDs on GaAs epilayers with different carrier densities as function of electron beam deposition rate, as indicated in the inset. Also shown are the values of n and φ_e for control samples (shielded from the filament during evaporation) [150].

0.1 Å s^{-1} to 1.05 at a rate of 10 Å s^{-1}, which compares favourably with the ideality factor of 1.03 for the control samples. The corresponding values for φ_e for these conditions are 0.83 eV and 0.97 eV, respectively. Again, φ_e for the highest deposition rate compares favourably with that of the control sample. Thirdly, the larger the free carrier density of the epilayer, the more severe the degrading effect of the stray electrons on the diode characteristics. This trend will be particularly important in the case of metal-semiconductor field effect transistor (MESFET) fabrication, for which the carrier densities below the gate are typically 10^{17}–10^{18} cm^{-3}.

Similar results to these above have been reported for EB deposition of Ti SBDs [88]. In this case, however, the effects of stray electron irradiation was significantly lower than for Pt SBDs, the reason being that EB evaporation of Ti requires a significantly lower electron beam current density. Consequently, the quality (in terms of ideality factor) of the Ti SBDs was higher than that of the Pt SBDs.

In summary, I-V measurements performed on Pt and Ti SBDs deposited with and without shielding the n-GaAs epitaxial layer from stray electrons during EB metallization, revealed that the use of a shield is essential for the fabrication of high quality devices. The poorer properties of devices formed in the presence of stray electron irradiation can be understood in terms of the results in Section 4, which showed that EB deposition introduces defects at and close to the GaAs surface.

9.2. Low Energy He-ions

It was found that, although the range of 5 keV He-ions in GaAs is less than 500 Å (Figure 3), the defects which these ions produce decreases from a maximum near the surface to below the DLTS detection limit at a depth of 1 μm below the surface [116]. Being located within the depletion layer of the SBD, these defects can significantly influence its characteristics. In this section the effects of low energy He-ion bombardment induced defects in GaAs on the properties of SBDs fabricated thereon are elucidated. The results presented here, were obtained after bombarding Si-doped (to 10^{16} cm^{-3}) n-GaAs with 1 keV He-ions at doses ranging from 5×10^{10} cm^{-2} to 5×10^{14} cm^{-2} [163]. After bombardment, Pd Schottky contacts were resistively deposited on the bombarded surfaces. The DLTS spectra obtained from these diodes were qualitatively the same as those for 5 keV He-ion bombarded GaAs, as depicted in Figure 18.

Figure 33 shows typical I-V curves of control (unbombarded) SBDs, as well as of SBDs fabricated on surfaces bombarded with doses of 5×10^{13} and 5×10^{14} He$^+$cm^{-2}. I-V measurements show that control diodes have ideality factors, n, of 1.01–1.02, effective barrier heights, φ_e, of 0.92 eV and reverse currents, I_R, (at 1 V reverse bias) of 2×10^{-11} A. In contrast, SBDs fabricated on bombarded surfaces all have a higher n, a lower φ_e and a higher I_R. Figure 33 further reveals that the I-V curves of SBDs on bombarded surfaces exhibit no typical generation-recombination current component. This, together with the increase in n and decrease in φ_e, suggest that the barrier modification is the result of surface and near-surface disorder which inhibits pure thermionic emission and results in an adjustment of the Fermi level position at the surface.

The trends observed in this experiment are summarized in Figure 34. Firstly, this figure shows that n remains below 1.05 for doses up to 5×10^{13} cm^{-2}, but for higher doses it increases sharply to 1.9 for a dose of 5×10^{14} cm^{-2}. Secondly, Figure 34 shows that φ_e decreases from 0.93 eV to 0.64 eV–0.67 eV for doses of 5×10^{10} cm$^{-2} - 5 \times 10^{14}$ cm^{-2}. Although not shown, it was found that this barrier reduction is accompanied by an increase in the reverse current for increasing doses. C-V measurements indicated that He-ion bombardment reduced the free carrier concentration at and closely below surface, but the carrier reduction was too close to the surface to allow the construction of accurate depth profiles.

In summary, low energy He-ion bombardment has a profound influence on the GaAs surface and near-surface properties, and consequently on the rectification properties of SBDs fabricated on the bombarded surfaces.

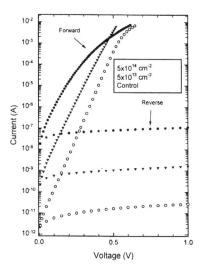

Fig. 33. Typical I-V curves of control (unbombarded) SBDs on 1×10^{16} cm^{-3} Si-doped OMVPE grown GaAs (open circles), as well as of SBDs on surfaces bombarded with doses of 5×10^{13} He$^+$cm^{-2} (filled triangles) and 5×10^{14} He$^+$cm^{-2} (filled circles) [163].

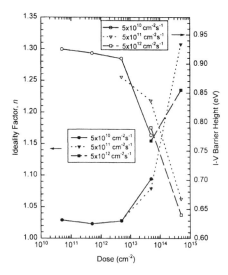

Fig. 34. Ideality factor, n, and effective barrier height, φ_e, of SBDs on 1×10^{16} cm^{-3} Si-doped OMVPE grown GaAs as function of He-ion dose for different dose rates, as indicated in the inset [163].

However, one of the main defects introduced during He-ion bombardment of GaAs, EHe3, is metastable (Section 6), and such defects may be significant in realizing optical non-linearity in compound semiconductors [11].

9.3. High Energy Electrons

Numerous studies of defect introduction rates and of the effect of these defects on materials and device properties have been reported since the extensive study of Auckerman [153]. Kol'chenko and Lomako [164] showed that the carrier removal rate in n-GaAs irradiated at room temperature with 2.5 MeV and 28 MeV electrons is independent of both the initial carrier concentration and the chemical nature of the donor. The carrier removal rate also does not depend strongly on the conductivity type [56]. It is significant to note that there is a rather wide spread in the carrier removal rate reported by different researchers. Closer inspection reveals that the larger carrier removal rates were for rather larger beam fluxes: 2 μA cm^{-2} [165] and 48 μA cm^{-2} [166].

Rezazadeh and Palmer [71] have reported that the carrier removal rate of n-GaAs irradiated with 1 MeV electrons is 0.65 cm^{-1} and 2.9 cm^{-1} at 293 K and 92 K, respectively, which are in good agreement with the 0.6 cm^{-1} measured at room temperature by Lang and Kimerling [58]. The effect of electron irradiation on the mobility profile in GaAs FATFETs have been discussed by Szentpali et al. [154] who found that irradiation with 0.65 MeV electrons at doses of 6 × 10^{15} to 1.2 × 10^{16} e$^-$cm^{-2} changed the drift mobility profiles in the active layer. These changes were attributed to radiation induced defects, acting as scattering centers, which they detected by DLTS and DLOS. Shaw et al. [155] demonstrated that 1 MeV electron irradiation caused increases in the dark currents of In$_{0.53}$Ga$_{0.47}$As photodiodes measured at various temperatures. Their calculations showed that the increase in dark current is the result of a generation current due to a single defect level at 0.29 eV below the conduction band, which they also detected by DLTS.

9.4. High Energy Protons and α-particles

Rezazadeh and Palmer [71] reported that 4.1 MeV α-particle irradiation resulted in a carrier removal rate of (1.2 ± 0.4) × 10^4 cm^{-1} at 293 K and (2.0 ± 0.6) × 10^4 cm^{-1} at 85 K. For 1 MeV protons they found the carrier removal rates to be (2.75 ± 0.5) × 10^3 cm^{-1} at 293 K and 4.5 × 10^3 cm^{-1} at 85 K. Several other reports followed (e.g., [109]), in which the focus was mainly on irradiation induced carrier reduction. Umemoto et al. [156] reported on the influence of alpha-particle induced damage on the performance of MESFETs. Galashan and Bland [111] reported that 52 MeV proton irradiation caused changes in resistor and FET threshold voltages, which are nearly an order of magnitude higher than those due to 1 MeV neutron irradiation. Recently, de Souza *et al.* [162] reported a comparison of the thermal stability of the electrical isolation layers in *n*-GaAs introduced by H, He, and B irradiation. These authors also reported the H, He, and B irradiation doses required for a complete removal of carriers by traps, as well as for significant hopping conduction to take place [162].

In the following section some detailed examples of the influence of the defects introduced during MeV proton and He-ion bombardment on the free carrier density of epitaxially grown GaAs, as well as on the properties of SBDs fabricated thereon, are illustrated and discussed.

Goodman *et al.* [99] compared the defects, change in SBD properties and free carrier removal in proton and He-ion (α-particle) bombarded epitaxially grown, *n*-type GaAs films, doped with Si to 1.2×10^{16} cm^{-3}. Alpha-particle irradiation was performed at doses of 10^{10}–$10^{12}\alpha$ cm^{-2} using an ^{241}Am source, and at doses ranging from 10^{12}–$10^{14}\alpha$ cm^{-2} in a Van de Graaff accelerator, while proton (2 MeV) irradiation was performed in the Van de Graaff accelerator at doses ranging from 10^{11}–10^{14} H^{+}cm^{-2}. From the results depicted in Figure 35, the carrier removal rate (at 298 K) was calculated as $(1.37 \pm 0.05) \times 10^4$ cm^{-1} for α-particles and $(7.1 \pm 0.2) \times 10^2$ cm^{-1} for protons. The results for α-particle irradiation are in good agreement with those of Rezazadeh and Palmer [71], but the carrier removal of protons is significantly lower than that reported by Rezazadeh and Palmer [71]. This may be due to the fact that Goodman *et al.* used 2 MeV protons for which the average defect introduction rate is about 1.5–2.2 times lower than for 1 MeV (Figure 14). In addition, Goodman *et al.* probed only the first 0.5 μm below the surface, and it is known (Section 2) that as the protons are slowed down by penetrating deeper into the materials, nuclear stopping becomes more important and therefore the rate of defect (carrier trap) creation is increased.

The radiation induced carrier removal is due to trapping of carriers by the radiation induced defects with energy levels at various positions in the band-gap. Blood [157] argued that the traps responsible for carrier removal at 77 K are the E3 and E4 irradiation-induced defects, the equivalents of Eα4 and Eα5 in α-particle irradiated GaAs (Section 6) and Ep4 and Ep5 in proton irradiation GaAs (Section 5). At room temperature, however, the emission rates of these defects are too high to trap and retain carriers. It is instructive to note that if the introduction rates of α-particle induced hole traps Hα1 - Hα5 (of which the energy levels lie below the Fermi level in *n*-GaAs at room temperature [67]) are summed, it approximately amounts to the carrier removal rate in *n*-type GaAs, calculated from Figure 35.

Figure 35 also shows that the increase in reverse current (ΔI_R) is almost directly proportional to the particle dose, both for α-particles and protons. The ΔI_R for α-particles is about an order of magnitude higher than

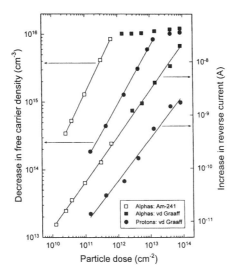

Fig. 35. Decrease in free carrier density (carrier removal) of 1.2×10^{16} cm^{-3} Si-doped OMVPE grown GaAs, and increase in reverse current of SBDs fabricated on it, as function of proton and α-particle (He-ion) irradiation dose [99].

for protons. From I-V measurements it was seen that the ideality factor (n) for the irradiated samples did not exceed 1.05, indicating that thermionic emission is the dominant current transport mechanism in the forward bias region (0.15–0.40 V). Further, there was no significant decrease in the barrier height, so the increase in reverse current cannot be accounted for by a reduction in the barrier height, but rather by a generation-recombination contribution to the thermionic emission current. To determine the barrier height, series resistance and the recombination current after α-particle irradiation, a modified version of the model proposed by Donoval *et al.* [158] was used. According to this model, which avoids the use of the ideality factor n, the current through the diode is the sum of thermionic emission, generation-recombination and shunt leakage:

$$
\begin{aligned}
I = I_S &\left\{ \exp\left(\frac{[q(V - I R_S)]}{kT} \right) - 1 \right\} \\
&+ I_G \left\{ \exp\left(\frac{[q(V - I R_S)]}{2kT} \right) \right\} + \frac{(V - I R_S)}{R_{sh}}
\end{aligned}
\tag{42}
$$

The thermionic emission saturation current pre-factor (I_S) can be expressed as:

$$
I_S = AA^*T^2 \exp(-\varphi_e/kT)
\tag{43}
$$

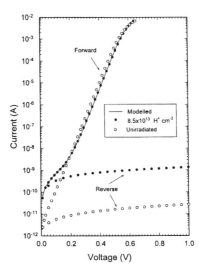

Fig. 36. Measured (symbols) and modelled (line) I-V characteristics of Si-doped (to 1×10^{16} cm^{-3}) OMVPE grown n-GaAs irradiated with a dose of 8.5×10^{13} H$^+$cm^{-2} [99].

with A the area of the diode, A^* the experimentally measured Richardson constant (equal to 4.8×10^4 A m^{-2}K^{-2} [159], φ_e the effective barrier height with image force lowering taken into account, R_{sh} being the shunt leakage resistance (important in low voltage regions, $qV \ll kT$), R_s the series resistance and I_G the generation-recombination pre-factor. The barrier height and series resistance are obtained by fitting the experimentally obtained forward I-V characteristics to Eq. (41).

The results of this modelling for α-particle irradiated n-GaAs is depicted in Figure 36 which shows that, after irradiation, the recombination current dominates the current characteristics in the low forward bias region (0–0.2 V). It was found that the recombination component increased with increasing irradiation dose. Although the most effective recombination centers are those with energy levels located near the middle of the band-gap [55], they are not the only centers which can contribute to recombination. For example, it was shown that, although the H(0.17) level of the divacancy in electron irradiated Si lies far below the middle of the band gap, it is the dominant recombination center in solar cells with p-type substrates [54]. The recombination current seen in Figure 36 is due to the presence of defects introduced during irradiation, of which the concentration increases with increasing particle dose, as discussed in Sections 5 and 6. This analysis further showed that the control sample has the lowest series resistance, whereas the diode that received the highest particle dose has the highest series resistance. The increase in series resistance is a result of the reduction in free carrier concentration in the depletion region.

9.5. Summary

I-V measurements performed on Pt and Ti SBDs, deposited with and without shielding the n-GaAs epitaxial layer from stray electrons during EB metallization, revealed that the use of a shield is essential for the fabrication of high quality devices. The poorer properties of devices formed in the presence of stray electrons can be understood in terms of the results in Section 4, which showed that EB deposition introduces defects at and close to the GaAs surface. Low energy He-ion bombardment was shown to have a profound influence on the surface and near-surface properties of GaAs, and consequently on the rectification properties of SBDs fabricated on the bombarded surfaces. Despite the adverse effects of He-ion bombardment on SBD performance, it does, however, introduce a prominent metastable defect in Si-doped n-GaAs which may be useful in advanced applications.

It was also shown that high energy α- and H$^+$-particle irradiation removes free carriers at much higher rates than electrons. For the 10^{16} cm^{-3} doped GaAs, the onset of a detectable carrier reduction and generation current which exceeds the thermionic emission current (at 298 K), occurs for doses of about 10^{10} cm^{-2} and 10^{11} H$^+$cm^{-2}, respectively. This free carrier removal is essential for device isolation but can cause severe degradation in the performance of depletion layer based devices, such as MESFETs, in radiation environments.

A final important point that should be borne in mind is that one of the main defects introduced during proton and He-ion bombardment of GaAs is metastable (Section 6). The effect that such a defect has on the GaAs free carrier density may be of significance in realizing optical non-linear devices in compound semiconductors [11].

10. SUMMARY AND CONCLUSIONS

Radiation induced defects play an important role in several areas of semiconductor device fabrication. In order to avoid the deleterious effects of some of these defects and utilize the beneficial effects of others, depending on the application, it is imperative to understand the effect of radiation on electronic materials and devices fabricated on them. To achieve this, it is essential that the electronic properties and concentration of radiation induced defects should be known, allowing calculation of their effect on the properties of electronic materials and devices. In addition, the structure, introduction rate, introduction mechanism and thermal stability of the defects should be determined so that they can be reproducibly introduced, avoided or eliminated, depending on the application. The spectroscopic nature of DLTS, which allows independent studies of different defect species in the same semiconductor, has rendered it a key characterization technique in providing most of this information.

Below, we summarize the most recent advances pertaining to the introduction defects in GaAs by electron-, proton- and He-ion irradiation, and the results obtained from DLTS characterization of these defects.

1. Radio-nuclides (e.g., ^{241}Am and ^{90}Sr) are useful, yet inexpensive, sources for providing α- and β-particles (He-ions and electrons) to introduce irradiation induced defects, since irradiation from these sources create essentially the same set of defects as electron, proton, and He-ion irradiation in Van de Graaff accelerators. The first radiation induced metastable defects in epitaxially grown GaAs were detected after α-particle irradiation from an ^{241}Am source.

2. Electrons, protons and He-ions produce essentially the same set of defects, with the relative defect concentration depending on the particle type. A general trend that was observed, is that the ratio of the concentration of the deep defect centers, located near mid-gap, to that of the shallower centers, increases with particle mass. This is important for applications such as device isolation based on carrier reduction due to defect trapping because deep defect levels are more effective than shallow centers in retaining carriers at room temperature.

3. The E5 defect, which was previously reported to be introduced by 1 MeV electrons, protons and He-ions, has been shown to have different signatures for 1 MeV electrons (E5a), and for 12 MeV electrons, protons and α-particles (E5b). The E5a defect is speculated to be created by collision cascades produced by knock-on Ga and As atoms. For electron irradiation, the abundance and energy of these knock-on atoms increase with increasing electron energy, explaining why 1 MeV electrons result mainly in E5b, whereas 12 MeV electrons introduce mainly the E5a. Collision cascades occur easily for protons and α-particle irradiation because these particles transfer much more energy during collisions (owing to their larger mass) than electrons with the same energy. Hence, E5a is the dominant of the two E5 defects for MeV proton and α-particle irradiation.

4. Metastable defects have been detected and characterized in epitaxially grown n-GaAs irradiated with electrons, protons and He-ions. The metastable defects detected in n-GaAs doped with Si- and S are different, suggesting the involvement of these dopants, or impurities specific to their introduction during doping, in radiation induced metastable

defect complexes. Metastable defects are important because they can be instrumental in realizing optical non-linear devices in compound semiconductors.

5. Contrary to what was reported in previous reviews, recent investigations showed that some irradiation induced electron traps in n-GaAs involve impurities. This was clearly demonstrated for the metastable irradiation induced defects in Si- and S-doped epitaxially grown GaAs. In p-GaAs, several impurity related irradiation introduced defect complexes have been known to exist. The most recently reported of these is a prominent radiation induced hole trap which is introduced in Be-doped MBE p-GaAs, but not in identically irradiated Zn-doped OMVPE grown p-GaAs. Impurity-specific radiation induced defects present an additional element of defect engineering to tailor the properties of electronic materials for specific applications. This is particularly important when the impurity related defects are metastable.

6. Whereas the E1 and E2 have for a long time been assumed to be two charge states of the isolated arsenic vacancy, V_{As}, this assignment has recently been proved to be incorrect by uniaxial stress measurements combined with DLTS. This experiment revealed that E1 and E2 have C3v symmetry, and can therefore not be simple arsenic vacancies, V_{As}, with Td symmetry. However, for E3, the uniaxial stress data are consistent with its original assignment as an As Frenkel pair, e.g., V_{As}–As_i, provided that the interstitial-vacancy direction is along the $\langle 111 \rangle$ direction. Similar studies on metastable radiation defects should be undertaken to cast more light on their structure.

7. All radiation induced defects in n- and p-GaAs thus far investigated, exhibit electric field enhanced emission, with the degree of emission enhancement depending on the individual defect. No radiation in-duced defect in n-GaAs has yet been reported to exhibit the classical Poole-Frenkel emission enhancement, eliminating the possibility of establishing their donor or acceptor nature using the field assisted emission. The electric field enhanced emission of carriers from several radiation induced defects in n-GaAs can be explained by phonon assisted tunnelling, which facilitated the determination of their Frank-Condon shifts. For p-GaAs, field assisted emission provided information re-garding the range of the potentials of irradiation induced defects. Since electric field assisted emission can cause incorrect or non identification of defects, it is important to note the field conditions (determined by the free carrier concentration, reverse bias and pulse amplitude) when comparing DLTS defect signatures.

8. The defects introduced in epitaxially grown GaAs by sub-threshold energy electron irradiation have similar, but not identical, signatures as the defects introduced by high energy (above-threshold) electron

irradiation. The important role that low energy electrons play in the microelectronics manufacturing industry, and the harmful influence that the defects they produce have on SBD properties, necessitate further investigations to determine the structure and introduction mechanisms of sub-threshold electron irradiation induced defects.

9. We have presented some examples as to how radiation induced defects influence the GaAs free carrier density and the properties of SBDs fabricated on it. In particular, it has been shown that stray electrons impinging on GaAs during electron beam metallization of Schottky contacts have a pernicious influence on the quality of SBDs. It is therefore imperative to shield the GaAs from these electrons during fabrication of high quality devices. Low energy He-ion bombardment was shown to have a profound influence on the surface and near-surface properties of GaAs, and consequently, on the rectification properties of SBDs fabricated on the bombarded surfaces. It was also shown that high energy α-particles and H^+-ions remove free carriers at much higher rates than electrons. This free carrier removal is essential for device isolation but can cause severe degradation of the performance of depletion layer based devices in radiation environments. The defects introduced during high energy particle irradiation also result in high recombination currents in rectifying devices, which are detrimental in detector applications.

10. Finally, although the electronic properties of irradiation induced defects in GaAs, determined by DLTS and similar junction spectroscopic techniques, have been reported in significant detail, no indisputable evidence has yet been presented for the structure of any of these defects as determined by, for example, EPR. The converse is also true. Clearly, this is an area that needs to be addressed.

Acknowledgements

I express my sincere gratitude to Prof. G. Myburg for the fabrication of numerous SBDs and ohmic contacts, and for initiating studies of defect introduction in semiconductors using radio-nuclides in our laboratory. I thank Prof. S.A. Goodman for his inestimable contribution to the DLTS measurements of radiation induced defects and for the low temperature irradiation studies that he initiated. I am grateful to the Physics Department of the University of Port Elizabeth for supplying the sulphur-doped GaAs. I would also like to express my sincere gratitude to Prof. J.B. Malherbe for the low energy ion bombardment, M. Hayes and T. Hauser for proton and He-ion implantation in the Van de Graaff accelerator at the University of Pretoria, Dr. M.C. Ridgway of the Australian National University for the ion-implantation of samples in their Pelletron implanter, the National Accelerator center

(NAC) at Faure (South Africa) for proton irradiation, and the H.F. Verwoerd and Bloemfontein National Hospitals for electron irradiation in their linear accelerators. I am is deeply indebted to W.E. Meyer for invaluable assistance with computer hickups, numerical modelling, and variable temperature I-V, C-V and TSCAP measurements, P.N.K. Deenapanray for the low temperature irradiation and DLTS of p-type GaAs and to P. de Villiers for priceless technical assistance in preparing the manuscript. Finally, I gratefully acknowledge the financial assistance of the South African Foundation for Research Development (FRD) and the Carl and Emily Fuchs Institute for Microelectronics (CEFIM).

References

1. D.E. Davies, J.F. Kennedy and A.C. Yang, *Appl. Phys. Lett.*, **23**, 615 (1973).
2. A.G. Foyt, W.T. Lindley, C.M. Wolfe and J.P. Donnelly, *Solid State Electronics*, **12**, 209 (1969).
3. J.C. Dyment, J.C. North and L.A. D'Asaro, *J. Appl. Phys.*, **44**, 207 (1973).
4. R.D. Rung, C.J. Dell'occa and L.G. Walker, *IEEE Trans. Electron. Devices*, **28**, 1115 (1991).
5. J.L. Benton, J. Michael, L.C. Kimerling, D.C. Jacobson, Y-H. Xie, D.J. Eaglesham, E.A. Fitzgerald and J.M. Poate, *J. Appl. Phys.*, **70**, 2667 (1991).
6. E.A. White, K.T. Short, R.L. Dynes, J.P. Garno and J.M. Gibson, *Appl. Phys. Lett.*, **50**, 95 (1987).
7. A. Grob, J.J. Grob, P. Thevenin and P. Siffert, *Nucl. Instrum. Methods*, **B58**, 236 (1991).
8. A. Mogro-Campero, R.P. Love, M.F. Chang and R.F. Dyer, *IEEE Trans. Electron Devices*, **33**, 1667 (1986).
9. D.C. Sawko and J. Bartko, *IEEE Nucl. Sci.*, **30**, 1756 (1983).
10. C. Jagadish, B.G. Svensson and N. Hauser, *Semicond. Sci. Technol.*, **8**, 481 (1993).
11. K. Wada and H. Nakanishi, *Materials Science Forum*, **258–263**, 1051 (1998).
12. H.G. Grimmeiss, Deep Levels in Semiconductors. *Ann. Rev. Mater. Sci.*, **7**, 341–376 (1977).
13. D.V. Lang, *J. Appl. Phys.*, **45**, 3014 (1974).
14. D.V. Lang, *J. Appl. Phys.*, **45**, 3023 (1974).
15. D. Pons and J.C. Bourgoin, *J. Phys.*, **C18**, 3839 (1985).
16. D. Stievenard, X. Boddaert, J.C. Bourgoin and H.J. von Bardeleben, *Phys. Rev.*, **B41**, 5271 (1990).
17. L.C. Kimerling, *Materials Research Society Proceedings, Vol 2: Defects in Semiconductors* (Edited by Narajan, J. and Tan, T.Y. North Holland, New York), p.85, (1981).
18. J. Lindhard, V. Nielsen, M. Scharff and P.V. Thompsen, *Kgl. Dan. Vid. Selsk. Mat. Fys. Medd.*, **33**, 10 (1963).
19. O.B. Firsov, *Zh. Eksp. Teor. Fiz.*, **34**, 447 (1958).
20. J.P. Biersack and L.G. Haggmark, *Nucl. Instrum. Methods*, **174**, 257 (1980).
21. J. Lindhard and M. Scharff, *Phys. Rev.*, **124**, 128 (1961).
22. H.Z. Bethe, *Phys.*, **76**, 293 (1932).
23. J. Lindhard, *Kgl. Dan. Vid. Selsk. Mat. Fys. Medd.*, **34**, 14 (1965).
24. J. Lindhard and M. Scharff and H. Schiott, *Kgl. Dan. Vid. Selsk. Mat. Fys. Medd.*, **33**, 14 (1963).
25. W. Eckstein, *Computer Simulation of Ion-Solid Interactions*, Springer Ser. Mat. Sci., Vol. 10 (Springer, Berlin, Heidelberg), (1991).
26. M.T. Robinson, *Nucl. Instrum. Methods in Phys. Res.*, **B67**, 396–400 (1992).
27. M.T. Robinson, *Nucl. Instrum. Methods in Phys. Res.*, **B48**, 408–413 (1990).
28. M.T. Robinson, *Phys. Rev.*, **B40**, 16, 10717–10726 (1989).
29. G.H. Kinchin and R.S. Pease, *Rep. Prog. Phys.*, **18**, 1 (1955).

30. D. Pons, *J. Appl. Phys.*, **55**, 2839 (1984).
31. P. Sigmund, *Appl. Phys. Lett.*, **14**, 114 (1969).
32. P. Marmier and E. Sheldon, in "Physics of Nuclei and Particles, Volume I" (Academic Press, New York), p.314 (1971).
33. A. Hallén and M. Bakowski, *Solid-State Electron.*, **32**, 1033 (1989).
34. G.L. Miller, D.V. Lang and L.C. Kimerling, Capacitance Transient Spectroscopy. *Ann. Rev. Mater. Sci.*, **7**, 377–448 (1977).
35. D.V. Lang and C.H. Henry, *Phys. Rev. Lett.*, **35**, 1525 (1975).
36. C.H. Henry and D.V. Lang, *Phys. Rev.*, **B15**, 989 1977.
37. J.A. Van Vechten and C.D. Thurmond, *Phys. Rev.*, **B14**, 3539 (1976).
38. C.O. Almbladh and G.J. Rees, *J. Phys. C: Solid State Phys.*, **14**, 4575–4601 (1981).
39. D.V. Lang, Space-Charge Spectroscopy in Semiconductors. In "Thermally Stimulated Relaxation of Solids" (P. Braunlich, ed., Springer-Verlag. Berlin), pp.93–133 (1979).
40. L.C. Kimerling, *J. Appl. Phys.*, **45**, 1839 (1974).
41. J.J. Shiau, A.L. Fahrenbruch and R.H. Bube, *J. Appl. Phys.*, **59**, 2879 (1986).
42. F.D. Auret and M. Nel, *J. Appl. Phys.*, **63**, 973 (1988b).
43. P.D. Kirchner, W.J. Schaff, G.N. Maracas, L.F. Eastman, T.I. Chappell and C.M. Ransom, *J. Appl. Phys.*, **52**, 6462 (1981).
44. A. Cola, M.G. Lupo and L. Vasanelli, *J. Appl. Phys.*, **69**, 3072 (1991).
45. L. Dobaczewski, P. Kaczor, M. Missous, A.R. Peaker and Z.R. Zytkiewicz, *J. Appl. Phys.*, **76**, 194 (1994).
46. Y. Zohta and M.O. Watanabe, *J. Appl. Phys.*, **53**, 1890 (1982).
47. J. Frenkel, *Phys. Rev.*, **54**, 647 (1938).
48. D. Pons and S. Makram-Ebeid, *J. de Physique*, **40**, 1161 (1979).
49. S. Makram-Ebeid and M. Lannoo, *Phys. Rev.*, **B25**, 6406 (1982).
50. S.A. Goodman, F.D. Auret and W.E. Meyer, *Jpn. J. Appl. Phys.*, **33**, 1949 (1994).
51. F.D. Auret, S.A. Goodman and W.E. Meyer, *Semicond. Sci. Technol.*, **10**, 1376 (1995).
52. W.E. Meyer, F.D. Auret and S.A. Goodman, *Jpn. J. Appl. Phys.*, **35**, L1 (1996).
53. H. Lefevre and M. Schultz, *Appl. Phys.*, **12**, 45 (1977).
54. T. Markvart, D.P. Parton, J.W. Peters and A.F.W. Willoughby, *Materials Science Forum*, **143–147**, 1381 (1994).
55. S.M. Sze, *Physics of Semiconductor Devices 2 Ed*, (John Wiley & Sons), p.37 (1981).
56. D.V. Lang, Inst. Phys. Conf Ser. No. 31, p.70 (1977).
57. J.C. Bourgoin, H.J. von Bardeleben and D. Stievenard, *Phys. Stat. Sol. (a)*, **102**, 499 (1987).
58. D.V. Lang and L.C. Kimerling, *Lattice Defects in Semiconductors 1974*, (Inst. Phys. Conf. Ser. 23), p.581 (1975).
59. D. Pons and J.C. Bourgoin, *Phys. Rev. Lett.*, **47**, 1293 (1981).
60. D. Pons, A. Mircea and J.C. Bourgoin, *J. Appl. Phys.*, **51**, 4150 (1980).
61. H. Lim, H.J. von Bardelben and J.C. Bourgoin, *J. Appl. Phys.*, **62**, 2738 (1987).
62. G.A. Baraff and M. Schlüter, *Phys. Rev. Lett.*, **55**, 1327 (1985).
63. B. Ziebro, J.W. Hemsky and D.C. Look, *J. Appl. Phys.*, **72**, 78 (1992).
64. H.J. Von Bardeleben, C. Delerue and D. Stievenard, *Materials Science Forum*, **143–147**, 223 (1994).
65. S. Loualiche, A. Nouilhat, G. Guilliot, M. Gavand, A. Laugier and J.C. Bourgoin, *J. Appl. Phys.*, **53**, 8691 (1982).
66. D. Stievenard, X. Boddaert and J.C. Bourgoin, *Phys. Rev.*, **B34**, 4048 (1986).
67. S.A. Goodman, F.D. Auret and W.E. Meyer, *J. Appl. Phys.*, **75**, 1222 (1994).
68. H. Lim, K.N. Kang and H.Y. Park, *J. Korean Phys. Society*, **23**, 38 (1990).
69. K. Thommen, *Radiat. Eff.*, **2**, 201 (1970).
70. D.V. Lang, L.C. Kimerling and S.Y. Leung, *J. Appl. Phys.*, **47**, 3587 (1976).
71. A.A. Rezazadeh and D.W. Palmer, *Defects and Radiation Effects in Semiconductors 1980* (Inst. Phys. Conf. Ser. 59) p.317 (1981).
72. D. Stievenard, J.C. Bourgoin and D. Pons, *Physica*, **116B**, 394 (1983).
73. H. Lim, H.S. Kim, B. Choe and H.Y. Park, *J. Korean Phys. Society*, **23**, 45 (1990).
74. F.D. Auret, S.A. Goodman, G. Myburg and W.E. Meyer, *Appl. Phys.*, **A56**, 547 (1993).
75. F.D. Auret, S.A. Goodman and W.E. Meyer, *Appl. Phys. Lett.*, **67**, 3277 (1995d).

76. F.D. Auret, S.A. Goodman, W.E. Meyer, R.M. Erasmus and G. Myburg, *Jpn. J. Appl. Phys.*, **32**, L974 (1993).
77. F.D. Auret, S.A. Goodman and W.E. Meyer, unpublished, (1998).
78. J.C. Slater, *J. Appl. Phys.*, **22**, 237 (1951).
79. F.L. Vook, *Phys Rev.*, **135A**, 1742 (1964).
80. F. Seitz and J. Koehler, *Solid State Phys.*, **2**, 305 (1965).
81. R. Kleinhenz, P.M. Mooney, C.P. Schneider and O. Paz, *13th International Conference on Defects in Semiconductors*, edited by L.C. Kimerling and J. Parsey, Colorado, CA. p.627 (1984).
82. F.D. Auret and P.M. Mooney, *J. Appl. Phys.*, **55**, 984 (1984).
83. F.D. Auret and P.M. Mooney, *J. Appl. Phys.*, **55**, 988 (1984).
84. M. Nel and F.D. Auret, *J. Appl. Phys.*, **64**, 2422 (1988).
85. M. Nel and F.D. Auret, *Jpn. J. Appl. Phys.*, **28**, 2430 (1989).
86. F.D. Auret, L.J. Bredell, G. Myburg and W.O. Barnard, *Jpn. J. Appl. Phys.*, **30**, 80 (1991).
87. K. Kanaya and S. Okayama, *J. Phys. D (Appl. Phys.)*, **5**, 43 (1972).
88. F.D. Auret, G. Myburg, L.J. Bredell, W.O. Barnard and H.W. Kunert, *Materials Science Forum*, **83–87**, 1499 (1992).
89. J.W. Corbett and J.C. Bourgoin, "Defect Creation in Semiconductors", in *Point Defects in Solids*, Vol 2, Edited by Crawford, J.H. and Slifkin, L.M. (Plenum Press, New York) p.1 (1975).
90. J.C. Bourgoin and J.W. Corbett, *Radiation Effects*, **36**, 157 (1978).
91. C. Christensen, J.W. Petersen and A.N. Larsen, *Appl. Phys. Lett.*, **61**, 1426 (1992).
92. S.S. Li, D.W. Schoenfeld, F. Llevada and K.L. Wang, *Solid State Electron.*, **21**, 1616 (1978).
93. G. Guillot, S. Loualiche and A. Nouailhat, *Defects and Radiation Effects in Semiconductors 1980* (Inst. Phys. Conf. Ser. 59) p.325 (1981).
94. Y. Yuba, K. Gamo, K. Murakami and S. Namba, *Defects and Radiation Effects in Semiconductors 1980* (Inst. Phys. Conf. Ser. 59), p.329. (1981).
95. D. Pons, A. Mircea, A. Mitonneau and G.M. Martin, *Defects and Radiation Effects in Semiconductors 1978* (Inst. Phys. Conf. Ser. 46), p.352 (1979).
96. S.S. Li, K.L. Wang, P.W. Lai, L.Y. Loo, G.S. Kamath and R.C. Knechtli, *IEEE Trans. Electron Devices ED*, **27**, 857 (1980).
97. F.D. Auret, M. Nel and H.C. Snyman, *Radiation. Effects*, **105**, 225 (1988).
98. F.D. Auret, S.A. Goodman, M. Hayes, G. Myburg, W.O. Barnard and W.E. Meyer, *S. Afr. J. Phys.*, **16**, 153 (1993).
99. S.A. Goodman, F.D. Auret and W.E. Meyer, *Nucl. Instr. and Meth. in Phys. Res.*, **B90**, 349 (1994).
100. W.O. Siyanbola and D.W. Palmer, *Semicond. Sci. Technol.*, **5**, 2038 (1990).
101. W.O. Siyanbola and D.W. Palmer, *Solid State Communications*, **74**, 209 (1990).
102. W.O. Siyanbola and D.W. Palmer, *Phys. Rev. Lett.*, **66**, 56 (1991).
103. A.C. Irvine and D.W. Palmer, *Phys. Rev.*, **B49**, 5695 (1994).
104. S.J. Hartnett and D.W. Palmer, *Materials Science Forum*, **258–263**, 1027 (1998).
105. S.A. Goodman, F.D. Auret, M.C. Ridgway and G. Myburg, *Nuclear Instr. and Meth. in Phys. Res. Sect. B*, **148**, 446 (1999).
106. F.D. Auret, A. Wilson, S.A. Goodman, G. Myburg and W.E. Meyer, *Nucl. Instr. and Meth. in Phys. Res.*, **B90**, 387 (1994).
107. G.M. Martin, E. Esteve, P. Langlade and S. Makram-Ebeid, *J. Appl. Phys.*, **56**, 2655 (1984).
108. B.G. Svensson, B. Mohadjeri, S. Hallen, J.H. Svensson and J.W. Corbett, *Phys. Rev.*, **B43**, 2292 (1991).
109. M.A. Hopkins and J.R. Srour, *IEEE Trans. Nucl. Sci.*, **NS-30**, 4457 (1983).
110. A.R. Knudson, A.B. Cambell, A.B. Stapor, P. Shapiro and G.P. Mueller, *IEEE Trans. Nucl. Sci.*, **NS-32**, 4388 (1985).
111. A.F. Galashan and S.W. Bland, *J. Appl. Phys.*, **67**, 173 (1990).
112. L.F. Zakharenkov, V.K. Kozlovskii and B.A. Shustrov, *Phys. Status Solidi (a)*, **117**, 85 (1990).
113. V. Raineri, G. Fallica and S. Libertino, *J. Appl. Phys.*, **79**, 9012 (1996).
114. I.L. Singer, J.S. Mudray and J. Comas, *J. Vac. Sci. Technol.*, **18**, 161 (1981).

342   F.D. AURET

115. N.J. Devlin, *Electron. Lett.*, **16**, 138 (1980).
116. F.D. Auret and S.A. Goodman, *Appl. Phys. Lett.*, **68**, 3275 (1996).
117. F.D. Auret, R.M. Erasmus, S.A. Goodman and W.E. Meyer, *Phys. Rev.*, **B51**, 17 521 (1995).
118. F.D. Auret, S.A. Goodman, R.M. Erasmus, W.E. Meyer and G. Myburg, *Nucl. Instr. and Meth. in Phys. Res.*, **B106**, 323 (1995).
119. S.A. Goodman and F.D. Auret, *Jpn. J. Appl. Phys. Lett.*, **32**, L1120 (1993).
120. A.A. Rezazadeh and D.W. Palmer, *J. Phys.*, **C18**, 43 (1985).
121. F.D. Auret, P.N.K. Deenapanray and S.A. Goodman, unpublished, (1998).
122. G. Hoffman, J. Madok, N.M. Haegel, G. Roos, N.M. Johnson and E.E. Haller, *Appl. Phys. Lett.*, **61**, 2914 (1992).
123. D. Pons, *Physica*, **B116**, 388 (1983).
124. W.R. Buchwald, N.M. Johnson and L.P. Trombetta, *Appl. Phys. Lett.*, **50**, 1007 (1987).
125. A.W.R. Leitch, Th. Presca and J. Weber, *Phys. Rev.*, **B44**, 1375 (1991).
126. A.W.R. Leitch, Th. Presca and J. Weber, *Phys. Rev.*, **B45**, 14400 (1992).
127. A. Chantre and L.C. Kimerling, *Appl. Phys. Lett.*, **48**, 1000 (1986).
128. M. Levinson, M. Stavola, J.L. Benton and L.C. Kimerling, *Phys. Rev.*, **B28**, 5848 (1983).
129. T.I. Kol'chenko and V.M. Lomakoi, *Semiconductors*, **28**, 501 (1994).
130. F.K. Koschnick, M. Hesse, K. Krambrock and J.-M. Spaeth, *Proceedings of the 8th Conference on Semi-insulating III–V Materials*, Warsaw (1994), (edited by M. Godlewski, World Scientific, Singapore), p.229 (1994).
131. F.D. Auret, R.M. Erasmus and S.A. Goodman, *Jpn. J. Appl. Phys.*, **33**, L491 (1994).
132. J.L. Benton and M. Levinson, In *Defects in Semiconductors II.*, ed. by S. Mahajan and J.W. Corbett (North-Holland, New York) p.95 (1983).
133. D.L. Scharfetter, *Solid-State Electronics*, **8**, 299 (1964).
134. F.D. Auret and M. Nel, *J. Appl. Phys.*, **64**, 2422 (1988).
135. A. Chantre, *Appl. Phys.*, **A48**, 3 (1989).
136. S.A. Goodman, F.D. Auret and G. Myburg, *Appl. Phys.*, **A59**, 305 (1994).
137. M.J. Legodi, F.D. Auret and S.A. Goodman, *J. Physics B*, **273–274**, 762 (1999).
138. J.C. Bourgoin and M. Lannoo, *Point Defects in Semiconductors II, Experimental Aspects* (Springer, New York), pp.191–202 (1983).
139. K.L. Wang and G.P. Li, *Solid State Commun.*, **47**, 233 (1983).
140. S.J. Pearton, A.J. Tavendale and A.A. Williams, *Electronics Letters*, **16**, 483 (1980).
141. J.L. Hartke, *J. Appl. Phys.*, **39**, 4871 (1968).
142. G. Vincent, A. Chantre and D. Bois, *J. Appl. Phys.*, **50**, 5484 (1979).
143. G. Pensl, G. Roos, C. Holm and P. Wagner, *Proceedings of the Fourteenth International Conference on Defects in Semiconductors*, edited by von Bardeleben, H.J., Materials Science Forum (Transworld Technology, Switzerland), Vol. 10–12, pp.911–916 (1986).
144. G.P. Li and K.L. Wang, *Appl. Phys. Lett.*, **42**, 838 (1983).
145. D. Pons and J.C. Bourgoin, *Phys. Rev.*, **B43**, 11840 (1991).
146. Q.S. Zhu, K. Hiramatsu, N. Sawaki, I. Akasaki and X.N. Liu, *J. Appl. Phys.*, **73**, 771 (1993).
147. W. Csaszar and A.L. Endros, *Phys. Rev. Lett.*, **73**, 312 (1994).
148. G.A. Sai-Halasz and J. Gazecki, *Appl. Phys. Lett.*, **45**, 1067 (1984).
149. K. Bhattacharya, A. Reisman and M.C. Chen, *J. Electron. Mater.*, **17**, 273 (1987).
150. G. Myburg and F.D. Auret, *J. Appl. Phys.*, **71**, 6172 (1992).
151. F.D. Auret, G. Myburg, H.W. Kunert and W.O. Barnard, *J. Vac. Sci. Technol.*, **B10**, 591 (1992).
152. S.W. Pang, *J. Electrochem. Soc.*, **133**, 784 (1986).
153. L.W. Aukerman and D.D. Graft, *Phys. Rev.*, **127**, 1576 (1962).
154. B. Szentpali, B. Kovacs, D. Huber, C.D. Kourkoutas, P.C. Euthymiou and G.E. Zardas, *Solid State Communications*, **80**, 321 (1991).
155. G.J. Shaw, S.R. Messenger, R.J. Walters and G.P. Summers, *J. Appl. Phys.*, **73**, 7244 (1993).
156. Y. Umemoto, N. Masuda and K. Mitsusada, *IEEE Elect. Dev. Lett.*, **EDL-7**, 396 (1986).
157. P. Blood, *Inst. Phys. Conf. Ser.*, **56**, 251 (1981).
158. D. Donoval, J. de Sousa Pires, P.A. Tove and R. Harman, *Solid-State Electron.*, **32**, 961 (1989).

159. R.M. Erasmus, W.E. Meyer, F.D. Auret and G. Myburg, *S. Afr. J. Phys.*, **16**, 89 (1993).
160. E.M. Verbitskaya, V.K. Eremin, A.M. Ivanov, E.S. Ignatenko, N.B. Strokan, U.Sh. Turebekov, J. von Borany and B. Schmidt, *Sov. Phys. Semicond.*, **25**(5), 516 (1991).
161. D. Pons, P.M. Mooney and J.C. Bourgoin, *J. Appl. Phys.*, **51**, 2038 (1980).
162. J.P. de Souza, I. Danilov and H. Boudinov, *J. Appl. Phys.*, **81**, 650 (1997).
163. M.J. Legodi, F.D. Auret, S.A. Goodman, J.B. Malherbe and M. Swart, *Nuclear Instr. and Meth. in Phys. Res. Sect. B*, **148**, 441 (1999).
164. T.I. Kol'chenko and V.M. Lomakoi, *Sov. Phys. Semicond.*, **9**, 1153 (1975).
165. P.L. Pegler, J.A. Grimshaw and P. Banbury, *Radiat. Effects*, **15**, 183 (1972).
166. A. Kahan, L. Bouthillette and H.M. DeAngeles, *Radiation Effects in Semiconductors* (New York: Gordon and Breach), pp.281–285 (1971).
167. P.N. Brunkov, V.S. Kalinovsky, V.G. Nikitin and M.M. Sobolev, *Semicond. Sci. Technol.*, **7**, 1237 (1992).

CHAPTER 8

Instability and Defect Reaction

YUZO SHINOZUKA

Department of Material Science and Chemistry
Faculty of Systems Engineering, Wakayama University
Sakaedani 930, Wakayama 640-8510, Japan
yuzo@sys.wakayama-u.ac.jp

1. ELECTRONIC-ATOMIC PROCESSES IN SEMICONDUCTORS

The present semiconductor technology is based on a physical hypothesis where carriers (conduction electron and valence hole) move in an almost static lattice and sometimes interact with static (vacancy, impurity, ...) and dynamical (lattice vibrations) crystal imperfections. It has long been considered that the electron-lattice interaction in semiconductors can be treated as a small perturbation. Recently, however, there have been found various phenomena which break this hypothesis. A carrier trapping at a defect sometimes induces large atomic displacements, such as structural change of a point defect, bond breaking, creation of a defect, desorption of an atom from a surface and so on [1]. One of the most striking phenomena might

345

be the off-center instability of a substitutional impurity, which has been found at several donors in III–V semiconductors [2, 3]. This phenomenon invalidates the traditional common sense of the valence control using shallow impurity doping, and is considered as a key mechanism of defect reactions in semiconductors. Degradation of light emitting semiconductor diodes and lasers is caused by the climb and glide motions of dislocations [4]. The control of defect reactions is an important key to achieve the high efficiency and reliability of devices.

The purpose of this chapter is to give an overview on various electronic-atomic processes at defects in semiconductors. First, I will discuss the strong localization of a carrier in semiconductors caused by interactions with the lattice in connection with the shallow-deep instability. Next I will discuss the symmetry breaking structural changes at point defects recalling the Jahn-Teller effect. Because of a strong electron localization at deep-level defects, the effects of the electron-electron interaction as well as the electron-lattice interaction becomes important. The cooperation and competition of two interactions at point defects will be precisely discussed with a tetrahedral T-U-S model. Finally, I will discuss mechanisms of electronically induced (enhanced) defect reactions. Proposed mechanisms so far are reexamined including the structural instability and the recombination enhanced. These two mechanisms can be treated in a unified scheme by using the correct configuration coordinate diagram, which enable us to treat correctly the correlation in successive captures of electrons and holes. The mechanism of the energy conversion during the reaction is discussed paying attention to the relation between the lattice relaxation mode and the symmetry breaking reaction coordinate.

2. ELECTRON LOCALIZATION: SHALLOW-DEEP INSTABILITY

In condensed matter, a crystalline structure is realized when the electrons and the atoms are in the ground state — the absolute minimum of the total energy. If a carrier (conduction electron, valence hole) is created in a semiconductor, the balance of interatomic forces mediated by valence electrons is disturbed and then the atomic configuration is rearranged. This is the so-called electron-lattice interaction [5]. Since carriers move almost freely in normal semiconductors, it has been considered that strong electron-lattice interaction is absent there and is realized only in ionic materials. Recent experimental studies have shown that this is not true. If the electronic excitation is spatially localized by a point defect, impurity, or due to self-trapping as the secondary process, and if the change of the electronic charge density is sufficiently large, then the surrounding atoms are affected by the induced force and are displaced significantly, as has been found in some deep-level defects in covalent semiconductors [1].

Let us estimate the strength of the electron-lattice interaction using a continuum model proposed by Toyozawa [6]. Suppose a point defect in a semiconductor, which attracts a conduction electron (valence hole) with an effective mass m. The strength of the attractive potential is U_l for the long-range part and U_s for the short-range part. The lattice is regarded as a superposition of the elastic medium and the dielectric medium, representing the acoustic phonon field and the optical phonon field, respectively. If an electron is localized around a point defect with a radius a of the wave function,

$$\varphi(r) \sim \exp\left(-\pi \frac{r^2}{a^2}\right), \tag{1}$$

both mediums deform so as to relax the elastic strain and induced polarization. The lattice relaxation energies caused by two interactions are given by

$$E^{ac}_{LR}(a) = E_d^2/2Ca^3 \quad \text{and} \quad E^{op}_{LR}(a) = e^2/\varepsilon a, \tag{2}$$

where E_d is the deformation potential, C is the elastic constant, $\varepsilon^{-1} = \varepsilon_\infty^{-1} + \varepsilon_0^{-1}$ with ε_0 and ε_∞ are the static and high frequency dielectric constants, respectively. Both $E^{ac}_{LR}(a)$ and $E^{op}_{LR}(a)$ increase as the electron radius a decreases. Their maximum values are realized at $a = d$, where d is the lattice constant of the crystal, since the wave function (1) is an envelope function of Wannier orbitals. For a typical semiconductor with $E_d = 6$ eV, $C = 4 \times 10^{10}$ N/m^2, $\varepsilon = 4$ and $d = 3$ Å, the maximums are $E^{op}_{LR}(d) = 1.2$ eV and $E^{ac}_{LR}(d) = 2.7$ eV. Adding the increase of the electron kinetic energy by localization and the potential gain by a defect, the total energy is given by

$$E(a) = B(d/a)^2 - (U_s + E^{ac}_{LR})(d/a)^3 - (U_l + E^{op}_{LR})(d/a), \tag{3}$$

where $B = 3\pi\hbar^2/2md^2$ [6]. The stable electron radius $\bar{a}$ is determined so as to give the minimum of $E(a)$ in the region of $0 \le d/a \le 1$. The value of $\bar{a}$ depends on two strengths for electron localization: intrinsic $\vec{E}_{LR}/B = (E^{ac}_{LR}/B, E^{op}_{LR}/B)$ and extrinsic $\vec{U}/B = (U_s/B, U_l/B)$. Figure 1(a) shows typical behaviors of $E(a)$ and Figure 1(b) shows the phase diagram of an electron in the deformable lattice. In the region Sh, an electron is shallowly localized around a point defect ($\bar{a} \gg d$) with small lattice distortion. In the region De, it is deeply localized ($\bar{a} = d$) with strong lattice distortion. If there is a metastable state, it is indicated in the parenthesis.

For a typical donor in nonpolar semiconductors such as P in Si, $E^{op}_{LR}(d) \sim 0$, $U_s \sim 0$, $\varepsilon \gg 1$ and m is light, then $\bar{a} \gg d$ and $E^{ac}_{LR}(\bar{a}) \sim 0$ in spite of large $E^{ac}_{LR}(d)$. As a result, the effect of the electron-lattice interaction can be negligible for shallow donor (acceptor) impurities. On the other hand, it is possible that even if both $\vec{E}_{LR}/B$ and $\vec{U}/B$ locate in the region

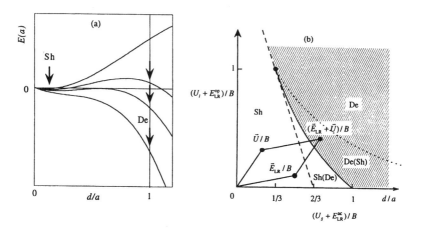

Fig. 1. (a) The total energy and (b) the phase diagram of an electron interacting with a point defect and the lattice.

Sh but the cooperation locates in the region De or De(Sh) as is shown in Figure 1(b). Then the ground state is a deep state with strong lattice distortion. This mechanism is called the extrinsic self-trapping [7, 8]. Here, a strong electron-localization and a strong lattice distortion are the cause and the result each other. A nitrogen impurity in Si [9, 10] might be an example of this extrinsic self-trapping (De(Sh)), which accompanies a metastable shallow state with small lattice distortion. Thus we have seen that strong electron-lattice interaction, mainly due to the short-range interaction by acoustic phonons (see the a dependence in eq. (2)), induces a shallow-deep instability at point defects in semiconductors.

3. SYMMETRY BREAKING

So far we have assumed a spherical symmetry in electron localization. When an electron is strongly localized ($a \sim d$), it feels the lattice structure and also interacts with other valence electrons. Then we must explicitly consider the lattice structure and its symmetry to discuss the lattice distortion in real space. In this section we discuss symmetry breaking structural changes recalling the Jahn-Teller (J-T) instability, which has been first proposed to discuss the stability of molecular structures [11]. Let us consider a point defect in a semiconductor, whose point symmetry is characterized by a symmetry group denoted by **G**. The adiabatic Hamiltonian of the system is given by

$$H = H_e(\{r\}) + V(\{r\}, \{R\}) + U_L(\{R\}). \tag{4}$$

Here $\{r\}$ denotes the set of the position of electrons $\{r_1, r_2, \ldots\}$ and $\{R\}$ represents the displacement $\{R_0, R_1, \ldots\}$ of atoms around the defect. We must distinguish the relevant electrons and the other valence electrons that are combined with ions to form the lattice medium [5]. $U_L(\{R\})$ is the elastic energy for the displacements of atoms. The set of displacements $\{R\}$ can be transformed into the set of normal modes $\{Q_\Gamma\}$ in the irreducible representation Γ of $\mathbf{G}$ [12]. The electron-lattice interaction term $V(\{r\}, \{R\}) = V(\{r\}, \{Q_\Gamma\})$ can be expanded as a power series of R_j's or Q_Γ's.

$$V(\{r\}, \{R\}) = V(\{r\}, \{R\} = 0) + \sum_j \frac{\partial V}{\partial R_j} R_j + \ldots$$

$$= V(\{r\}, \{Q_\Gamma\} = 0) + \sum_\Gamma \frac{\partial V}{\partial Q_\Gamma} Q_\Gamma + \ldots. \quad (5)$$

In the rigid lattice $\{R\} = 0$, all electronic states $\Psi(\{r\})$ which obey the Schrödinger equation

$$[H_e(\{r\}) + V(\{r\}, \{R\} = 0)]\Psi(\{r\}) = E\Psi(\{r\}), \quad (6)$$

are classified by the irreducible representation Γ_e of $\mathbf{G}$ as $\Psi_{\Gamma_e}(\{r\})$.

Next, we allow the displacements of atoms. The generalized force $F_{\Gamma_e}^\Gamma$ with respect to Q_Γ for the electronic state $\Psi_{\Gamma_e}(r)$ is given by

$$F_{\Gamma_e}^\Gamma = \left\langle \Psi_{\Gamma_e} \left| \frac{\partial V}{\partial Q_\Gamma} \right| \Psi_{\Gamma_e} \right\rangle. \quad (7)$$

If $\Psi_{\Gamma_e}(\{r\})$ is degenerate, the symmetric representation $[\Gamma_e \times \Gamma_e]$ always contains at least one non-identity representation Γ [12]. This means $F_{\Gamma_e}^\Gamma \neq 0$ and then atoms are *always* displaced to lower the symmetry. Figure 2 shows an example of the J-T distortion at a square molecule of D_{4h}. When the electronic state is doubly degenerate E_g, it distorts to D_{2h}. It should be noted that there are several equivalent stable configurations: in Figure 2 there are two equivalent J-T distortions ($Q_{B1g} > 0$ and $Q_{B1g} < 0$), thus the total symmetry remains conserved in some sense. Table 1 shows typical examples of the relation between electronic states and the J-T distortion modes for some point symmetries. A vacancy (V) in the diamond structure has the T_d symmetry. In Si crystal when a vacancy is positively charged V_{Si}^+, the electronic ground state is triply degenerate ($\Gamma = T$) then the E-type J-T distortion takes place to lower the symmetry to D_{2d}, which has triple equivalent minimums. This J-T distortion at V_{Si}^+ has been observed experimentally by EPR measurements [13] and the potential barrier height between equivalent minimums is estimated as 0.07 eV [14].

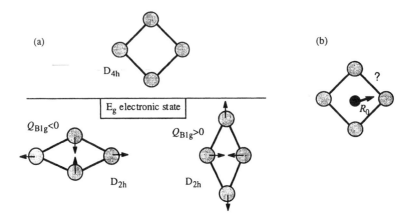

Fig. 2. An example of the Jahn-Teller effects. (a) D_{4h} structure is unstable for the E_g electronic states and distorts to D_{2h}. (b) Does the J-T distortion contain the displacement of the central atom?

So far we have treated the atomic motion classically. As has been stated in the original paper [10], the J-T theorem only gives us the initial instability but does not give the magnitude of the atomic distortion and hence that of E_{LR}. When the lattice relaxation energy E_{LR} is smaller than the average phonon energy, $\hbar\omega$, quantum transitions occur rapidly between the equivalent minimums of the distorted atomic configuration. If the mean transition time is shorter than the characteristic time of the experiment, the average of the symmetry lowering is zero (the dynamic J-T effect). This explains why the J-T effect is absent at shallow-level impurities which are affected by the point symmetry in some degree. On the other hand, if the electron wave function is localized as is in deep-level defects, electrons interacts strongly with the lattice, then $E_{LR} \gg \hbar\omega$ and the present J-T instability really takes place and lowers the symmetry for degenerate electronic states.

4. STABILITY OF THE POSITION OF AN IMPURITY

Let us next apply the above Jahn-Teller theorem to the stability of positions of an impurity atom in semiconductors, either *interstitial* or *substitutional*. In the former case, an impurity atom may occupy a position with larger empty space, and hence with high point symmetry. Let us denote the symmetry of the position as **G**. If the electronic state is degenerate, atoms around the impurity are always displaced to lower the symmetry due to the J-T effect as shown above. *The point now is whether these J-T displacements*

TABLE I
Typical examples of the Jahn-Teller distortion modes and the type of the displacement
R_0 of the center atom. E states are doubly degenerate and T states are triply degenerate.

T_d electronic state Ψ_Γ	J-T distortion mode Q_Γ	D_{3d} electronic state Ψ_Γ	J-T distortion mode Q_Γ	C_{3v} electronic state Ψ_Γ	J-T distortion mode Q_Γ
E	E				
T_1	E, T_2	E_g, E_u	A_{2g}, E_g	E	E
T_2	E, T_2				
R_0 belongs to T_2		R_0 belongs to A_{2u} and E_u		R_0 belongs to A_1 and E	
off-center instability T_d position is unstable for T_1 and T_2 electronic states.		*off-center instability* D_{3d} position does not show the instability caused by J-T effect.		*off-center instability* C_{3v} position is unstable for E electronic states.	

contain the displacement R_0 of the impurity atom located at the center. The number of atoms which couple the electronic states is not limited, then the transformation of $\{R_j\}$ into $\{Q_\Gamma\}$ is practically impossible. However it is much easy to find to which representation $\Gamma('s)$ the central displacement R_0 belongs. The criterion of the instability of the positions of an impurity for several point symmetries has been discussed [15] and some of the results are summarized in Table I. Thus, we have a useful information on the stability of an impurity position without any elaborate calculations. It should be noted that *an impurity atom at any positions that have the inversion symmetry does not show the instability caused by the J-T effect.* This is because the irreducible representations for any system with the inversion symmetry are classified into *gerade* and *ungerade*. All the J-T distortion modes are *gerade* while the displacement of the central atom always belongs to *ungerade* modes. Examples are the interstitial hexagonal position (D_{3d}) in the diamond structure, the body-centered position (O_h) and the face-centered position (D_{4h}) in the simple cubic lattice.

If electronic states $\Psi_{\Gamma_e}(r)$ contain a spin freedom, the condition of the Jahn-Teller instability depends on the number of bound electrons. The criterion for even number electrons is whether the symmetric representation $[\Gamma_e \times \Gamma_e]$ contains the displacement R_0 of the impurity, while that for odd number is whether the antisymmetric representation $\{\Gamma_e \times \Gamma_e\}$ contains R_0 or not [15]. The present J-T instability for *substitutional* impurities corresponds to off-center instability. Examples are N in Si [9, 10], EL2 in GaAs [2, 3], and DX center in AlGaAs [16, 17]. It is surprising that although the T_d is the most popular symmetry met in the diamond and the zincblende structures, it shows an inherent instability as shown here. Thus the diamond structure is not so stable as we have been considered with respect to local electronic excitations.

5. STRUCTURAL CHANGE OF A TETRAHEDRAL POINT DEFECT

In case of strong electron localization we must also take account of the electron-electron interaction. Even if we start with a single electron in the conduction band, as it becomes strongly localized around a particular bond, it interacts with another valence electrons which have been combined with ions to form the lattice medium in the first place. Then we have to extend the concept of the electron-lattice interaction, which can be consistently treated with the many-electron scheme (see a review article [5]). In this section, let us study the structural instability of a "defect molecule", the opposite scheme to "one electron in a crystal solid" discussed in Section 2.

Most of covalent semiconductors have either the diamond, zincblende or wurtzite structures, which are constructed by the four-fold coordination network with sp^3 hybridization of valence electrons. Then the tetrahedral symmetry (T_d) is the most frequent and important point symmetry met in point defects in covalent semiconductors. Several point defects have been found to show symmetry breaking structural changes from T_d to D_{2d}, C_{2v} or C_{3v} when the electronic occupation is changed. The stable structure is governed by the electron-electron interaction and the electron-lattice interaction. The simplest model to discuss the cooperation and competition between them would be the T-U-S model [18], which consists of several interacting electrons coupled with a deformable lattice. Recently El-Maghraby and Shinozuka [19, 20] have done a comprehensive study on a tetrahedral (T_d) four-site system to discuss the competition between these interactions and the metamorphosis of the stable structures. Hereafter I will review their results to discuss the symmetry breaking processes of point defects in covalent semiconductors.

Figure 3 schematically shows (a) a vacancy and (b) a substitutional impurity in the diamond (zincblende) lattice. As a first step of approximation we cut them out of the crystal and form them into the T_d T-U-S model shown in Figure 3(c). It is characterized by the electronic transfer energy, $-T$, describing the quantum hopping of an electron from site to site, the on-site Coulomb repulsion energy, U, between two electrons, the lattice relaxation energy, S, due to the short range interaction of an electron with the lattice mode, Q_ℓ, at the site ℓ ($= 1 \sim 4$) and the number n ($= 0 \sim 8$) of the occupied electrons. The site ℓ represents either atomic or molecular orbital: four dangling bonds for a vacancy and four bonding (or antibonding) orbitals for a substitutional impurity.

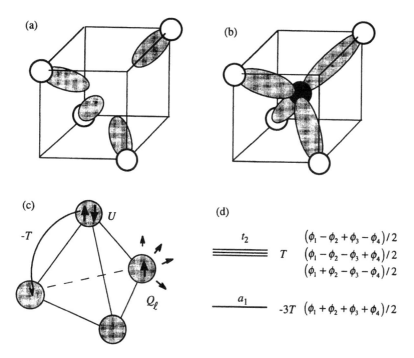

Fig. 3. (a) a vacancy and (b) a substitutional impurity in semiconductors. (c) the four-site T_d T-U-S model. Hatching in (c) represents (a) sp^3 dangling bond and (b) bonding or antibonding orbitals. (d) One-electron scheme for U=S=0, where ϕ_ℓ is one-electron orbital at the ℓ-th site.

Treating the system adiabatically, the Hamiltonian is given by

$$H = -T\sum_\ell \sum_{\ell'} \sum_\sigma a^+_{\ell\sigma} a_{\ell'\sigma} + U\sum_\ell n_{\ell+} n_{\ell-} - c\sum_\ell (n_{\ell+} + n_{\ell-}) Q_\ell + \frac{1}{2}\sum_\ell Q^2_\ell.$$

(8)

Here, $a^+_{\ell\sigma} (a_{\ell\sigma})$ is the creation (annihilation) operator of an electron with spin σ (+ (up), − (down)) at the site $\ell = 1, 2, 3, 4$ and $n_{\ell\sigma} \equiv a^+_{\ell\sigma} a_{\ell\sigma}$. The site energy at ℓ is assumed to change linearly with the lattice distortion Q_ℓ around the site with c the lattice-coupling constant. The Hamiltonian commutes with the number n of electrons and the total spin S_σ. Then they are the constants of motion. Diagonalizing the Hamiltonian matrix for each electronic subspace (n, S_σ), one obtains the adiabatic potential $W_i(\{Q_\ell\})$ of both the ground $(i = g)$ and excited states $(i = ex)$. The minimum of $W_g(\{Q_\ell\})$ in four-dimensional $\{Q_\ell\}$ space corresponds to the stable configuration. Applying the Feynman-Hellmann theorem, the lattice

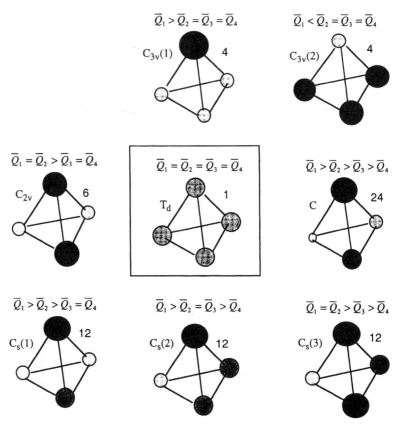

Fig. 4. Schematic presentation of the stable structures, which are characterized by the point symmetries. The volume of a sphere is proportional to the distortion and the local electron density. The number indicates the multiplicity of equivalent configurations.

distortion at the stable configuration is shown to be proportional to the local electron density as $\overline{Q}_\ell = \langle n_{\ell+} + n_{\ell-} \rangle c$ and then the sum rule holds, $\sum_\ell \overline{Q}_\ell = nc$.

In the rigid lattice ($Q_\ell = 0$), if the electron-electron interaction is neglected ($U = 0$), one electron eigenstates are one a_1 state with an eigenenergy $-3T$ and three t_2 states with T. Then the system does *not* have the electron-hole symmetry. If we allow the lattice to deform ($Q_\ell \neq 0$) and if the electron-electron interaction is switched on ($U \neq 0$), however, eigenstates of n electrons become complicated and structural changes may take place lowering the symmetry, as will be shown below.

Possible stable structures characterized by point symmetries are illustrated in Figure 4. The obtained stable configurations for the electronic ground state

TABLE II
Metamorphosis of the stable structure of the ground state of T_d T-U-S model for different electron occupation.

parameters	$n=0$	$n=1$	$n=2$	$n=3$	$n=4$	$n=5$	$n=6$	$n=7$	$n=8$
$U/T = 0.4$, $S/T = 0.8$	T_d	T_d	T_d	$C_{3v}(1)$	$C_{3v}(1)$	C	$C_{3v}(2)$	$C_{3v}(2)$	T_d
$U/T = 1$, $S/T = 0.5$	T_d	T_d	T_d	$C_{3v}(1)$	T_d	${}^*T_d$	C_{2v}	$C_{3v}(2)$	T_d
$U/T = 1$, $S/T = 2.4$	T_d	T_d	T_d	C_{2v}	C_{2v}	C	$C_{3v}(2)$	$C_{3v}(2)$	T_d
$U/T = 1$, $S/T = 8$	T_d	$C_{3v}(1)$	$C_{3v}(1)$	$C_s(1)$	C_{2v}	C	$C_{3v}(2)$	$C_{3v}(2)$	T_d
$U/T = 4$, $S/T = 2.6$	T_d	T_d	T_d	C_{2v}	T_d	${}^*T_d$	C_{2v}	$C_{3v}(2)$	T_d
$U/T = 10$, $S/T = 8$	T_d	$C_{3v}(1)$	C_{2v}	$C_{3v}(2)$	T_d	$C_{3v}(1)$	C_{2v}	$C_{3v}(2)$	T_d
S_σ of the lowest configuration	0	1/2	0	1/2	0	1/2, ${}^*3/2$	0	1/2	0

of each subspace (n, S_σ) are summarized in the phase diagrams in Figure 5 $(a \sim n)$ [20]. Along the T-U side of each phase diagram, the character of the lowest electronic state for the rigid lattice $(Q_\ell = 0)$ is indicated. We can see that if it is triply degenerate (T_1 and T_2) the stable structure is always C_{3v} for $S \neq 0$, as has been discussed in Section 4 (Table I). For excited states the T_d configuration is more unstable because they contain large amplitude of the charge transfer electronic states. Thus the 4-site T_d "defect molecule" shows various stable structures depending on the electron number, n, total spin, S_σ and the strengths of the electron-lattice interaction, S/T, and the electron-electron interaction, U/T. When the electron occupation is changed by an electron (hole) capture or a photoexcitation, a structural change takes place: some typical examples are listed in Table II.

Let us apply the results to discuss stable structures of typical point defects: a vacancy and a substitutional impurity. Table III shows the relation between the possible charge state of typical point defects in IV and III–V materials and the electron occupation n in the T_d T-U-S model.

5.1. Vacancy

The one-electron energy at a dangling bond is located in the middle of the band gap, and then it is a good approximation to separate electrons at dangling bonds with other host valence electrons. The lattice distortion

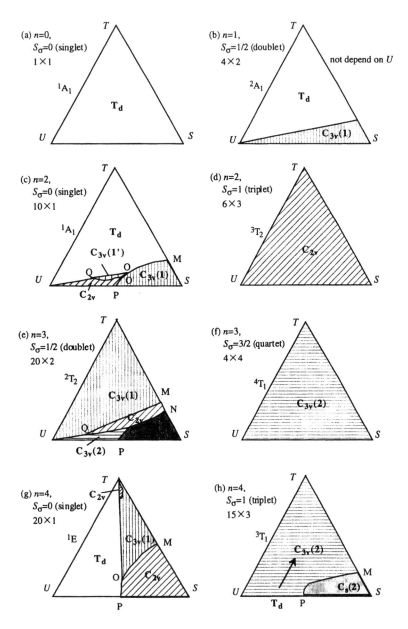

Fig. 5a-h.   Phase diagrams of the 4-site T_d T-U-S model for each electronic subspace (n, S_σ). The coordinates of (T, U, S) at any point in the triangle are proportional to the lengths of the perpendiculars from that point to the US, ST, and TU side respectively [20].

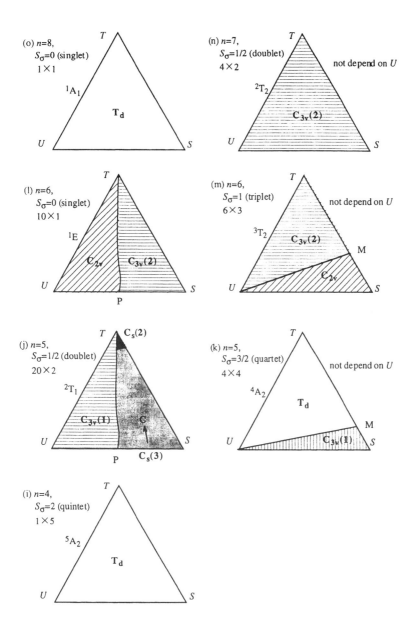

Fig. 5i-o. Phase diagrams of the 4-site T_d T-U-S model for each electronic subspace (n, S_σ). The coordinates of (T, U, S) at any point in the triangle are proportional to the lengths of the perpendiculars from that point to the US, ST, and TU side respectively [20].

TABLE III
The relation between T_d T-U-S model and typical point defects.

vacancy in IV	charge	V^{2+}	V^+	V^0	V^-	V^{2-}
	$n =$	2	3	4	5	6
III vacancy in III–V	charge	V_{III}^{2+}	V_{III}^+	V_{III}^0	V_{III}^-	V_{III}^{2-}
	$n =$	3	4	5	6	7
V vacancy in III–V	charge	V_V^{2+}	V_V^+	V_V^0	V_V^-	V_V^{2-}
	$n =$	1	2	3	4	5
	charge	D^{2+}	D^+	D^0	D^-	D^{2-}
V impurity in IV	antibonding		$n = 0$	1	2	3
	bonding	$n = 7$	$n = 8$			
	charge	A^{2+}	A^+	A^0	A^-	A^{2-}
III impurity in IV	antibonding				$n = 0$	1
	bonding	$n = 5$	6	7	8	

for bonding $Q_\ell > 0$ for antibonding $Q_\ell > 0$

Q_ℓ here represents a linear combination of the displacements of three host atoms next to the host atom at ℓ (Figure 3(a)). The vacancy (V) in Si has been found to have different stable structures depending on its charge [13, 14]: D_{2d} for V^0 and V^+, C_{2V} for V^-, and T_d for V^{++}. The neutral V^0 corresponds to $n = 4$ in the present model, the parameters for V_{Si} may fall into the third case $(U/T = 1, S/T = 2.4)$ in Table II. Since only the diagonal type electron-lattice interaction has been taken into account, the D_{2d} type structure does not appear here. It is stabilized by the off-diagonal electron-lattice interaction: modulation of the electron transfer $T(Q_\ell)$.

5.2. Substitutional Impurity

As a first step of approximation we only take account of electrons occupying the bonding and the antibonding orbitals between an impurity and four nearest neighbor host atoms. If four bonding orbitals are partially filled by electrons, we concentrate the electrons in the bonding orbitals and neglect the antibonding orbitals. The lattice distortion Q_ℓ (Figure 3(b)) then represents a contraction mode between the impurity and one of four host atoms (Table III). On the other hand, if four bonding orbitals are completely filled by eight electrons, we will combine them with ions into the lattice and concentrate

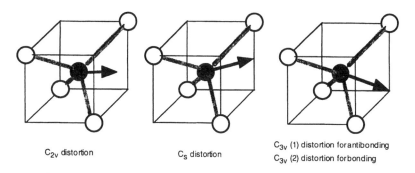

C$_{2v}$ distortion   C$_s$ distortion   C$_{3v}$ (1) distortion for antibonding
                                                                           C$_{3v}$ (2) distortion for bonding

Fig. 6. Several types of the displacement of an impurity at T$_d$ site.

only electrons in the antibonding orbitals. Then the lattice distortion Q_ℓ (Figure 3(c)) represents a stretching mode between the impurity and one of four host atoms.

Since singly ionized donor (D$^+$) and acceptor (A$^-$) correspond to $n = 0$ or 8, then the T$_d$ structure is stable there. But when the electronic occupation is changed, a structural instability may take place, which is different for donor and acceptor, because there is no electron-hole symmetry. The displacement of an impurity atom in real space are illustrated in Figure 6. For D^0 donor, T$_d$ is stable for $S/T < 6$ and C$_{3v}$(1) is stable for $S/T > 6$. Nitrogen impurity in Si, which has been found to show an off-center configuration for a neutral state, may correspond to the latter case. For D$^-$ donor, C$_{3v}$(1) type off-center configuration is more stable when $S/T > 3$ and $S > U$. DX center in Al$_{1-x}$Ga$_x$As may correspond to this case: T$_d$ structure for neutral donor and C$_{3v}$(1) off-center structure for DX$^-$ state. On the other hand, for neutral acceptor A^0, T$_d$ is always unstable and C$_{3v}$(2) type off-center is always stable. Then we can conclude that the on-center type deep-acceptor A^0 is absent in tetrahedrally bonded semiconductors. This is not the case for a shallow acceptor A^0, which should be regarded as A$^-$ + one shallowly bound hole in the present approximation. It is also noted that we have used a crucial approximation where a "defect molecule" has been cut out of the crystal.

6. DEFECT REACTION

So far, we have seen that the stable structure of a point defect will change when the electronic state has changed. Then a carrier capture or an electronic excitation can induce atomic motions around a defect towards a new stable

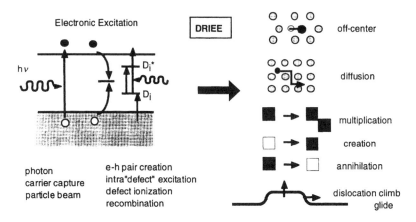

Fig. 7. Defect reaction induced by electronic excitation (DRIEE).

structure. During the lattice relaxation an amount of the electronic energy is converted to the kinetic energy of atoms, which may induce defect reaction. In this section, we will discuss the dynamics of defect reactions.

Figure 7 schematically shows typical examples of defect reactions [21]: impurity diffusion, structural change of a defect (creation, annihilation and multiplication), and climb and glide motions of a dislocation. It has been shown that defect reactions are induced or enhanced by photoexcitation, electron beam excitation and carrier injection [22]. Hereafter we will call these phenomena DRIEE in abbreviation and indicate the reaction symbolically as $D_i \rightarrow D_f$.

Several models have been proposed so far, which can be classified into two categories as schematically shown in Figure 8. One is the structural instability mechanism [23, 24] (Figure 8(a)) where a particular electronic state of a defect induces a symmetry breaking atomic motion such as the off-center distortion. For example, if the relaxation mode Q_1 is nothing but the reaction coordinate Q_R ($D_i \rightarrow D_f$) itself, an electronic excitation to ex_1 promptly induces the reaction or generates a precursor of a new defect configuration. An excitation to ex_2 decreases the potential barrier height ($E_t \rightarrow E_t^*$) of the reaction. The other mechanism is the recombination enhanced (or phonon kick) mechanism (Figure 8(b)), where an electron-hole recombination at a defect induces a transient atomic motion in the reaction coordinate Q_R [25, 26]. The first mechanism (a) stresses why and how the instability takes place toward a "defect" configuration. The second mechanism (b) stresses how the electronic energy converts to the lattice vibration energy and the motion in the reaction coordinate to overcome the potential barrier.

(a) *Structural Instability Mechanism*   (b) *Recombination Enhanced Mechanism (Phonon Kick)*

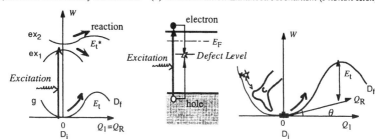

Fig. 8. Two mechanisms for DRIEE.

Though two mechanisms look different, they can be treated in a unified scheme as follows.

6.1. Annihilation of an Impurity Complex

A hydrogen atom can passivate or activate donors, acceptors and other impurities in semiconductors. In silicon, a hydrogen impurity can occupy several stable positions in the diamond lattice. If it occupies a bond center position between a substitutional carbon (isoelectronic) impurity and a silicon atom, this H-C complex acts as a shallow donor whose thermal depth is $E_{th}^e = 0.15$ eV. The positively charged H-C complex is stable above the room temperature, while the neutral state is unstable at room temperature and the activation energy for the annihilation is estimated as $0.5 \sim 0.7$ eV [27]. Since the "united atom" of a H-C complex is a nitrogen atom in the molecular chemistry, the effective potential seen by a conduction electron away from the $(H$-$C)^+$ complex is similar to those by the N^+ impurity. Then we can derive an analogy between the stability of a N impurity and that of a H-C complex in Si. For a shallow donor state or a positively charged state (an electron is away) the central position is stable for a N impurity, and the H-C complex is stable. If an electron comes close to them, on the other hand, it induces the off-center displacement of a N impurity, and the off-center displacement of a hydrogen atom. That is, a hydrogen atom is pushed away from a carbon: the annihilation of a H-C complex. This would be a typical example of a defect reaction caused by the structural instability mechanism.

6.2. Impurity Diffusion

Let us next discuss the impurity diffusion, especially the interstitial mechanism, where an impurity atom diffuses via interstitial positions. An interstitial

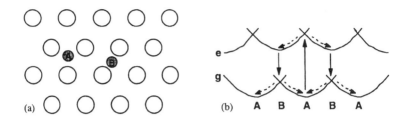

Fig. 9.   (a) Interstitial positions in a lattice. (b) Electronically enhanced impurity diffusion.

impurity may occupy a larger vacant site such as **A** and **B** in Figure 9. Then they have a high point symmetry and we can apply the results in Section 4. If a localized electronic state which shows the off-center J-T instability at position **A** continuously changes into a state which does not show at position **B** and *vice versa* (Figure 9(b)), the diffusion is enhanced by the electronic excitation and deexcitaion. The self-interstitial in Si is shown to be this type [17]. This mechanism is similar to the Bourgoin-Corbett mechanism [23] (Figure 10), but the number of bound electrons (holes) here remains constant. The B-C mechanism is proposed for an enhancement of an impurity diffusion, where the stable position of an impurity is unstable for another charged state and *vice versa*.

6.3.   Successive Carrier Captures and Induced Transient Lattice Vibration

Carrier captures by a defect as well as localized electronic excitations take place via a localized electronic state whose level is located in the band gap. Therefore it is convenient to introduce a configuration coordinate diagram (CCD) [5]. A typical example is illustrated in Figure 11: for simplicity the defect is assumed to have D_i^+ and D_i^0 charged states. It has been suggested that a carrier capture induces a transient lattice vibration, which then enhances the following capture of an opposite carrier. These coherent captures are considered as a key mechanism for recombination enhanced defect reactions, which are observed in many optical semiconductor devices [4].

Let us study the transient lattice vibrations around a deep-level defect in a semiconductor and its role in the defect reaction [28]. The normal mode of the lattice vibration is denoted by q_k with an angular frequency ω_k. The suffix k runs over the modes and the wave numbers for the bulk mode and also over localized modes if any. Without loss of generality we assume that the system is a p-type, and the equilibrium positions are $q_k = 0$ when the defect is positive (D_i^+) and $q_k = \bar{q}_k$ when the defect is neutral (D_i^0).

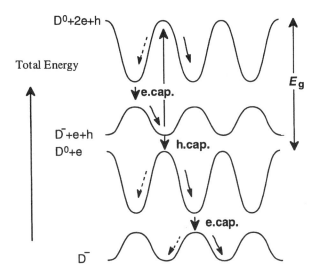

Fig. 10. The Bourgoin-Corbett mechanism for impurity diffusion.

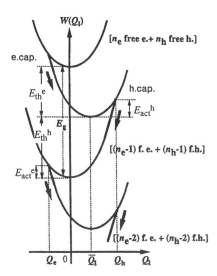

Fig. 11. The configuration coordinate model for a deep-level defect with many carriers.

The interaction mode Q_1 defined by

$$Q_1 = \frac{1}{\overline{Q}_1} \sum_k \omega_k^2 \bar{q}_k q_k, \tag{9}$$

represents the total effect of all the lattice distortions ($q_k = 0 \rightarrow \bar{q}_k$) induced by the electronic change. Before the capture of an electron, the lattice q_k vibrates thermally around each equilibrium $q_k = 0$ and is sometimes activated to distort $Q_1 = Q_e$, at which a nonradiative electron capture can takes place [29, 30]. If we assume the minimum activation energy for each mode q_k so as to reach $Q_1 = Q_e$ [26], the time evolution of the interaction mode after the capture, which has occurred at $t = 0$, is given by,

$$Q_1(t) = \frac{1}{\overline{Q}_1} \sum_k \omega_k^2 \bar{q}_k^2 \left[1 - \left(\frac{Q_e}{\overline{Q}_1} + 1 \right) \cos \omega_k t \right]. \tag{10}$$

Since the interaction mode Q_1 is a linear combination of many normal modes q_k with different frequencies ω_k, $Q_1(t)$ shows a damping oscillation, which lasts in a period $\tau \sim 2\pi/\Delta\omega$ where $\Delta\omega$ is the width of the frequency distribution. While in the normal modes, $q_k(t)$ vibrate harmonically (no damping and no energy dissipation). If the electronic system couples mainly with the localized lattice vibrational mode, the transient oscillation lasts very long.

Let us simulate the dynamics of the transient vibration and a series of carrier captures. We use following assumptions:

(1) Each normal mode q_k vibrates around either $q_k = 0$ or $q_k = \bar{q}_k$ depending on the charge of the defect (D_i^+ or D_i^0).
(2) The time dependence of $Q_1(t)$ is determined by $\{q_k(t)\}$ through the equation (9).
(3) The first electron capture takes place by a thermal activation $Q_1 \rightarrow Q_e$.
(4) If the motion of the interaction mode, $Q_1(t)$, crosses the point Q_e (Q_h) in the adiabatic potentials, there arises a large probability P_e (P_h) per time to capture an electron (hole) nonradiatively.
(5) If the motion $Q_1(t)$ has been damped out before the next carrier capture, the coherence in successive captures is quitted. The next capture is again thermally activated process.

Figure 12 shows typical examples of the time evolution of $Q_1(t)$ with parameters $P_e/\omega_0 \sim P_h/\omega_0 \sim 0.05$, where ω_0 is the mean frequency [28]. A small black circle indicates the time of a carrier capture. Figure 13 shows the time evolution of the adiabatic potential $W(Q_1)$. It is readily seen that the vibronic energy, i.e. the kinetic energy of the vibration of $Q_1(t)$, increases every time when the defect captures a carrier. If $P_i \tau < 1$, it is damped out

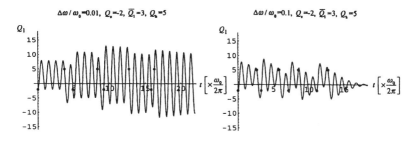

Fig. 12. Time evolution of $Q_1(t)$ during a series of captures [28].

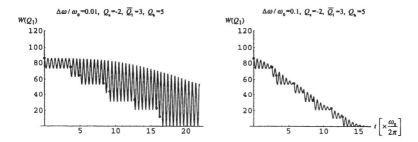

Fig. 13. Time evolution of $W(Q_1)$ during a series of captures [28].

before the next capture. The amount of the thermal depth E_{th}^i is dissipated by a capture of $i = e, h$. On the other hand, if $P_i\tau > 1$ the energy dissipation is weak and the amplitude of $Q_1(t)$ increases as time goes. The induced vibration in turn enhances the following carrier capture process. If $P_e\tau > 1$ and $P_h\tau > 1$ are satisfied, coherent captures can take place. Thus the carrier captures and the transient lattice vibration are highly correlated processes (capture enhanced capture), which depend on the carrier densities n_e and n_h, the capture cross sections σ_e and σ_h, the activation energies E_{act}^e and E_{act}^h, and the frequency distribution $\Delta\omega$ of phonons.

We have assumed that an electron (hole) can be athermally captured just when $Q_1(t)$ crosses the intersection $Q_e(Q_h)$ of two adiabatic potentials. The probability P_i ($i = e, h$) is related to the prefactor of the capture cross section $\sigma^i = \sigma_\infty^i \exp(-E_{act}^i/k_B T)$ for the nonradiative multiphonon process as $P_i = n_i v_T^i \sigma_\infty^i$ [29]. Here n_i is the carrier density, v_T^i the thermal velocity and σ^i the capture cross section. If the transient vibration enables $Q_1(t)$ to cross Q_e (Q_h), it is theoretically shown that $P_i \sim \omega_0/2\pi$ [26]. Then the condition for the coherent captures turns out to be $P_e\tau \sim P_h\tau \sim \omega_0/\Delta\omega > 1$. The

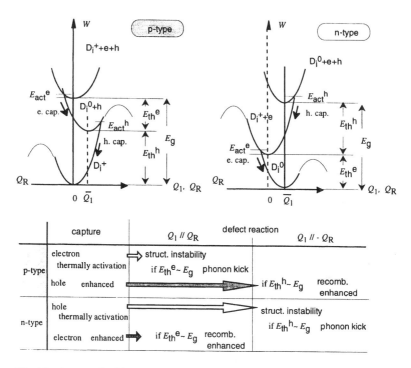

Fig. 14. An example of the configuration coordinate diagrams for the same defect in p-type and n-type semiconductors. Thick adiabatic potentials are for the interaction mode Q_1 (relaxation mode) and thin adiabatic potentials for the reaction coordinate Q_R. The table summarizes the most probable path for DRIEE.

details depend on the parameters: Q_e, Q_h, $\overline{Q}_1$, ... If several pairs of electrons and holes are captured within a short period $\sim 2\pi/\Delta\omega$, the amplitude of the interaction mode $Q_1(t)$ increases remarkably, and it may overcome the potential barrier E_t for a defect reaction ($D_i \rightarrow D_f$). More than the band gap energy $E_g = E_{th}^e + E_h^h$ can be supplied for this phonon kick mechanism.

6.4.   Recombination Enhanced Defect Reaction

There are many situations depending on the variety of charged states of a defect, the magnitude of E_{th}^e and E_{th}^h, the relation between the relaxation mode Q_1 and the reaction coordinate Q_R, the magnitude of capture cross sections, and so on (Figure 14). If the relaxation mode Q_1 is a symmetric mode such as a breathing mode, Q_1 is then orthogonal to Q_R that is a symmetry breaking mode by definition. Then the symmetrical relaxation process does not affect defect reactions. When Q_1 is not orthogonal to Q_R, Sumi [26] has shown theoretically that the transient lattice vibration by a capture reduces the activation energy E_t for a defect reaction to $E_t - (E^i_{act} + E^i_{th})$ or $\left(\sqrt{E_t} - |\cos \theta| \sqrt{E^i_{act} + E^i_{th}} \right)^2 /(1 - \cos^2 \theta)$ where θ is the angle between Q_1 and Q_R.

When the off-center instability takes place by the charge change of a defect, it is important which corresponds to the off-center configuration, $Q_1 = 0$ or $\overline{Q}_1$? If the reaction coordinate Q_R is almost parallel to Q_1, a defect reaction in p-type semiconductors occurs by the structural instability mechanism induced by a minority electron capture. If E^e_{th} is large and $Q_1//Q_R$ but there is a potential barrier, the phonon kick mechanism takes place with the aid of E^e_{th} induced by an electron capture. If E^h_{th} is large and $Q_1//-Q_R$, the phonon kick mechanism takes place by a hole capture which has been enhanced by an electron capture. If a capture of a majority carrier occurs very quickly within the local transient vibration, the band gap energy $E_g = E^e_{th} + E^h_{th}$ can be used for an excitation of a defect reaction. This could be the most effective path to the defect reaction and nothing but the so-called recombination enhanced mechanism. In an n-type semiconductors, defect reactions can be discussed in a similar way as above, starting from $Q_1 = \overline{Q}_1$ and $D^0_i + e + h$. In contrast to a p-type, the first capture of a minority hole by a neutral center D^0_i is not accelerated by the Coulomb attraction while the second electron capture by D^+_i is accelerated. We have assumed that the initial defect has D^0_i and D^+_i charged states. If it has D^-_i and D^0_i, the above discussion can be also valid with respective translations $D^0_i \rightarrow D^-_i$ and $D^+_i \rightarrow D^0_i$. As we have shown, either in n-type and p-type, the capture of a majority carrier always enhanced by a minority carrier capture, and *vice versa*. But only a hole capture is accelerated by the Coulomb attraction of D^-_i.

References

1. Y. Shinozuka, "Relaxations of Excited States and Photo-Induced Structural Phase Transitions", *Proc. 19th Taniguchi Symposium* (Springer, 1997) ed. K. Nasu, 229.
2. J. Dabrowiski and M. Scheffler, *Phys. Rev. Lett.*, **60**, 2183 (1988).
3. D.J. Chadi and K.J. Chang, *Phys. Rev. Lett.*, **60**, 2187 (1988).
4. O. Ueda, "Reliability and degradation of III–V optical devices" (Artech House Publishers, Boston-London) 1996.
5. Y. Shinozuka, *Jpn. J. Appl. Phys.*, **32**, 4560 (1993).
6. Y. Toyozawa, *Physica*, **116B**, 7 (1983).
7. Y. Toyozawa, *Solid State Electron.*, **21**, 1313 (1978).
8. Y. Shinozuka and Y. Toyozawa, *J. Phys. Soc. Jpn.*, **46**, 505 (1979).
9. K.L. Brower, *Phys. Rev.*, **B26**, 6040 (1982).
10. K. Murakami, H. Kuribayashi and K. Masuda, *Phys. Rev.*, **B38**, 1589 (1988).
11. H.A. Jahn and E. Teller, *Proc. Roy. Soc.*, **A161**, 220 (1937).

12. See for example, Y. Onodera, "Group Theory and Application to Quantum Mechanics", (Springer, 1995).
13. G.D. Watkins, *Proc. Defects in Semiconductors*, 1972, 228; *Physica*, **117B** & **118B**, 9 (1983).
14. M. Sprenger, S. Muller, E. Sieverts and C. Ammerlaan, *Phys. Rev.*, **B35**, 1566 (1987).
15. Y. Shinozuka, *Materials Science Forum*, **83–87**, 527 (1992).
16. D.J. Chadi and K.J. Chang, *Phys. Rev. Lett.*, **61**, 873 (1988), *Phys. Rev.*, **B39**, 10063 (1989).
17. M. Saito, A. Oshiyama, and O. Sugino, *Phys. Rev.*, **B47**, 13205 (1993).
18. Y. Toyozawa, *J. Phys. Soc. Jpn.*, **51**, 1861 (1981).
19. M. El-Maghraby and Y. Shinozuka, *Proc. Materials Science Forum*, **258–263**, 647 (1997).
20. M. El-Maghraby and Y. Shinozuka, *J. Phys. Soc. Jpn.*, **67**, 3524 (1998).
21. Y. Shinozuka, *Proc. MRS 1996 Fall Meeting*, **442**, "Defects in Electronic Materials II", ed. J. Michel, T. Kennedy, K. Wada, and K. Thonke, p.225 (1996).
22. K. Maeda, M. Sato, A. Kubo and S. Takeuchi, *J. Appl. Phys.*, **54**, 161 (1983).
23. J.C. Bourgoin and J.M. Corbett, *Phys. Lett.*, **38A**, 135 (1972), and *Radiation Effects*, **36**, 157 (1978).
24. M.K. Sheinkman, *JETP Lett.*, **38**, 330 (1983), and M.K. Sheinkman and L.C. Kimerling, "Defect Control in Semiconductors", ed. K. Sumino (North-Holland, 1990) p.97 (1983).
25. J.D. Weeks, J.C. Tully and L.C. Kimerling, *Phys. Rev.*, **B12**, 3286 (1975).
26. H. Sumi, *Phys. Rev.*, **B29**, 4616 (1985), *J. Phys.*, **C17**, 6071 (1984).
27. Y. Kamiura, M. Tsutsune, M. Hayashi, Y. Yamashita and F. Hashimoto, *Materials Science Forum*, **196–201**, 903 (1995).
28. Y. Shinozuka and T. Karatsu, *Materials Science Forum*, **258–263**, 659 (1998), and *Physica* **B273–274**, 999 (1999).
29. C.H. Henry and D.V. Lang, *Phys. Rev.*, **B15**, 989 (1977).
30. Y. Shinozuka, *J. Phys. Soc. Jpn.*, **51**, 2852 (1982).

CHAPTER 9

Device Degradation

KAZUMI WADA[1] AND HIROSHI FUSHIMI[2]

[1]*Department of Materials Science and Engineering,
Massachusetts Institute of Technology, 77 Massachusetts Avenue,
Cambridge, MA 0.2139 USA*
[2]*Nippon Telephone and Telegraph Corporation, Atsugi, 243-0198, Japan*

Extremely high density of carriers is often injected into their microscopic structures in recent electron and optoelectronic semiconductor devices in order to achieve higher performances. This however often enhances device degradation. There are various modes in device degradation: Degradation of device extrinsic structures such as resistivity increase of metallic wires by electromigration, and degradation of device intrinsic structures such as deterioration of device performances. Main factors governing the degradation are semiconductor-material-related, but the device-structure-related factors are present and make phenomena complex. In the present chapter, material-related degradation of the device intrinsic structures is described. Devices discussed here are minority carrier injection devices, and materials are III–V compound semiconductors, especially GaAs.

It is well known that a long-term operation of semiconductor laser diodes (LDs) results in increase in non-radiative recombination between electrons and holes at elevated temperatures, leading threshold current increase, i.e., so called rapid degradation [1]. History of LDs development is a challenge how to suppress degradation. In course of such development, it became clear that dislocations in the active layers degraded LDs. All LDs have thus been subjected to initial screening tests prior to the market. In other words, those survived are being supplied to the market. It is reasoning why stable LDs are commercially available. However, slow degradation became noticeable. Micro dislocation loops were reported in light emitting diodes, LEDs. Thus, it can be said that point defect aggregation occurs in these degradation during minority carrier injection. However, physics behind slow degradation as well as rapid degradation have not been clarified. Accordingly, no fundamental solution has been prepared for suppression of possible degradation of next generation devices, except for recipe for dislocation suppression.

Recently, similar phenomena have again been reported in high performance transistors, i.e., heterojunction bipolar transistors (HBTs) and not suppressed yet [2]. No dislocations have been observed in the base layers of degraded HBTs but micro-defects have been reported in the base layer. These features are similar to the slow degradation in LDs. However, there is a essential difference between slow degradation of LDs and degradation of HBTs: The former occurs in a nominally undoped luminescent layer, whereas the latter occurs in a heavily doped base layer. In other words, Fermi level effect or chemical effect by impurities would take part in degradation of HBTs.

What should be clarified for suppression of device degradation? Is it impurity or intrinsic point defect? Is it on As sublattices or Ga sublattices? How they diffuse and react under thermal ground states or under electronic excitation? Which is enhanced, diffusivity or reaction? Is the mechanism recombination enhanced defect reaction (REDR) or charge state effect (CSE) [3]? We have plenty of questions to answer.

In this chapter degradation of HBTs is described, especially GaAs/AlGaAs HBTs with heavily carbon doped base layer. Here, two types of the device degradation are found, i.e., hydrogen-related degradation and carbon related degradation. The mechanisms governing the degradation are discussed from the frame work of REDR and CSE.

1. HBT STRUCTURES AND THEIR DEGRADATION

HBTs are bipolar transistors whose emitter/base structures are heteto-junctions. Figure 1 shows a typical structure of n-p-n HBTs based on

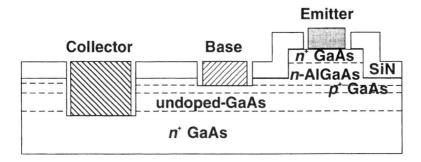

Fig. 1. Typical structure of *n-p-n* HBTs based on GaAs/AlGaAs systems.

GaAs/AlGaAs systems, where the emitter is *n*-AlGaAs:Si, the base is p^+-GaAs:C, and the collector is undoped GaAs. Employment of the widegap emitter provides the following advantages over homojunction bipolar transistors:

(1) Suppression of hole injection from the base layer into the emitter.
(2) High current gains even at heavily doping of the base layer for decrease in the base resistance.

This should allow high current gain at high current density.

Recently, it has been reported that HBTs show degradation [2]. In the current voltage characteristics, so called Gummel plot, the base current increases and the collector current decreases by a high current injection for a long time. In Figure 2 typical examples of changes in Gummel plot is shown. This leads current gain decrease with time, i.e., degradation. It is interesting to note that the base current increase at low voltage was accompanied with current gain decrease at high voltage. In the following section, for simplifying the degradation behavior to extract fundamental essence of degradation physics, emitter/base junction diodes are used instead of real HBTs structures. In this diodes the base current increase is induced by forward biasing at elevated temperatures as shown later. Thus, the bias current increase at low bias voltage will be monitored for quantitative analysis of HBTs degradation.

It is found that there are two factors dominating the degradation, hydrogen and carbon. In the next two sections, these two degradation factors are described more in detail.

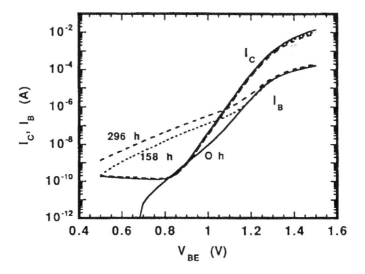

Fig. 2.   Gummel plots measured for C-doped GaAs/AlGaAs HBTs before and after minority-carrier injection. Injection minority carrier density is 50 kA/cm^2 at 250°C. The base current increase at low voltage is clearly shown.

2. HYDROGEN RELATED DEGRADATION OF HBTS

There are two kinds of states of hydrogen in heavily doped p-GaAs:C. The first origin of hydrogen is undissociated methyl radicals formed during epitaxial growth at low temperatures [4–9]. The second is hydrogen decomposed from AsH$_3$ during cooling process after epitaxial growth [10, 11]. It is shown in this section that hydrogen is involved as C-H complex in the first case and isolated donor in the second case. Accordingly, carbon acceptors are inactivated by hydrogen induced by these two mechanisms, i.e., neutralization by C-H bond formation and charge compensation with hydrogen donors [12–14]. In the following section, these two states of hydrogen and their roles for HBT degradation are described.

3. THERMAL BEHAVIOR OF HYDROGEN IMPURITY

3.1. Isolated Hydrogen Interstitials

Figure 3 shows the structure and the growth sequence of p^+-GaAs:C epilayers described in this chapter. The growth method is metalorganic vapor

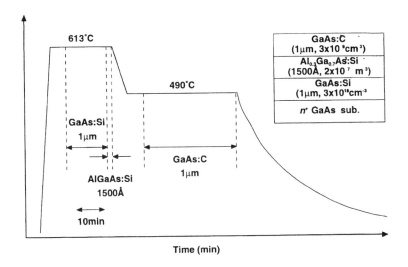

Fig. 3. The structure and the growth sequence of p^+-GaAs:C epilayers described in this chapter. The structure is a p^+-n heterojunction diode which imitate the base/emitter heterojunction structure of HBTs.

phase epitaxy (MOVPE). The structure is a p^+-n heterojunction diode, which imitates the base/emitter heterojunction structure of HBTs. In other words, a GaAs:Si buffer layer and the succeeding $Al_{0.3}Ga_{0.7}As$ emitter layer was grown on the Horizontal Bridgman (HB) grown n^+-GaAs:Si substrate. Then, p^+-GaAs:C base layer was grown at a fairly low temperature on the emitter layer, and cooled down to room temperature. C-H complexes are introduced during GaAs:C growth and isolated hydrogen interstitials are during cooling, as will be shown later.

Figure 4 shows the depth profiles of hole and hydrogen concentrations [15]. They are measured by electrochemical etching hole concentration profiler (Polaron) and secondary ion mass spectrometry (SIMS). Figures 4(a) and (b) show these profiles before and after annealing at 450°C for 5 min. Before annealing the hydrogen profile is inclined: higher near the surface and lower in the deeper region, while it becomes flat after annealing. Hole concentration is inclined as well, but the profile is inverse to the one of hydrogen. After annealing, the profile becomes flat as well as the hydrogen profile. This behavior can be fully explained by assuming that hydrogen is mobile at 450°C and acts as a donor to compensate a carbon acceptor. The depth profiles of H donors and holes quantitatively indicates that H is a single donor.

It is important to note that this hydrogen does not form a C-H complex but an isolated interstitial. This is verified by various methods as follows.

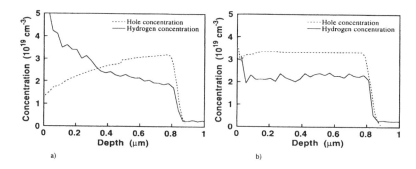

a)   b)

Fig. 4. (a) Depth profiles of both hydrogen and hole concentrations in the as-grown state. The concentration profiles of both the hydrogen impurities and holes are slope-like. The hole concentration profile, however, is inverted from the hydrogen profile. This suggests that hydrogen play a dominant role in carbon deactivation; (b) Depth profiles of both hydrogen and hole concentrations after annealing at 450°C for 5 min. Both slope-like profiles, hydrogen and hole concentration, clearly disappear. This shows that hydrogen deactivates carbon.

Concentration of C-H complexes measured by Localized Vibrational Mode (LVM) Infrared Absorption Spectroscopy is not changed before and after 380°C annealing (Figure 5). Diffusivity of hydrogen estimated by employing simple square root Dt relation is in a good agreement with the data reported so far [16–19]. This is shown in Figure 6. This is further confirmed in Section 3.2.

The fact that interstitial hydrogen donors do not form complexes with carbon acceptors suggests that there should be a reaction barrier between interstitial hydrogen donors and carbon acceptors. The barrier is high enough at temperatures below 450°C where interstitial hydrogen donors are to be introduced.

Recently Stockman and others reported annealing temperature dependence of hole concentration of p^{+}-GaAs:C [20]. Their samples were post-growth annealed in the MOVPE furnace at temperatures between 280°C and 440°C for 5 min. Hole increased starts at 320°C and finished at 400°C. It is strongly suggestive that hydrogen behavior they observed is of interstitial hydrogen donors, not of C-H complexes.

3.2. C-H Complexes

In GaAs:C used in this chapter, carbon concentration is 7×10^{19} cm^{-3}, as shown in Figure 7. However, hole and hydrogen concentration after 450°C annealing in Figure 4 is 3.5×10^{19} cm^{-3} and 2.5×10^{19} cm^{-3}. This indicates that hydrogen still remains in the epilayer. Indeed, further annealing at 530°C dramatically increases hole concentration to 7×10^{19} cm^{-3} which corresponds to carbon concentration. This increase is accompanied with

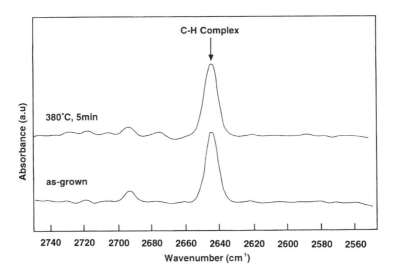

Fig. 5. LVM spectra relate to C-H complex in GaAs:C at 2643 cm^{-1} before and after annealing. The concentration of C-H complexes is not changed before and after 380°C annealing.

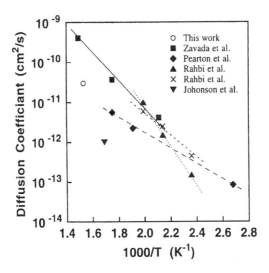

Fig. 6. The diffusion coefficients of hydrogen in GaAs. The data are replotted from the previous reports [17–19].

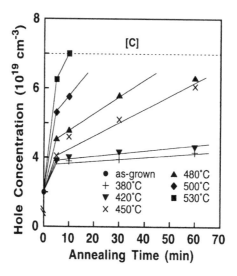

Fig. 7. Hole concentration changes due to annealing at temperatures between 380°C and 530°C obtained by Hall effect measurements. The hole concentration abruptly increases within 5 min and then the increasing rate becomes slow; the rapid process and the slow process. There should be at least two mechanisms dominating the hole concentration.

decrease in LVM peak (2643 cm^{-1}) intensity of C-H complexes, indicating dissociation of C-H complexes. Thus, the hole increase is due to hydrogen dissociation. However, there are two-hole recovery stages in Figure 7: slow and rapid recovery.

The slow recovery is due to decomposition of C-H complexes and the fast recovery is to isolated hydrogen interstitials. This is confirmed by Hall effect measurements. Figure 8 shows hole mobility dependence on hole concentration with theoretical prediction by Walukiewicz [22]. It is clear that the samples without annealing and annealed at low temperatures have mobility lower than the prediction. These samples are in a state of rapid recovery. The samples annealed at temperatures higher than 450°C have mobility corresponding to the prediction. These samples are in the slow recovery stage. Thus, it is evidently shown that the sample in the rapid recovery stage contains ionized hydrogen donors, while those in a slow recovery stage do not. This is confirmed by LVM FTIR. The absorbance intensity due to C-H complexes exponentially decreases with annealing time at various temperatures, as shown in Figure 9. The rate constants are identical to those of hole concentration increase shown in Figure 7. In other words, carrier reduction mechanism by hydrogen is neutralization of carbon acceptor, i.e., C-H complex formation in the samples in the slow recovery

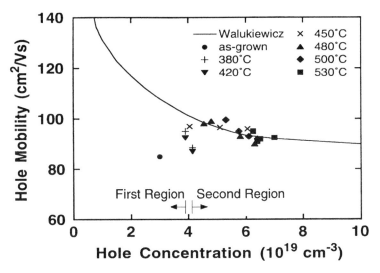

Fig. 8. Relation between hole concentration and mobility at room temperature. Solid curve is the theoretical calculation by Walukiewicz [23]. This calculation was done under assumption of no compensation. These data are divided into two regions. The samples with the flat profiles of hydrogen concentration are in the slow process and the samples with the slope-like profiles of hydrogen concentration are in the rapid process. This shows that hydrogen showing the slope-like profiles acts as a donor.

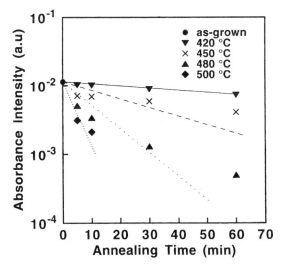

Fig. 9. Absorbance intensity due to annealing at temperatures between 420°C and 500°C obtained by FTIR. The absorbance intensity due to C-H complexes exponentially decreases with annealing time.

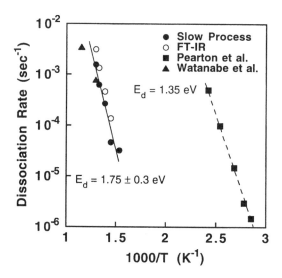

Fig. 10. Arrhenius plot of the rate constants of the hole increase in the slow process. Previously reported dissociation rate by Pearton *et al.* [21] and Watanabe *et al.* [28] are also analyzed and plotted in this figure. An activation energy is 1.75 ± 0.3 eV in this study. The dissociation rate in this study is about 5 orders of magnitude smaller than reported by Pearton *et al.* [21]. The possible mechanism is discussed in the text.

stage. On the other hand, charge compensation due to isolated hydrogen donors is additionally taking place in the samples in the rapid recovery stage. Therefore, it is clear those hydrogen interstitial donors and C-H complexes are present in this particular GaAs:C.

Figure 10 shows the temperature dependence of the rate constants of hole increase and LVM intensity decrease. This indicates that the activation energy of the dissociation rate of hydrogen from C-H complexes is around 1.75 eV. Pearton and others reported the dissociation rate of hydrogen and its activation energy. Their data are replotted in Figure 10. It is found that the activation energy is almost the same but the rate is several orders of magnitude smaller in the present measurements. This can be understood in the following way. The dissociation rate of C-H complexes can be written by

$$\frac{d[\text{C-H}]}{dt} = 4 \cdot \pi \cdot R \cdot D_H \cdot [\text{C}] \cdot [\text{H}] - \nu \cdot [\text{C-H}], \tag{1}$$

$$\approx -\left(-4 \cdot \pi \cdot R \cdot D_H \cdot [\text{C}] + \nu \right) \cdot [\text{C-H}] \tag{2}$$

where [ ] denotes concentration, R reaction radius, D_H interstitial hydrogen diffusivity, and ν dissociation rate constant. Here, [H]/[C-H] is assumed to be one. When carbon concentration is low enough, the first term of Eq. (1), i.e., association term, can be ignored, which has been employed by Pearton *et al.* However, carbon concentration is too high to ignore the association term in the present case. In other words, interstitial hydrogen dissociated from the C-H complexes simultaneously associate with carbon acceptor to form the complex again at temperature where dissociation occurs. This substantially reduces "apparent" dissociation rate, i.e., $-4 \cdot \pi \cdot R \cdot D_H \cdot [C] + \nu$, resulting in the slower dissociation rate than those reported by Pearton.

4. ELECTRONIC EXCITED HYDROGEN BEHAVIOR — HYDROGEN INDUCED DEGRADATION

Since degradation of HBTs was reported [2], microscopic analyses have been energetically performed. One of degradation factors is hydrogen as follows. It has been reported that planar defects on {111} planes were present in degraded HBTs. The planar defects are quite similar to the defects formed in the surface layer of hydrogen plasma irradiated Si [22]. Burgers vector of the plasma irradiation induced defects cannot be defined, indicating these planar defects are not bounded by dislocation but simply extend lattice constant perpendicular to the plane. It is suggestive that the defects would be hydrogen platelets. Because of similarity and the fact that hydrogen is present in the base layer of HBTs, the planar defects observed in degraded HBTs may be hydrogen platelets. This will be further supported in the next paragraph.

4.1. Isolated Hydrogen Related Degradation

The effect of hydrogen on HBTs degradation is examined by *p-n* hetero-junction diodes whose structure is shown in Figure 11. In other words, the diode is 1 μm thick p^+-GaAs:C ($[p] = 3 \times 10^{19}$ cm^{-3}) and 150 nm thick n-Al$_{0.3}$Ga$_{0.7}$As:Si ($[n] = 2 \times 10^{18}$ cm^{-3}) and this imitates the emitter and base structure of HBTs. Hereafter, the p^+ layer is referred to be "Base layer" and n layer is to be "Emitter layer", for convenience. Three types of diodes are examined, as shown in Table I. Here, the diode type A is explained. First, without any annealing, the epilayer is mesa-etched to fabricate *pn* diodes 80 $\times$ 80 μm^2. Then, AuGe/Ni/Ti/Pt/Au is deposited on the back surface of n^+-GaAs substrate and Ti/Pt/Au is deposited on the surface of epilayer. Finally, they are annealed at 380°C for 30 seconds to make ohmic contacts. This device fabrication processing is all wet processing and does not include any plasma processing, which prevents process-induced damage. Additionally, the annealing for ohmic contacts is short enough not to evacuate interstitial hydrogen from the GaAs:C layer. Thus, in this layer there should be two types of hydrogen:

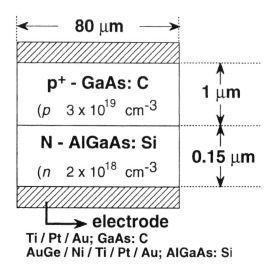

Fig. 11. The p^+-n junction diode structure with electrode. The diode is $80 \times 80 \ \mu m^2$ and is fabricated by mesa etching.

(1) interstitial hydrogen donors ($1 \times 10^{19} \ cm^{-3}$)

(2) C-H complexes ($3 \times 10^{19} \ cm^{-3}$)

Before the fabrication process, type B and C are subjected from annealing at 450°C and 530°C for 5 min. These annealing evacuate interstitial hydrogen donors from the type B diode and interstitials hydrogen donors and hydrogen bonded to carbon from the type C.

Minority-carrier injection into the GaAs:C layer is performed at 175°C to 200°C in these diodes. Current injected is 0.08 to 1.6 kA/cm^2.

Figure 12 shows one of typical examples of the Base current increase after injection to the diode B. It is clearly found that current leakage is increased by injection. In contrast, no leakage current increased without current injection. Figure 13 summarizes leakage current increase measured

TABLE I

	Hydrogen Donor	C-H Complex	Annealing Conditions
A	$1 \times 10^{19} \ cm^{-3}$	$3 \times 10^{19} \ cm^{-3}$	as- grown
B	–	$3 \times 10^{19} \ cm^{-3}$	450°C, 5 min.
C	–	–	530°C, 5 min.

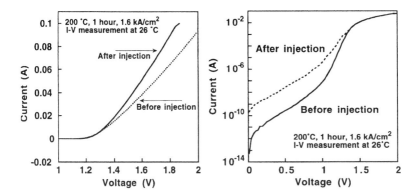

Fig. 12. The I-V characteristics before and after minority-carrier injection for sample B at room temperature. Injection minority carrier density is 1.6 kA/cm² at 200°C for 1 hour. Hole concentration increases to 6.39×10^{19} cm^{-3} by minority-carrier injection. This shows that carbon acceptors were reactivated to loose isolated hydrogen donors.

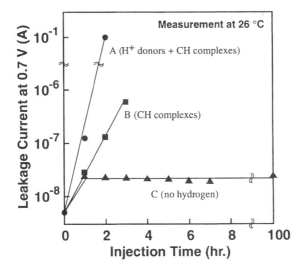

Fig. 13. Injection time dependence of the leakage current at 0.7 V at room temperature. Injection minority carrier density is 1.6 kA/cm² at 200°C. Hydrogen induces an increase in the leakage current, under minority-carrier injection. Isolated hydrogen donors (H⁺) induce rapid degradation and C-H complexes induce slow degradation. The diode currents without minority-carrier injection do not change.

at 0.7 V with injection time. This indicates that

(1) hydrogen induces leakage current.
(2) interstitial hydrogen donors induce rapid leakage current increase.
(3) hydrogen bonded to a carbon acceptor induces slow increase.
(4) carbon acceptor itself is not a degradation factor.

4.2. CH Complex Related Degradation

It is well known that hydrogen passivation eliminates states in bandgap. Thus, C-H complexes should not generate any states within the bandgap. Despite of that, diode degradation has been induced by the complexes. In order to understand what is going on behind leakage current increase, degradation kinetics due to C-H complexes are studied in the next paragraph.

If C-H complexes really decomposed by minority carrier injection, there should be increase in hole concentration, as shown in Figure 7. This indeed occurs by current injection at temperature as low as 200°C, as shown in Figure 14. Here hole concentration is estimated by using a resistive component in current-voltage (I-V) characteristics shown in Figure 12, based on the following equation.

$$R = \left(\frac{W_n}{q \cdot \mu_n \cdot n} + \frac{W_p}{q \cdot \mu_p \cdot p} \right) \cdot \frac{1}{S}, \qquad (3)$$

where q denotes elemental charge, μ_n electron mobility, μ_p hole mobility, n electron concentration, p hole concentration W_n the Emitter layer thickness, W_p the Base layer thickness and S diode area. Here it is assumed that resistivity change occurs only in the Base layer, not in the Emitter layer. This is rather appropriate assumption, since most holes are localized in the Base layer due to the presence of the Emitter barrier and minority-carrier (hole) injection to the Emitter layer hardly occurs. This hole increase shown in Figure 14 indicates that carbon acceptors are reactivated by minority-carrier injection. In other words, hydrogen dissociation occurs from the C-H complexes in terms of electron injection into the Base layer. The rate constant of carbon reactivation is 1.3×10^{-4} sec^{-1} at 200°C at 1.6 kA/cm^2. This rate constant is in good agreement with the one obtained in leakage current increase shown in Figure 13, i.e., 4.2×10^{-4} sec^{-1}.

These results clearly demonstrate that hydrogen dissociation from C-H complexes occurs under electronic excitation due to minority-carrier injection and causes leakage current increase. Furthermore, it is not difficult to imagine that resulting interstitial hydrogen donors could induce the diode degradation, as shown in Figure 13.

Figure 15 shows increase of leakage currents of the diode B under minority-carrier injection at various temperatures, 160°C, 175°C, 185°C,

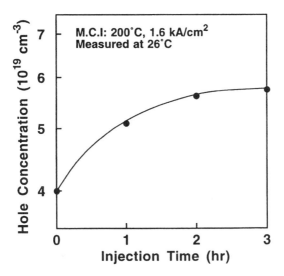

Fig. 14. Injection time dependence of hole concentration for the B-type diodes. Minority-carrier injection was carried out at 200°C at 1.6 kA/cm². The rate constant of the hole increase, estimated to be 1.3×10^{-4} sec^{-1}, is similar to that of leakage current in Figure 13.

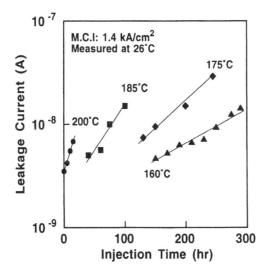

Fig. 15. Injection time dependence of leakage current for the B-type diodes measured at room temperature. Minority-carrier injection was carried out at temperature range between 160°C and 200°C at 1.4 kA/cm². Current was measured when the diodes were forward biased at 0.7 V.

and 200°C. The leakage current increases with injection time and the increase is faster at higher temperature.

Prior to analyzing the kinetics in Figure 15, it is required if the mechanism of leakage current increase is diffusion-limited or reaction-limited. Since the leakage current increase well corresponds to increase of hole concentration in a quantitative bases, the rate of leakage current increase in Figure 15 can be regarded to be the rate constant of hole increase. This means the rate-limiting event for leakage current increase is dissociation reaction of hydrogen from C-H complexes. Generated interstitial hydrogen donors would induce the current leakage, since interstitial hydrogen donor is most efficient leakage source shown in Figure 13. It can be thus said that leakage current increase results from C-H decomposition enhanced by minority-carrier injection. Therefore, an elemental step for current leakage is the following reaction:

$$C\text{-}H + me^- \Leftrightarrow C^- + H^{+1-m},   (4)$$

where m is the number of electrons involved in this reaction enhancement. The rate equation of C-H complex density decrease can be expressed by

$$\frac{d[C\text{-}H]}{dt} = -\kappa_f \cdot [C\text{-}H] \cdot n^m + \kappa_r \cdot [C] \cdot [H],   (5)$$

$$[C\text{-}H] = [C\text{-}H]_0 \exp(-\kappa_f \cdot n^m \cdot t),   (6)$$

κ_f and κ_r denote forward and reverse reaction rate constants of Eq. (5), $[CH]_0$ initial density of C-H complexes. Since current leakage occurs dissociated hydrogen from the complexes, resulting interstitial hydrogen should be proportional to $\exp(-\kappa_f \cdot n^m \cdot t)$ in concentration. Figure 15 indeed shows exponential increases of leakage current at various temperatures. Figure 16 shows decomposition rates of complexes, i.e., dissociation rates of hydrogen from the complexes. One data set is obtained from thermally stimulated decomposition shown in Figures 7 and 9. The other set is from decomposition under minority-carrier injection (MCI) shown in Figure 15. It is found that the decomposition under minority-carrier injection is enhanced by nearly 6 orders of magnitude around 200°C. Activation energies under MCI and thermal conditions are 0.91 and 1.75 eV. The mechanism of enhancement is charge state effect, as discussed next.

As expressed in Eq. (6), minority carriers, electrons, are involved in the decomposition reaction. Thus, it is worth studying the decomposition rate dependence on injection current density. Square dependence on injection current density is clearly observed when the injection to the diode B is performed at 200°C for 48 h, as shown in Figure 17. This indicates that the number of electrons, m, involved in the decomposition reaction is two. Thus, coulomb repulsion between C- and H- enhances the decomposition

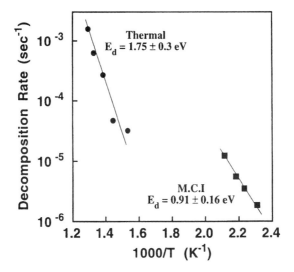

Fig. 16. Arrhenius plot of the rate constants of the decomposition of C-H complexes replotted from Figure 15. Previously reported thermal decomposition rate is also shown [16]. The numbers in the figure are activation energies of each decomposition.

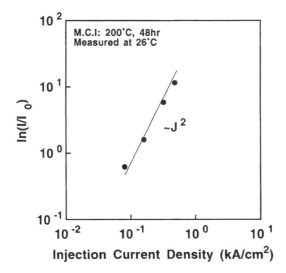

Fig. 17. Leakage current dependence of injection current density for the B-type diodes measured at room temperature. Minority-carrier injection was carried out at 200°C for 48 hours. This shows that the exponential decay is proportional to the square of injection current density.

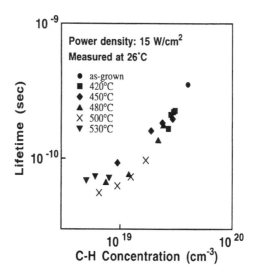

Fig. 18. Luminescence lifetime versus the concentration of C-H complexes at room temperature. C-H concentration is varied by the annealing at temperatures between 420°C and 530°C. C-H complexes do not act as recombination centers, since the luminescence lifetime increases with an increase in the concentration of C-H complexes.

reaction of C-H complexes, indicating that the mechanism of decomposition enhancement is so-called structural instability due to charge state effect described in Section 2.1.

Contribution of recombination enhanced defect reaction REDR is considered in the following. Time correlated single photon counting method is used to measure photoluminescence lifetime. Figure 18 shows photoluminescence lifetime dependence on C-H density. The density was controlled by thermal annealing of the samples shown in this figure. The samples were excited by pyridine 2 dye laser at 740 nm excited by mode-locked Nd YAG laser at power of 15 W/cm^2. The measurement was performed at room temperature.

It is clear that C-H complexes do not act as recombination centers, since lifetime becomes longer with C-H complex density. If the complexes do act as recombination centers, then lifetime should be shorter with the complex density. The reason why lifetime becomes longer with the complex density is shown in Figure 19. Here lifetime is replotted against hole concentration. Here hole concentration increases with decrease of the C-H complexes. The lifetime is inversely proportional to hole concentration. It is well known that lifetime in the heavily doped GaAs used in the present captor is determined by Auger recombination, in which p^{-2} dependence is obtained. This concludes that the decomposition of C-H complexes is not REDR.

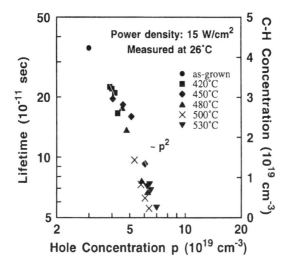

Fig. 19. Luminescence lifetime against hole concentration. The relation between the concentration of C-H complexes and hole concentration is also shown. A decrease in luminescence lifetime occurs with a hole concentration decrease.

These findings clearly demonstrate that decomposition of C-H complexes is induced by "Charge State Effect", not by "Recombination Enhanced Defect Reaction".

Regarding decomposition of C-H complexes, ab-initio calculation has been reported. Figure 20 shows the results reported by Breuer, Jones, Öberg, and Briddon (BJÖB) [23]. According to their report, thermally stimulated dissociation of H^+ requires 1.88 eV, which is in good agreement with our experimental data 1.76 eV. H^+ stays at the position of bond center between Ga and C. However, one electron capturing by a C-H complex tends to change the charge of hydrogen from H^+ to H^0. Since H^0 stays at the position of anti-bonding, H^0 tends to move from the bond center to anti-bonding position. This substantially reduces the saddle point energy of the hydrogen dissociation energy, resulting in reduction of its dissociation energy. Although their calculation predicts one electron capturing leads decomposition of C-H complexes, this can be extended to two electron capturing, since both H^0 and H^- are supposed to favor the anti-binding state.

Finally, hydrogen related degradation is discussed. There are two kinds of hydrogen in heavily carbon-doped p^+ GaAs:C; interstitial hydrogen donors and C-H complexes. Interstitial hydrogen donors are most efficient degradation factor. Defects observed after HBTs degradation are platelets, which are quite similar to those, observed in GaAs after hydrogen plasma

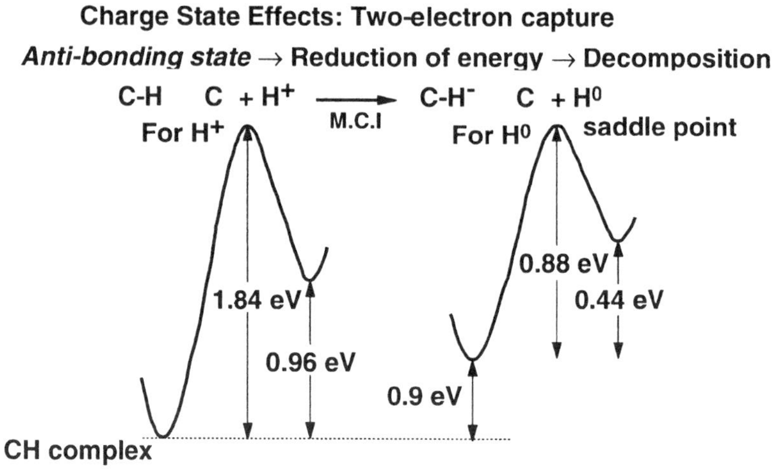
Charge State Effects: Two-electron capture
Anti-bonding state → Reduction of energy → Decomposition
C-H C + H+
M.C.I
C-H- C + H0
For H+
For H0
saddle point
1.84 eV
0.96 eV
0.9 eV
0.88 eV
0.44 eV
CH complex

Fig. 20. Regarding decomposition of C-H complexes, *ab-initio* calculation has been reported by Breuer, Jones, Öberg and Briddon (BJÖB).

irradiation. This strongly suggests that hydrogen in GaAs:C aggregates into the platelets under minority-carrier injection. Interstitial hydrogen is quite mobile to aggregate into the platelets. C-H complexes could decompose to release hydrogen under minority-carrier injection, resulting in interstitial hydrogen acceptor H^- formation. Hydrogen acceptors make molecules 2H with interstitial hydrogen donors H^+. 2H molecules become hydrogen precipitates by agglomeration of hydrogen interstitials. It is strongly suggestive that such precipitates would be microplatelets being observed by TEM [2].

In order to verify the model for hydrogen molecule formation as degradation factors of HBTs, further efforts are definitely required. In order to suppress HBTs degradation due to hydrogen it is essential not to incorporate hydrogen in the GaAs:C base layer. Hydrogen is efficiently incorporated into GaAs:C at low temperature, thus, it should be expected to grow GaAs:C layers at relatively high temperature. Recently, Kohda *et al.* reported that heavily carbon doped p^+ GaAs can be grown at high temperature under high V/III ratio [24]. This is quite promising to achieve highly stable HBTs based on GaAs/AlGaAs systems.

5. CARBON RELATED DEGRADATION OF HBTS

Carbon is an excellent acceptor for the base layer of HBTs, since it is a slow diffuser and has a high activation ratio in contrast to Be acceptor [25–27].

Indeed, Uematsu, *et al.* reported that HBTs with the Be doped base layers are degraded quite easily in terms of recombination enhance Be diffusion [28–29]. The difference is related to the difference of lattice occupation sites: C sits on the As sublattices and proceeds As vacancy diffusion, while Be sits on the Ga sublattices and proceeds Ga interstitial kick-out diffusion. This suggests that V_{As} and/or I_{Ga} would be most probable point defects in heavily p-GaAs where the Fermi level is within the valence band [30] and thus they would mediate diffusion of these acceptors.

Even carbon shows instability when the layers are annealed at high temperature; Decrease in hole concentration and PL intensity [31–32]. Phenomenologically such instabilities are induced by generation of compensation centers, scattering centers, and/or recombination centers [33–37]. However, the defect and defect reaction dominating such defect generation are still open question. In HBTs material processing, the epilayers are annealed at 500°C or above for device isolation [38–40]. Although such annealing duration is not very long, it is possible that such defects would be induced and enhance the degradation. Indeed, Figure 21 shows slow increase of leakage current in the annealed diodes. Therefore, it is quite important to understand the nature of annealing induced defects and its formation mechanism, as described in the next paragraph.

Before discussing any further, we must address how to measure increase of defect density responsible for current leakage. As is observed in Figure 21,

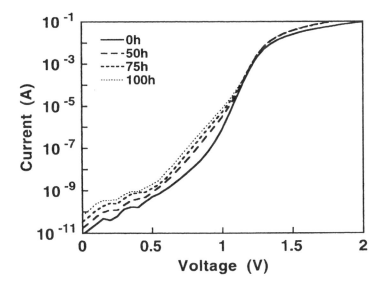

Fig. 21. The increase of leakage current in the annealed diodes.

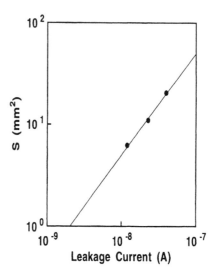

Fig. 22. Relationship between leakage current and integrated leakage current.

the ideality factor of I-V characteristics is more than 2, i.e., 3 to 4, which is similar to the hydrogen-induced degradation. This cannot be explained by simple assumption that recombination occurs in the depletion layer between the Emitter and Base layer of this diodes. This large ideality factor is supposed to be explained by tunneling assist recombination, but the analytical method is not clearly understood. Thus, defect density increase may be expressed by integrated leakage current. Here, we further simplify it by using leakage current not integrated leakage current to estimate degree of degradation. Figure 22 indicates the integrated leakage current S against leakage current. A linear relationship is observed between integrated leakage current and leakage current at a given voltage, here 0.7 V. Thus, the leakage current can phenomenologially be used for estimation of degree of degradation, although the mechanism governing current leakage is as yet unclear.

5.1.   Thermal Behavior of Carbon Related Defects

Figure 23 shows hole concentration reduction by high temperature annealing. The hole concentration was measured by the Hall effect measurements. The annealing temperatures were 530, 600, 700, and 750°C, and the duration was 10 min. Hole concentration decreases with annealing temperature increase. As shown in Figure 24, mobility becomes lowered with hole decrease, indicating that some kind of scattering centers are formed. Since hydrogen in

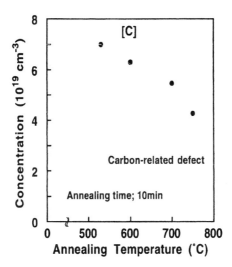

Fig. 23. Hole concentration changes due to annealing at temperatures between 500°C and 750°C for 10 min obtained by Hall effect measurements. There are at least two possible mechanisms responsible for the reduction in the number of holes during annealing at temperatures higher than 600°C: One is deactivation of carbon acceptors and the other is charge compensation.

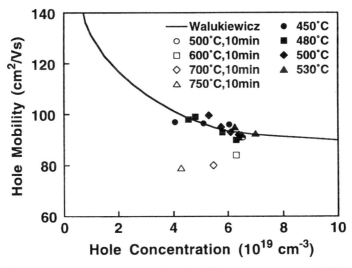

Fig. 24. Relation between hole concentration and mobility at room temperature. Solid curve is the theoretical calculation by Walukiewicz (see Ref. 18). This calculation was done under the assumption of no compensation. Our previously reported results are also plotted in this figure (see Ref. 17). High-temperature annealing generates compensation centers and scattering centers.

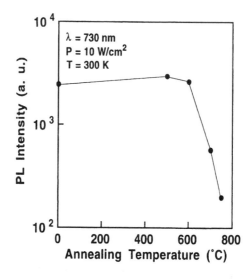

Fig. 25. Annealing temperature dependence of the PL intensity of near-band-edge emission at room temperature. The PL intensity decreases rapidly after annealing at higher than 600°C. The PL intensity decrease induced by high-temperature annealing is cause by nonradiative center increase.

the annealed samples was below detection limit of SIMS, hydrogen is not a cause of hole and mobility decreases. It is suggested that carbon would play roles in these phenomena.

Figure 25 shows photoluminescence intensity reduction by high temperature annealing. Annealing at 600°C or higher strongly reduce PL intensities. This is clear evidence that the high temperature annealing introduces nonradiative recombination centers as well. The introduction of the centers becomes efficient at higher temperature. Figure 26 shows PL lifetime dependence on hole concentration with data shown in the previous section on hydrogen. The lifetime decrease with hole concentration is governed by the Auger recombination in this regime. However, in these epilayers annealed at higher temperature PL lifetime becomes reduced despite that hole concentration becomes reduced. This cannot be explained by the Auger recombination, but clearly shows that high temperature annealing generates nonradiative recombination centers.

Based on these isochronal annealing, it is now clear that recombination centers as well as compensation centers are generated by high temperature annealing. However, it is not yet clear whether or not these centers are related to carbon. In the following, behavior of carbon during high temperature annealing is focused.

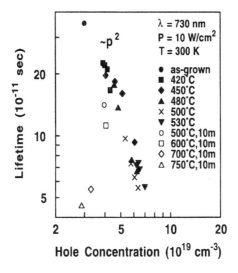

Fig. 26. Luminescence lifetime against hole concentration. Previously reported results on luminescence lifetime are replotted in this figure (see Ref. 19). The decrease in luminescence lifetime is induced by the formation of nonradiative recombination centers.

First, change in the lattice constant of GaAs:C is studied. As far as carbon acceptors sit on the As sub-lattice, the lattice constant should be constrained. Once carbons leave to interstitial sites, sizable lattice shrinkage is relieved. Thus, the lattice constant measurements give a clue for improving deeper understanding of carbon behavior upon annealing. Figure 27 shows the dependence of the lattice constant on annealing temperature. Additionally the lattice constant of undoped GaAs is shown as reference. At 500°C the lattice constant decreases where hydrogen is evacuated. This is because hydrogen at the bond-center site forms complexes with carbon to relax the lattice strain around carbon at the As sublattices [41]. Thus, the lattice constant becomes further reduced at the temperature by means of hydrogen decrease. On the other hand, the lattice constant increases with the higher annealing temperature toward undoped GaAs. This increase is quite important findings, because no hydrogen plays any roles on it and no carbon loss is observed by SIMS.

Hole decreases observed after high temperature annealing might be explained by formation of carbon donors on the Ga sublattices, since carbon is believed an amphoteric impurity. However, this figure clearly denies this possibility, since the lattice constant should not be so changed between carbon on the Ga sublattices and on the As sublattices. This suggests that carbon moves from the As sublattices to interstitial sites. The formation of carbon

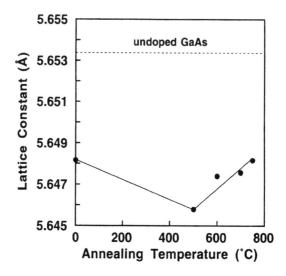

Fig. 27. Effect of annealing on the lattice parameter of heavily carbon-doped GaAs. The lattice constant increases as the temperature is raised from 600°C to 750°C.

interstitials has also been implied [42]. Figure 28 shows schematics of carbon interstitial formation.

The reaction of carbon interstitial formation can be explained by extrinsic Frenkel defect formation mechanism or kick-out mechanism by As interstitials. In other words, the reaction of carbon interstitial formation can be expressed by

$$C_{As} \Leftrightarrow C_i + V_{As} \tag{7}$$

$$C_{As} + I_{As} \Leftrightarrow C_i, \tag{8}$$

Here the reaction (7) may be referred as the extrinsic Frenkel defect formation mechanism. In the reaction (8) As interstitials replacing carbon on the As sublattices should be provided by the intrinsic Frenkel defect formation and the Schottky defect formation mechanisms. It has been reported so far that the intrinsic Frenkel defect formation mechanism on the Ga sublattices works during n-type epilayer growth because of the Fermi level pinning at the growth surface [5, 18]. This is also described in Section 1.1.2. Here we assume that the Frenkel defect formation mechanism on the As sublattices would occur during p-type epilayer growth due to the same reason. By considering charge states of carbon and point defects [37], the reaction of

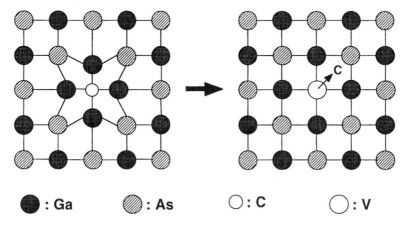

● : Ga   **◎** : As   ○ : C   ○ : V

Fig. 28. Schematics of carbon interstitial formation. The reaction of carbon interstitial formation can be explained by extrinsic Frenkel defects formation mechanism or kick-out mechanism by As intertitial.

carbon interstitial formation can be expressed by

$$C_{As}^- \Leftrightarrow C_i^{++} + V_{As}^+ + 4e^-, \tag{9}$$

$$As_{As} \Leftrightarrow I_{As}^{+++} + V_{As}^+ + 4e^-, \tag{10}$$

$$C_{As}^- + I_{As}^{+++} \Leftrightarrow C_i^{++} + As_{As}. \tag{11}$$

Here, the reaction (9) shows extrinsic Frenkel defect formation, and the reaction (10) intrinsic Frenkel defect formation. Carbon interstitials are generated by the reactions (9) and (11). This suggests that formation of one carbon interstitial consume 4 holes.

Based on the Vegard's law, the lattice constant can be converted to concentration of carbon on the As sub-lattice. There have been reported two conversion factors, relaxed and strained cases [42]. The relaxed case is valid for thick epilayers, while the strained case for thin layers. So, the conversion factor for relaxed case is used.

$$\Delta a_\perp = \frac{4}{\sqrt{3}} \frac{\Delta r_{As} \cdot C_{As}}{2.22 \times 10^{22}}, \tag{12}$$

Here, Δr_{As} is the difference of covalent bond length between C-Ga and As-Ga, i.e., -0.043 nm. In addition, the difference in the strained case is 1.9 times of the relaxed case. Figure 29 shows interrelationship between the carbon concentration on the As sublattices and hole concentration after high temperature annealing. It is experimentally clear that formation of one carbon interstitial consumes around 3 holes. This indicates that carbon

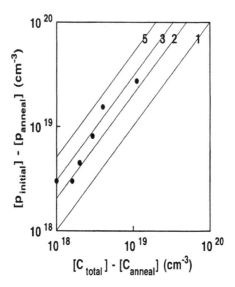

Fig. 29. Relation between the amount of hole decrease and the amount of CAs decrease analyzed by X-ray diffraction measurements. $[p_{initial}]$ and $[C_{total}]$ are equal to 7×10^{19} cm^{-3}. This shows that the proportional decay factor is 3. This result suggests that As vacancies always take part of the hole decrease caused by high-temperature annealing.

interstitials formed by high temperature annealing cause hole decrease. It is strongly suggestive that defects formed in these reactions should play as dominant scattering centers. It is not yet clear that these defects also act as recombination centers formed by high temperature annealing. This is discussed in the next paragraph.

First, the formation kinetics of these centers are discussed. Figure 30 shows hole decrease with annealing time. The annealing temperature ranged from 550 to 700°C. The hole decrease saturates around 5.8×10^{19} cm^{-3}. Concentration of carbon related defect can be expressed by the difference between initial hole concentration 7×10^{19} cm^{-3} and reduced hole concentration. The defect concentration is shown by Fig. 31. In other words, Frenkel defect formation saturates with annealing time. It is also clear from this figure that the saturation occurs faster at higher annealing temperature.

Here, the reaction (9) is assumed to analyze the kinetics of hole decrease.

$$\frac{d[C_i^+]}{dt} = \kappa_f^C \cdot [C_{As}^-] - \kappa_r^C \cdot [V_{As}^+] \cdot [C_i^+] \cdot n^3, \tag{13}$$

$$[C_i^+] = \frac{\kappa_f^C \cdot [C_{As}^-]}{\kappa_r^C \cdot [V_{As}] \cdot n^3} \left(1 - \exp(-\kappa_r^C \cdot [V_{As}^+] \cdot n^3 \cdot t)\right), \tag{14}$$

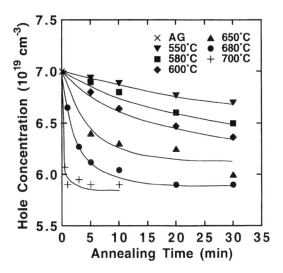

Fig. 30. The decrease in hole concentration by high-temperature annealing. Hole decrease with annealing time. This result suggests that the defects act as scattering centers formed by annealing.

According to these equations, the final concentration and the initial formation rate of C_i can be expressed

$$[C_i^+]_{t \to \infty} = \frac{\kappa_f^C \cdot [C_{As}^-]}{\kappa_r^C \cdot [V_{As}^+] \cdot n^3}, \tag{15}$$

$$[C^+]_{t \to 0} = \kappa_f^C \cdot [C^-]t \tag{16}$$

This shows the features of Figure 31: The formation of C_i is initially proportional to annealing time and saturates at a given concentration.

One the other hand, recombination centers behave quite differently. In order to compare these with that of recombination centers, here PL intensity dependence on annealing time is used. Prior to the comparison, the PL intensity is compared with the PL lifetime which is more fundamental data. As discussed in the previous section, PL lifetime is inversely proportional to concentration of recombination centers. Figure 32 shows the interrelationship. It is shown that PL intensity is clearly proportional to PL lifetime, indicating that PL intensity can be used to monitor the formation kinetics of recombination centers. Figure 33 shows PL intensity decrease with annealing time. The same samples shown in Figures 31 and 32 were used. In clear contrast to the defects reducing hole concentration, saturation of recombination center increase is not observed, as shown in Figure 34. In the initial stage, recombination

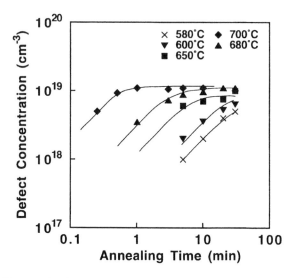

Fig. 31. The saturation in defect concentration by high-temperature annealing. This resul suggests that Frenkel defect formation saturate with annealing time.

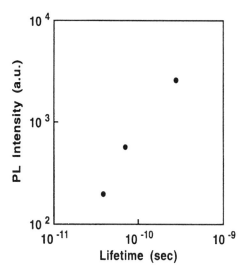

Fig. 32. Relation between the luminescence lifetime and the PL intensity.

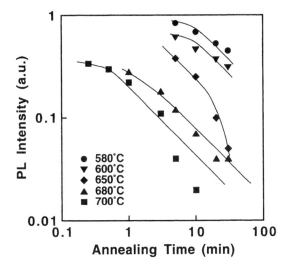

Fig. 33. PL intensity decrease with annealing time.

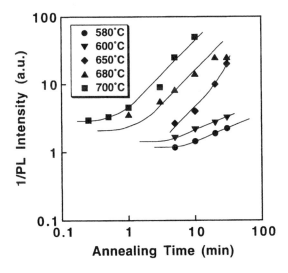

Fig. 34. The increase in recombination center by high-temperature annealing.

centers do not increase, which is probably due to pre-existing recombination centers. In addition, at 700°C hole concentration saturates within 1 min while PL intensity keeps decreasing after 1 min. This concludes that defects or defect reactions responsible for the hole decreases is not the same as those for recombination centers. In other words, the Frenkel defect formation is over within 1 min at 700°C, then, recombination centers are dominantly formed. This can be understood by assuming that defects through the Frenkel defect formation mechanism are used to form recombination centers. It can be said that recombination centers are secondary defects, not primary defects.

The candidates for the recombination centers are discussed as follows. As is described above, it can be assumed that using defects in terms of the Frenkel defect formation would form recombination centers. Following this assumption, the possible reactions are:

$$C_{As}^- + C_i^+ \Leftrightarrow (2C_i)^+, \tag{17}$$

$$C_{As}^- + V_{As}^+ \Leftrightarrow (V_{As}C_{As})^0, \tag{18}$$

$$C_i^{++} + V_{As}^+ + 4e^- \Leftrightarrow C_{As}^-. \tag{19}$$

The reaction (19) should be excluded, since this reaction is the reverse reaction of the reaction (9). If this is the case, carbon on the As sublattices must act as a recombination center, which is not plausible. Accordingly, di-carbon interstitials and/or complexes of carbon interstitials and As vacancies would be recombination centers. Indeed di-carbon interstitials has been experimentally confirmed [44, 45] and theoretically predicted to be favorable [35]. Figure 35 illustrates a di-carbon interstitial theoretically predicted, i.e., ⟨100⟩ split interstitial.

$$C_i + C_i \Leftrightarrow 2C_i. \tag{20}$$

$$2C_i + C_i \Leftrightarrow 3C_i. \tag{21}$$

Here, $2C_i$ denotes a di-carbon interstitial, $3C_i$ a tri-carbon interstitial. Formation rate and concentration of $2C_i$ can be expressed by

$$\frac{d[2C_i]}{dt} = \kappa_1 [C_i]^2 - (\kappa_{-1} + \kappa_2 \cdot [2C_i]) \cdot [2C_i]. \tag{22}$$

$$[2C_i] = \frac{\kappa}{(\kappa_{-1} + \kappa_2[C_i])} \cdot [C_i]^2 \cdot \left(1 - \exp(\kappa_{-1} + \kappa_2 \cdot [C_i] \cdot t)\right). \tag{23}$$

Based on this expression, the concentration of $2C_i$ saturates with time at the concentration of

$$[2C_i]_{t \to \infty} = \frac{\kappa_1 \cdot [C_i]^2}{(\kappa_{-1} + \kappa_2 \cdot [C_i])}, \tag{24}$$

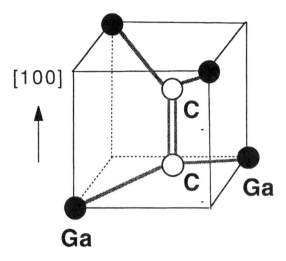

Fig. 35. Illustrates a di-carbon interstitial theoretically predicted. This illustrates is ⟨100⟩ split interstitial.

and the initial formation rate is proportional to time:

$$[2C_i]_{t \to 0} = \kappa_1 \cdot [C_i]^2 \cdot t, \tag{25}$$

In fact Figure 31 shows defect concentration is proportional to annealing time in the initial stage and shows its saturation as predicted by these equations. This is strongly suggestive that $2C_i$ would act as recombination centers.

In order to further confirm the difference between centers for hole decrease and for recombination centers, rate constants of hole and of PL decreases are compared each other. As shown in Figure 36, the activation energy is about 2 eV for the hole decrease. On the other, it is 3.8 eV for PL intensity decrease, as shown in Figure 37. This agrees well with the conclusion that decrease in PL intensity is not due to hole decrease. Here it is worth to compare degradation due to hydrogen and to carbon, in order to focus the minority carrier injection effect.

In the degradation due to hydrogen,

H1. Decomposition of C-H complexes in terms of charge state effect (CSE) due to Fermi level effect in terms of minority carrier injection.

H2. Aggregation of hydrogen interstitials released to form platelets, here, hydrogen diffusion should be enhanced most likely by charge state effect, less likely by recombination enhanced defect reaction (REDR). It has not been experimentally confirmed.

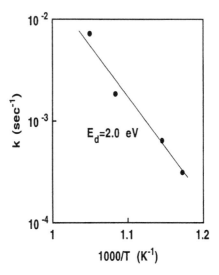

Fig. 36. The activation energy for the hole decrease. The activation energy is 2.0 eV.

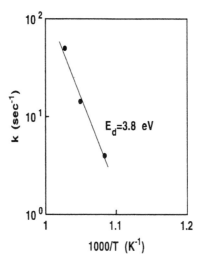

Fig. 37. The activation energy for PL intensity decrease. The activation energy is 3.8 eV.

H3. Formation of minority carrier recombination paths to degrade devices.

In the degradation due to carbon related defects as follows.

C1. Carbon interstitials are formed in terms of the Frenkel defect formation mechanism by high temperature annealing. This is purely thermal effect.
C2. Aggregation of carbon interstitials to form microprecipitates. In this process it is unclear whether or not diffusion of carbon interstitials is enhanced by minority-carrier injection and, if enhanced, what mechanism is working behind diffusion enhancement of carbon interstitials.
C3. Formation of minority carrier recombination paths to degrade devices.

Both $H3$ and $C3$ processes are unclear, but it is not possible to discuss the formation mechanism of minority carrier recombination paths in this section. Now, the $C2$ process is focused.

5.2. Carbon Related Degradation

Here, before we get into these topics, we propose new concept for device degradation which is related to point defects. The point defects generation can be written by

$$\frac{dN}{dt} = kN, \tag{26}$$

$$N = N_0 \exp(k \cdot t), \tag{27}$$

where N denotes point defect concentration, N_0 initial point defect concentration, k rate constant, t times. This idea has an essential problem. The defect density ceaselessly increases whereas in reality it saturates in most cases. This is obvious, because source of point defects is not unlimited. In the hydrogen case, hydrogen concentration is mid 10^{19} cm^{-3}. A new way to formulate the saturation kinetics is needed. In the present chapter, the following rate equation is employed to express the saturation.

$$\frac{dN}{dt} = k(N_\infty - N), \tag{28}$$

Under initial condition $t = 0$, $N(t) = 0$, the solution can be written by

$$N = N_\infty \{1 - \exp(-k \cdot t)\}, \tag{29}$$

denotes saturation of defect increase. N can be approximated to be in initial stage. Based on this model, we discuss the increase in the leakage current. Figure 38 shows leakage current increase by minority-carrier injection. This results shows that the increase in the leakage current saturate. Figure 39 shows

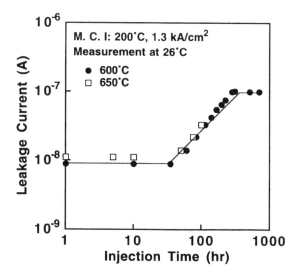

Fig. 38. Injection time dependence of the leakage current at 0.7 V at room temperature. Injection minority carrier density is 1.3 kA/cm² at 200°C. The increase in the leakage current saturate.

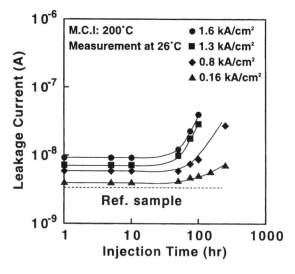

Fig. 39. Injection time dependence of the leakage current. Injection current density is 0.16 to 1.6 kA/cm² at 200°C. The parameter is injection current density.

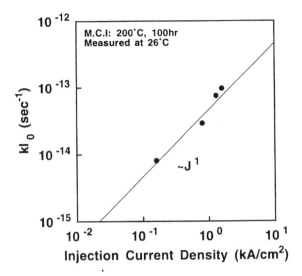

Fig. 40. Leakage current dependence of injection current density. Minority-carrier injection was carried out at 200°C for 100 h. This shows that the exponential decay is proportional to the injection current density.

leakage current increase by means of minority-carrier injection into the devices suffered from high temperature annealing. In this stage, isolated carbon interstitials have already been formed, but no degradation occurred without minority-carrier injection (see Ref. 2. for example). This indicates that the reaction shown by Eq. (20) would not occur at temperature as low as 200°C. Thus, formation of recombination centers would be enhanced under minority-carrier injection. It is likely that increase in the leakage current is proportional to injection time. This is also explainable by Eq. (25). The leakage current in the initial stage should be caused by pre-existing recombination centers. This suggests that carbon interstitials become mobile under minority-carrier injection. In order to clarify what is going on behind this diffusion enhancement, the leakage current increase is quantitatively studied. Figure 39 shows leakage current increase with injection time. The parameter is injection current. It is shown that leakage current is proportional to injection current. This is an evidence that recombination enhanced defect reaction (REDR) occurs.

Figure 40 has been replotted from Figure 39. The linear dependence of leakage current on injection time is clearly observed except very initial stage. In the very initial stage, leakage current does not depend on injection time, suggesting that there are other sources of current leakage.

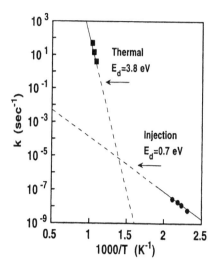

Fig. 41. Arrhenius plot of the rate constants of the carbon interstitial diffusion from Figure 39. The carbon interstitial diffusion would be enhanced under minority-carrier injection.

Now, the rate constants, which are not purely the rate constants but the product of N and k, can be obtained from Figure 39. The arrhenius plot is shown in Figure 41 with thermal decomposition rates. The activation energy of the carbon interstitial diffusion due to minority-carrier injection is 0.7 eV, while it is 3.8 eV for the thermal annealing. These results show that carbon interstitial diffusion would be enhanced under minority-carrier injection.

6. CONCLUSION

Hydrogen incorporated in p-GaAs:C decrease hole concentration by acting as isolated hydrogen donors (H^+) in addition to forming complexes with carbon acceptors (C-H). There are two stages, fast and slow, in thermal recovery of hole concentration in p-GaAs:C. Diffusion of Isolated hydrogen donors dominates the fast stage, while decomposition of complexes dominates the slow stage. Thermal dissociation energy of H from CH complexes in heavily doped p-GaAs is equal to that in lightly doped p-GaAs. However, the rate is several orders of magnitude lower than in heavily doped GaAs. This is due to hydrogen-retrap by high density of acceptors.

Hydrogen dissociation from the complexes is enhanced under electronic excitation even at low temperature. This leads to device degradation of minority carrier injection devices such as heterojunction bipolar transistors, HBTs. The mechanism of dissociation is not recombination enhanced defect reaction but charge state effect in terms of two-electron capture.

High temperature annealing decreases hole concentration in heavily doped p-GaAs:H. Hydrogen is not an origin of the hole reduction, since both isolated hydrogen donors and CH complexes have been rapidly eliminated by the thermal annealing. The defects decrease hole mobility and luminescence lifetime as well. They are induced most likely by the Frenkel pair formation mechanism at carbon on an As site, here named as "Extrinsic Frenkel defect mechanism". In other words, $C_{As}^- \Longleftrightarrow C_I^{++} + V_{As}^+ + 4e^-$. This reaction predicts consumption of 4 holes, although 3-hole consumption is experimentally confirmed. This suggests that charge density theoretically predicted might be reconsidered. It is shown that recombination center formation starts after hole density decrease is saturated. Based on this observation, it is proposed that recombination centers would be secondary defects with interstitial carbon and/or arsenic vacancies.

Under electronic excitation the extrinsic Frenkel defect reaction is not observed. However, recombination centers are generated, once the epilayer involves extrinsic Frenkel defects induced by high temperature annealing. This increase of recombination centers causes degradation. The defect reaction rate is proportional to injection carrier. This indicates that recombination enhanced defect reaction is working behind.

In device degradation, various factors including hydrogen and carbon are working on, such as device processing beside epilayer growth and high temperature annealing. These process-related factors must be considered. For example, lifetime of HBTs with passivation layer depends on the thickness of the layer.

References

1. See recent review, O. Ueda, "Reliability and Degradation of III–V Optical Devices", Artech House, Boston, (1996).
2. H. Sugahara, J. Nagano, T. Nittono and K. Ogawa, GaAs IC Symp., 1993, IEEE, New York, 1993, pp.115.
3. L.C. Kimerling, *Solid State Electron.*, **21**, 1391 (1978). Also see the chapter 2.1 in this book.
4. T.F. Kuech, M.A. Tischler, P.J. Wang, G. Scilla, R. Potemski and F. Cardone, *Appl. Phys. Lett.*, **53**, 1317 (1988).
5. B.T. Cunningham, L.J. Guido, J.E. Baker, J.S. Major, N. Holonyak and G.E. Stillman, *Appl. Phys. Lett.*, **55**, 687 (1989).
6. E. Tokumitsu, Y. Kudo, M. Konagai and K. Takahashi, *Jpn. J. Appl. Phys.*, **24**, 1189 (1985).
7. M. Konagai, T. Yamada, T. Akatsuka, K. Saito and E. Tokumitsu, *J. Cryst. Growth*, **74**, 292 (1989).
8. C.R. Abernathy, S.J. Pearton, R. Caruso, F. Ren and J. Kovalchik, *Appl. Phys. Lett.*, **55**, 1750 (1989).

9. N. Putz, E. Veuhoff, H. Heinecke, M. Heyen, H. Luth and P. Balk, *J. Vacuum Sci. Technol.*, **B3**, 671 (1985).
10. S.A. Stockman, A.W. Hanson, S.L. Jackson, J.E. Baker and G.E. Stillman, *Appl. Phys. Lett.*, **62**, 1248 (1993).
11. K. Woodhouse, R.C. Newman, R. Nicklin, R.R. Bradley and M.J.L. Sangster, *J. Cryst. Growth*, **120**, 323 (1992).
12. D.M. Kozuch, M. Stavola, S.J. Pearton, C.R. Abernathy and J. Lopata, *Appl. Phys. Lett.*, **57**, 2561 (1990).
13. B. Clerjaud, D. Cota and C. Naud, *Phys. Rev. Lett.*, **58**, 1755 (1987).
14. S.J. Pearton, C.R. Abernathy and J. Lopata, *Appl. Phys. Lett.*, **59**, 3571 (1991).
15. H. Fushimi and K. Wada, *J. Cryst. Growth*, **145**, 420 (1994).
16. J.M. Zavada, H.A. Jenkinson, R.G. Sarkis and R.G. Wilson, *J. Appl. Phys.*, **58**, 3731 (1985).
17. S.J. Pearton, W.C. Dautremont-Smith, J. Chevallier, C.W. Tu and K.D. Cummings, *J. Appl. Phys.*, **59**, 2821 (1986).
18. R. Rahbi, D. Mathiot, J. Chevallier, C. Grattepain and M. Razeghi, *Physica*, **B170**, 135 (1991).
19. N.M. Johnson and C. Herring, *Mater. Sci. Forum.*, **143–147**, 867 (1994).
20. S.A. Stockman, A.W. Hanson, S.M. Lichtenthal, M.T. Fresina, G.E. Hofler, K.C. Hsieh and G.E. Stillman, *J. Electron. Mater.*, **21**, 1111 (1992).
21. See the chapter 1.1.
22. S. Muto, S. Takeda and M. Hirata, *Mater. Sci. Forum.*, **143–147**, 897 (1993).
23. S.J. Breuer, R. Jones, S. Öberg and P.R. Briddon, *Mater. Sci. Forum.*, **196–201**, 951 (1995).
24. H. Kohda and K. Wada, *J. Cryst. Growth.*, **167**, 557 (1996).
25. Y.C. Pao, T. Hierl and T. Cooper, *J. Appl. Phys.*, **60**, 201 (1986).
26. P. Enquist, J.A. Hutchby and T.J. de Lyon, *J. Appl. Phys.*, **63**, 4485 (1988).
27. N. Kobayashi, T. Makimoto and Y. Horikoshi, *Appl. Phys. Lett.*, **50**, 1435 (1987).
28. M. Uematsu and K. Wada, *Appl. Phys. Lett.*, **58**, 2015 (1991).
29. M. Uematsu and K. Wada, *Appl. Phys. Lett.*, **60**, 1612 (1992).
30. G.A. Baraff and M. Schlüter, *Phys. Rev. Lett.*, **55**, 1327 (1985).
31. K. Watanabe and H. Yamazaki, *Appl. Phys. Lett.*, **59**, 434 (1991).
32. T.J. de Lyon, N.I. Buchan, P.D. Kirchner, J.M. Woodall, G.J. Scilla and F. Cardone, *Appl. Phys. Lett.*, **58**, 517 (1991).
33. G.E. Höfler and K.C. Hsieh, *Appl. Phys. Lett.*, **61**, 327 (1992).
34. M.C. Hanna, A. Majerfeld and D.M. Szmyd, *Appl. Phys. Lett.*, **59**, 2001 (1991).
35. A.J. Moll, E.E. Haller, J.W. Ager III and W. Walukiewicz, *Appl. Phys. Lett.*, **65**, 1145 (1994).
36. H.M. You, T.Y. Tan, U.M. Gösele, S.-T. Lee, G.E. Höfler, K.C. Hsieh and N. Holonyak, Jr., *J. Appl. Phys.*, **74**, 2450 (1993).
37. Byoung-Ho Cheong and K.J. Chang, *Phys. Rev.*, **B49**, 17436 (1994).
38. O. Nakajima, H. Ito, T. Nittono and K. Nagata, *1990 IEDM Tech. Dig.*, San Francisco, CA, pp.673–676 (1990).
39. K. Watanabe, K. Nagata, H. Yamazaki, S. Ishida and T. Ichijo, *Appl. Phys. Lett.*, **57**, 1892 (1990).
40. F. Ren, S.J. Pearton, W.S. Hobson, T.R. Fullowan, J. Lothian and A.W. Yanof, *Appl. Phys. Lett.*, **56**, 860 (1990).
41. K. Watanabe and H. Yamazaki, *J. Appl. Phys.*, **74**, 5587 (1993).
42. T.J. de Lyon, J.M. Woodall, M.S. Goorsky and P.D. Kirchner, *Appl. Phys. Lett.*, **56**, 1040 (1990).
43. W. Walukiewicz, *Mater. Res. Soc.*, **163**, 845 (1990).
44. R. Jones and S. Öberg, *Mater. Sci. Forum*, **143–147**, 253 (1994).
45. J. Wagner, R.C. Newman, B.R. Davidson, S.P. Westwater, T.J. Bullough, T.B. Joyce, C.D. Latham, R. Jones and S. Öberg, *Phys. Rev. Lett.*, **78**, 74 (1997).

INDEX